Synthesis of Biaryls

Synthesis of Biaryls

Ivica Cepanec

Director of Research & Development Department
BELUPO Pharmaceuticals, Zagreb, CROATIA

2004

ELSEVIER

Amsterdam • Boston • Heidelberg • London • New York • Oxford • Paris
San Diego • San Francisco • Singapore • Sydney • Tokyo

ELSEVIER B.V.	ELSEVIER Inc.	**ELSEVIER Ltd**	ELSEVIER Ltd
Sara Burgerhartstraat 25	525 B Street, Suite 1900	**The Boulevard, Langford Lane**	84 Theobalds Road
P.O. Box 211, 1000 AE Amsterdam	San Diego, CA 92101-4495	**Kidlington, Oxford OX5 1GB**	London WC1X 8RR
The Netherlands	USA	**UK**	UK

First edition 2004

Library of Congress Cataloging in Publication Data
A catalog record is available from the Library of Congress.

British Library Cataloguing in Publication Data
A catalogue record is available from the British Library.

ISBN: 0 08 044412 1

♾ The paper used in this publication meets the requirements of ANSI/NISO Z39.48-1992 (Permanence of Paper).
Printed in The Netherlands.

PREFACE

Organic chemistry is one of the most rapidly growing sciences. There is a wide variety of applications of organic compounds, for instance, pharmaceutical active substances, agrochemicals, optoelectronics, etc. Within this group there are hundereds and thousands of new compounds synthesized or isolated from natural sources. Such important organic chemistry developments are accompanied by the profound break-through of new reactions, increasingly efficient methodologies, reagents and catalysts. The chemistry of biaryls is one of the most interesting fields in organic chemistry. The aryl-aryl, C-C bond forming reactions are not only of academic interest, but also play an important role in a number of industrial processes, fine chemical industry, natural products synthesis, chiral catalyst chemistry, chiral stationary phases chemistry, environmental pollutant chemistry, electronic devices production, etc. The synthesis of biaryls has been impressively developed during the last century, starting from the classical Ullmann and Gomberg-Bachmann-Hey, generally known as high-temperature or low yielding reactions, to the modern Suzuki-Miyaura cross-coupling reactions, as well as various recent reactions involving diaryliodonium salts or miscellaneous organometallics, which cleanly proceed even at room temperature or below, tolerating several sensitive functional groups. Whereas the synthesis of complicated biaryl structures several decades ago was realized with aryl-aryl, C-C bond forming reaction in very early steps of synthesis, very powerful modern reactions allow to build the biaryl structure from highly functionalized substrates in the last step.

The book is organized through eight Chapters. In the first Chapter, a short introduction to the topic, and elementary structure facts are given. The old methods with newer improvements are presented in the second Chapter, whereas in Chapters 3-7 all important reactions of biaryl chemistry are described. The Suzuki-Miyaura, as probably the most important general reaction for the synthesis of biaryls, is described in particular detail. Not less interestingly, the Chapter 7, includes several specific approaches to biaryl synthesis, e.g. Meyers, Motherwell synthesis or palladium catalysed arylations of arenes with aryl halides. In the last, Chapter 8, the most relevant methods for the synthesis of axially chiral biaryls are included. The material is presented from both a synthetic and mechanistic point of view, whilst the representative synthetic procedures may be helpful for understanding the methodology of the given reaction. I hope that readers, advanced undergraduate and post-graduate students, research and industrial chemists and engineers, will find the book interesting and useful. In addition, lecturers may find it a helpful support to improve their presentations. The knowledge has been taken from numerous literature references, from early 1900's till the end of 2003, with appreciable additional experiences from the Belupo Pharmaceuticals Research Department.

It is now my pleasure to express profound gratitude to my coworkers and friends, especially Dr. Mladen Litvić, Mrs Štefica Vrščak for type-writting, Mrs Mihaela Farquhar, her husband Graeme, and Miss Lana Donaldson for English language editing, as well as to the persons who, each of them in different times, helped me to realize the beauty of chemistry - many thanks to: Dr. Franjo Kajfež, Dr. Zvonimira Mikotić-Mihun, Dr. Vladimir Vinković and Prof. Dr. Vitomir Šunjić.

Finally, I want to thank my wife Katarina and children Ivan and Jelena, who with love and understanding endured me for many days, whilst I spent hours in the library and on the computer. I dedicate this work to them.

Zagreb, Croatia, 2004.

Dr. Ivica Cepanec

CONTENTS

SYNTHESIS OF BIARYLS

ABBREVIATIONS

Ac	Acetyl
acac	Acetylacetonate
AIBN	Azobisisobutyronitrile
aq.	Aqueous
AsPh$_3$	Triphenylarsine
BHT	2,6-Di-*t*-butyl-4-methylphenol
Bn	Benzyl
BOC	*t*-Butoxycarbonyl
bpy	2,2'-Bipyridine
n-Bu	*normal*-Butyl
s-Bu	*secondary*-Butyl
t-Bu	*tertiary*-Butyl
n-BuLi	*n*-Butyllithium
s-BuLi	*s*-Butyllithium
t-BuLi	*t*-Butyllithium
Bz	Benzoyl
°C	Degrees Celsius
CAN	Ammonium cerium(IV) nitrate
cat.	Catalytic
CBz	Carbobenzyloxy
CDI	1,1'-Carbonyldiimidazole
COD	1,5-Cyclooctadienyl
Cp	Cyclopentadienyl
CTAB	Cetyltrimethylammonium bromide
Cy	Cyclohexyl
% d.e.	% Diastereomeric excess
DABCO	1,4-Diazobicyclo[2.2.2]octane
DDQ	2,3-Dichloro-5,6-dicyano-1,4-benzoquinone
DIBAlH	Diisobutylaluminum hydride

DMAc	Dimethylacetamide
DME	1,2-Dimethoxyethane
DMF	N,N-Dimethylformamide
DMI	N,N-Dimethylimidazolidinone
DMPU	N,N'-Dimethylpropyleneurea
DMSO	Dimethylsulfoxide
dppb	1,4-Bis(diphenylphosphino)butane
dppe	1,2-Bis(diphenylphosphino)ethane
dppf	Bis(diphenylphosphino)ferrocene
dppp	1,3-Bis(diphenylphosphino)propane
EDTA	Ethylenediaminetetraacetic acid
% e.e.	% Enantiomeric excess
Et	Ethyl
eq.	Molar equivalent
h	Hour, hours
HMPA	Hexamethylphosphoramide
hν	Irradiation with light
i-Pr	Isopropyl
LAH	Lithium aluminum hydride
LDA	Lithium diisopropylamide
LHMDS	Lithium hexamethyldisilazide
LIC-KOR	super base KOt-Bu / n-BuLi
LTA	Lead(IV) acetate
Me	Methyl
MeCN	Acetonitrile
MIBK	Methyl isobutylketone
min	Minute, minutes
m.p.	Melting point
Ms (OMs)	Methanesulfonyl
MS	Molecular sieves (3 or 4 Å)
NCS	N-Chlorosuccinimide
NMP	1-Methyl-2-pyrrolidinone
OCA	oxidative coupling of arenes
PBu$_3$	Tri-n-butylphosphine
Pt-Bu$_3$	Tri-t-butylphosphine
PCy$_3$	Tricyclohexylphosphine
PEG	Polyethylene glycol
PEt$_3$	Triethylphosphine
PFu$_3$	Tri(2-furyl)phosphine
Ph	Phenyl

PIDA	Iodosobenzene diacetate
PIFA	Iodosobenzene bis(trifluoroacetate)
PPh$_3$	Triphenylphosphine
PS	Polystyrene backbone
PTC	Phase transfer catalyst
PTol$_3$	Tri(2-tolyl)phosphine
Py	Pyridine
SM	Suzuki-Miyaura reaction
TBAB	Tetra-n-butylammonium bromide
TBAF	Tetra-n-butylammonium fluoride
TDAE	Tetrakis(dimethylamino)ethylene
TFA	Trifluoroacetic acid
Tf (OTf)	Triflate
THF	Tetrahydrofuran
TMS	Trimethylsilyl
TMU	1,1,3,3-Tetramethylurea
Tr	Trityl
Ts (OTs)	Tosylate
TTFA	Thallium(III) trifluoroacetate

CHAPTER
1

1. INTRODUCTION

1.1. Aryl-aryl bond forming reactions

The formation of an aryl-aryl bond is one of the most important goals in the field of organic chemistry. The methodology for performing the synthesis of biaryls has been a challenging focus for over a century. Since Ullmann's first reports about the coupling reactions of aryl halides to biaryls with copper bronze, a number of valuable reactions and methods have been published. The main reason for such a different approach to the aryl-aryl bond forming reactions from alkyl-alkyl bond are the distinguished properties of electrophilic aryl-compounds. Thus ordinarily aryl halides, in contrary to alkyl halides, are not suitable counterparts for classical nucleophilic substitution reactions. Exceptions are the aryl halides bearing an electron-withdrawing substituent in *ortho* and/or *para*-positions. The latter readily undergo nucleophilic aromatic substitutions (S_NAr), however, these reactions are rarely useful in the generation of aryl-aryl bonds, since several common electron-withdrawing substituents, e.g. nitro, cyano, or alkoxycarbonyl, are not well tolerated with aryl-carbanion donors, e.g. aryllithiums or aryl Grignard reagents. Alternatives to the nucleophilic substitution approach to the aryl-aryl bond formation are the free-radical arylation, or reductive elimination of various diarylmetallics such as diarylnickel, diarylpalladium or diarylcopper complexes. Whereas the former process, involving free-radicals, is the basis of classical Gomberg-Bachmann-Hey reaction and several older or modern alternative arylations, the latter reaction proceeds within the coordinative sphere of nickel, palladium or copper. This organometallic process is crucial for a number of reactions whose only difference is in the pathway of obtaining the unstable diaryl-nickel, -copper, or -palladium species. Some reactions involve the formation of transient diarylmetallic species, e.g. diarylcoppers in the Ullmann, or diarylnickel compounds in the homo-coupling reactions of aryl halides with nickel(0) complexes, whereas some other certain reactions include the separate preparation of an arylmetallic reagent, e.g. arylboronic acids in the Suzuki-Miyaura, arylzincs in the Negishi, aryl Grignard reagents in the Kharasch reaction, etc., which, upon transmetallation to arylpalladium(II) complexes, give the desired diarylpalladium species. The unstable diaryl-nickel, -copper or -palladium complexes are the crucial species which, by reductive elimination of the metallic compounds in lower oxidation state, generate the new aryl-aryl bond. Beside these, diaryl compounds of various metals were given the reductive elimination reaction to produce biaryls, however, only with nickel, copper, and palladium, the process can be accomplished catalytically. Synthesis of biaryls have been the theme of a great number of papers during more than a hundered years. Apart from the several recent successful general reactions which are very important and valuable alternatives to the older, classical reactions for synthesis of biaryls, a number of basic problems is still waiting for efficient solutions.

1.2. The importance of biaryls

Biaryl structures are wide-spread in many of naturally occuring products including alkaloids, lignans, terpenes, flavonoids, tannins, as well as polyketides, coumarins, peptides, glycopeptides, etc. For example, vancomycin (**1**) is a basic structure of several related glycopeptide antibiotics [1]: balhimycin, actinoidin A, ristocetin A, teicoplanin A_2-2, complestatin, etc which are important in medicinal chemistry or as a HPLC chiral stationary phases (vancomycin) [2].

1

A lot of natural pigments are biaryls; as gossypol (**2**), a major constituent of cottonseed pigment, is a binaphthyl structured polyphenol, having also a male antifertility action

[3]. Among lignans, an illustrative example is steganacin (**3**), a constituent of *Steganotaenia araliacea*, the respective synthetic target of several scientific papers, which possesses a significant antileukemic activity [4].

2

3

Unfortunately, some of the biphenyls such as polychlorobiphenyls (PCB's) are important environmental pollutants as a result of their practical uses. Their general stability to air and environmental conditions, similarly to DDT, is the reason for extremely slow degradation process to nontoxic substances, and will be matter of ecological and health debates for years. The biaryl structure is the basis of several successful chiral separations by crown ethers, inclusion complexes or by preparative chromatography on the chiral stationary phases. Chiral binaphthols and related chiral auxiliaries are widely employed chemicals in some impressive industrial processes. The simplest and still the most important chiral biaryl molecule is 1,1'-binaphthyl-2,2'-diol (**4**) which is used as a chiral ligand or starting material in the production of several valuable catalysts for certain enantioselective reactions. A few heterobiaryls are important ligands in the coordinative chemistry and in certain useful catalysts. For instance, 2,2'-bipyridine (**5**) is a widely employed ligand for nickel(II) salts acting as the oxidation catalysts, e.g. toluenes to benzoic acids with sodium hypochlorite as cheap terminal oxidant, or as aryl halide homo-coupling reaction catalyst, see Chapter 3. 2,2':6',2''-Terpyridine (**6**) is a versatile tridentate ligand which forms several important metal complexes, e.g. Ru(IV), acting as the oxidation catalysts, bleaching agents, oxygen-binding molecules, as well as having miscellaneous interesting applications [5].

4

5

6

Recently, in the medicinal chemistry, biphenyl motif became an important "spacer" within the structure of several modern antihypertenzive agents, so-called sartanes. For example, losartan (**7**) is one of the most prescribed block-buster drugs of today.

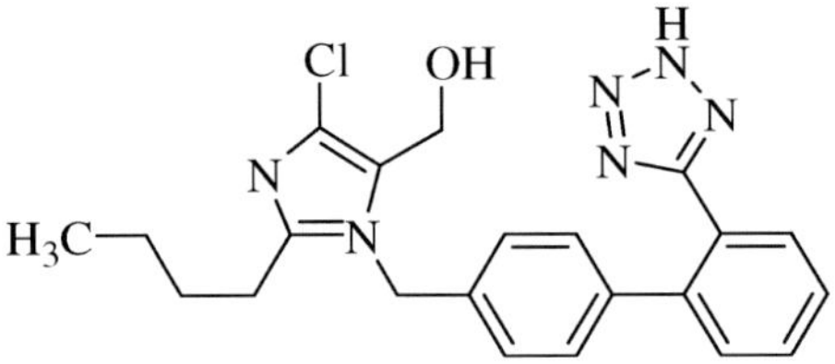

7

The aryl-aryl bond formation is the most important reaction in the synthesis of various polyaryl materials which possess valuable conducting properties. The polyaryls such as poly-*p*-phenylene (**PPP**), which by various doping procedures reach a conductive state, are in focus as a potential for use as electrode materials in light-weight rechargeable batteries, electrochemical cells, semi-conductor devices, solar cells, and in the several other possible electrochemical uses. Very comprehensive polyarene-chemistry and its practical applications are excellently rewieved by Takakazu Yamamoto [6].

1.3. Structure of biaryls. Atropisomerism

The structure formula of biphenyl (**8**) is usually written as two phenyl-rings connected by single C-C bond as a planar structure. However, a rotation about the aryl-aryl bond is normal, at room temperature this is a readily occuring process involving a planar and axial conformer (as well as all intermediate conformers). The first conformer is coplanar, all twelve carbon atoms lay in a single plane, whereas in the latter two planes which contain two phenyl-rings are axial, mutually under the angle of 90° (interplanar angle). Observing this two planes from the left- (or right-) side, a cross can be seen, Figure 1.

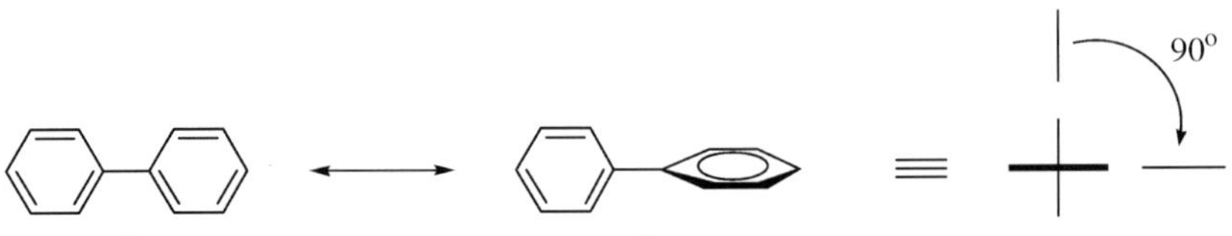

Fig. 1

The interplanar angle of the biphenyl itself in the ground state is actually 44°, however, in the crystalline state the rings are coplanar, presumably because of the packing forces. Whereas the planar conformation offers the maximum stabilization of the structure by π-electron overlap, the axial is imposed by the steric demands of the *ortho*-substituents. Practically, the UV spectra of the desired *ortho*-substituted biaryl is a simple and useful way of judging the content of axial conformer at the given temperature, since the strong conjugation band disappears upon more extensive *ortho*-substitution, or with more bulky *ortho*-groups at the same range of *ortho*-substitution. Thus in planar conformation, the π-orbital overlap is maximal and, as a consequence, the strong conjugation band in the 240-250 nm region appears, while in the opposite case, the complete absence of resonance stabilization occurs at an interplanar angle of 90° the above mentioned band becomes weaker or completely disappears.

Moreover, the biaryls substituted in the *ortho*-positions which are too hindered to rotate, under the standard conditions, mainly tetra-*ortho*-, but also tri- and, in some cases of extreme encumbered *ortho*-substituents, even di-substituted biaryl, can be principally resolved into the two isomeric forms named atropisomers [7].

The optical activity of the tri- and tetra-*ortho* substituted biaryls (atropisomers) is a consequence of the chiral axis, present in this structure. The configuration (*R*- or *S*-) of these compounds is matched according to the Cahn-Ingold-Prelog (CIP) system and an additional sequence rule: Near groups precede far groups. An example with 2-chloro-6-methoxy-2'-nitro-6'-carboxy-1,1'-biphenyl (**9**) is illustrative, Figure 2:

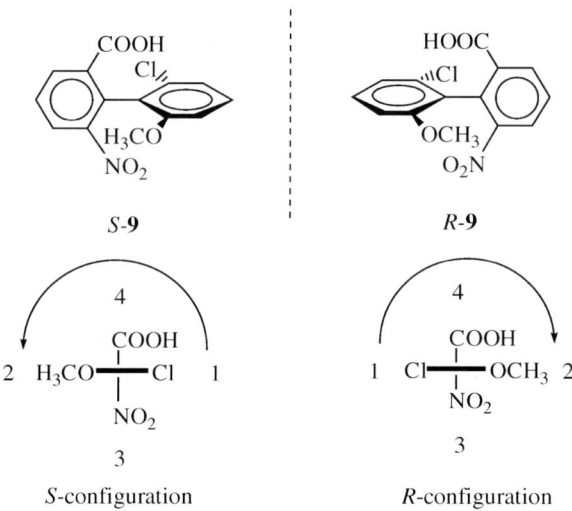

S-configuration   *R*-configuration

substituent priority: 1 > 2 >> 3 > 4

Fig. 2

However, the direct assignment of the absolute configuration at the biaryl axis is achieved by the measurement of a circular dichroism (CD) and the X-ray determination of crystal structure [7].

During more than a hundered years since the first biphenyls were obtained by synthetic methods, several excellent rewievs on this topic have appeared. Many of them were used during the preparation of this book and are cited throughout discussion in the following Chapters. Herein, two valuable rewievs are noteworthy [8,9]. First, published in 2002 by Lemaire's group [8], is the most detailed and extremely useful overwiev of all relevant aryl-aryl bond forming reactions. The second one, also excellently presented, focused mainly on the natural products synthetic problems and chiral biaryls, was written by Bringmann and his coworkers [9].

1.4. References

1. K. C. Nicolaou, H. Li, C. N. C. Boddy, J. M. Ramanjulu, T.-Y. Yue, S. Natarajan, X.-J. Chu, S. Bräse and F. Rübsam, Chem. Eur. J. 5 (1999) 2584.
2. K. C. Nicolaou, C. N. C. Boddy, S. Bräse and N. Winssinger, Angew. Chem., Int. Ed. Engl. 38 (1999) 2097.
3. M. C. Venuti, J. Org. Chem. 46 (1981) 3124.
4. S. M. Kupchan, R. W. Britton, M. F. Ziegler, C. J. Gilmore, R. J. Restivo and R. F. Bryan, J. Am. Chem. Soc. 95 (1973) 1335.
5. R.-A. Fallahpour, Synthesis (2003) 155.
6. T. Yamamoto, Synlett (2003) 425.
7. E. L. Eliel, S. H. Wilen and L. N. Mander, Stereochemistry of Organic Compounds, John Wiley & Sons, Inc., New York, 1994.
8. J. Hassan, M. Sévignon, C. Gozzi, E. Schulz and M. Lemaire, Chem. Rev. 102 (2002) 1359.
9. G. Bringmann, R. Walter and R. Weirich, Angew. Chem., Int. Ed. Engl. 29 (1990) 977.

CHAPTER
2

2. CLASSICAL METHODS FOR SYNTHESIS OF BIARYLS

2.1. The Ullmann and Related reactions

The formation of a biaryl by the condensation of two molecules of an aryl halide in the presence of finely divided copper is known as the Ullmann reaction [1-4]. Although several newer methods for synthesis of biaryls were developed, the Ullmann reaction still remained its importance in some specific cases. Due to its efficiency and simplicity, the reaction is very useful in the synthesis of simple symmetrical and some classes of unsymmetrical biaryls. More than a century old, the classical Ullmann reaction involves heating the mixture of the aryl halide(s) (**I**) with the copper powder or bronze at 100-360 °C in or without any solvent. Products of this reaction are the biaryl(s) (**II**) and the corresponding copper(I) halide, Scheme 1.

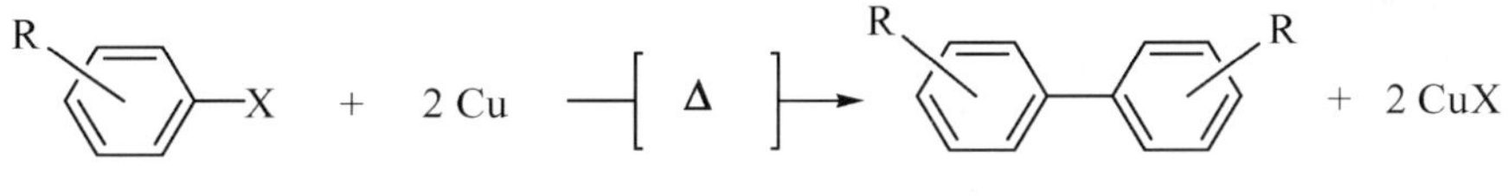

I   X = I, Br, Cl   **II**

Scheme 1

The yields of symmetrical biaryls in the Ullmann reaction may vary from approximately a few to more than 90%, depending on the reactant structure and reaction conditions [1-6]. However, in many examples the yields of biaryls were moderate to high. Theoretically, the Ullmann reaction requires two equivalents of the copper metal per one equivalent of the biaryl formed, thus two equivalents of copper(I) halide are generated. Practically, several equivalents, three to tenfold excess, of the copper metal are used in order to reach high conversion of the aryl halides [3,4,5]. The reactivity of the aryl halide with the copper metal in homo-coupling reactions is traditionally measured by the temperature required to initiate the reaction or by the yield of biaryl obtained. This two parameters greatly depend on the structure of the aryl halide. Some aryl halides, e.g. 2-nitro-iodobenzene, react very vigorously with the copper bronze [7]. This reaction is usually performed by adding the copper bronze or activated powder to the aryl halide in such rate which allows control of the exothermic reaction. In general, the order of reactivity of aryl halides is iodides > bromides > chlorides while in no instance is the fluorine atom eliminated [4-6]. Closely related substrates are aryl halogenoids, e.g. thiocyanates, and diaryl disulfides [4]. The Ullmann reactions have been performed without any solvent, in the presence of sand, or in boiling nitrobenzene, toluene, xylene, cymene, cumene, pyridine, but DMF proved to be the most efficient reaction solvent [4-6]. DMF permits the use of somewhat lower reaction temperatures, a lower proportion of copper and in many instances the biaryls were obtained in higher yields [6]. The reactivity of the selected aryl halides under the classical Ullmann's conditions are given in the Table 1.

Table 1. Comparative reactivities of aryl halides at 190-195 ^{o}C / 1.5 h [9], Scheme 1.

Aryl halide	Biaryl, yield (%)
Iodobenzene	30
2-Methoxy-iodobenzene	70
3-Methoxy-iodobenzene	65
4-Methoxy-iodobenzene	70
2-Nitro-chlorobenzene	50
2-Nitro-bromobenzene	75
3-Nitro-bromobenzene	2.5
4-Nitro-bromobenzene	2.5
4-Nitro-iodobenzene	25
Methyl 2-bromobenzoate	80
2-Cyano-bromobenzene	2.5

The substituents in the phenyl (aryl) ring have a remarkable influence on the reactivity of the aryl halide in the Ullmann coupling reaction [4,5,9,10]. They can be divided into three groups:

a) activating substituents: NO_2, CN, $COOCH_3$, etc in the *ortho*-position to halogen atom,

b) inhibiting substituents: free COOH, SO_3H, OH, NH_2, NHCOR groups,

c) bulky groups in *ortho*-position(s).

The activating effect in the Ullmann coupling reaction of aryl halides provides the electronegative groups such as nitro, alkoxycarbonyl or cyano in the *ortho*-position to the halogen atom. These activating substituents present in the *meta*- or *para*-positions do not significantly activate the aryl halide. For example, 2-nitro-bromobenzene reacts with the copper bronze more rapidly and at a lower temperature, while the

3- or 4-nitro-bromobenzenes are almost equally reactive as bromobenzene [9]. Electropositive groups such as methoxy or methyl in the *ortho-* and *para*-positions do not deactivate the aryl halide to the Ullmann coupling reaction as one could expect. However, substituents like phenols, amines and free-carboxylic acid functional groups strongly inhibit or prevent formation of biaryls by providing alternative reaction pathways [11-13]. Thus amines and phenols are arylated at *N-* or *O*-atoms affording triarylamines [11] and diarylethers [12,13]. Presence of the free-carboxylic acids causes either reduction of the parent halide to arene, or generation of the corresponding aryl esters [4,5]. Phenols and anilines have to be protected through the suitable ester or imide derivatives. Amides react with aryl halides under the Ullmann reaction conditions to give *N*-arylated products, this is known as the Goldberg reaction. Esterification is a useful method for protection of carboxylic and sulfonic acids [4,5]. Bulky groups in *ortho*-positions to halogen atom apparently influence the reactivity of the aryl halides. For example, 2,4,6-trimethylphenyl iodide gives a very poor yield of corresponding biaryl product, 2,4,6,2',4',6'-hexamethylbiphenyl [4].

The Ullmann reaction was used in the coupling of a wide variety of aryl halides, Ar-X, where the Ar includes: mono-, di-, poly-substituted phenyl [4,5,14], naphthalene [4,5,15,16], azulene [17], pyridine [18], pyrimidine [19], thiophene [4,5,20], carbazole [21], and even ferrocene [22], while X = I, Br, Cl. The reaction was successfully used in the intramolecular cyclization reactions affording four- [23], five- [24], six-membered [25], and some other larger rings [6]. Selected examples where halides **10-13** were converted to biaryls **14-17** are given in the Scheme 2 [14,16,17,19].

Since the first step of the Ullmann reaction obviously includes the heterogeneous reaction of the aryl halide at the metal surface, the form of copper used is crucial for yield and reaction temperature required for successful coupling [4,5,19,26]. In a number of cases the purchased copper bronze or precipitated copper powder ("for Ullmann synthesis") have been the most commonly employed forms of the metal. In some instances, the activation of copper bronze by treatment with the iodine in acetone, followed by washing with hydrochloric acid, acetone and drying in a desiccator have given much more reactive copper [4]. In most cases where the Ullmann coupling reactions were carried out at very high temperatures, the form of copper metal seems to be less important. However, the copper forms became important where the goal is coupling of less reactive halides or lowering of the reaction temperature. This latter parameter is actually the most important, as the Ullmann reaction usually requires rather harsh reaction conditions. The use of activated forms of copper metal permits the coupling reaction of sensitive substrates at much lower temperatures with higher yields. Older activation methods include treatment of copper powder with the complexing agents such as THF solution of biquinolyl, aqueous solutions of ethylenediamine, ammonia or disodium EDTA [26]. Performing the

Ullmann coupling reaction of 4-iodotoluene with the Na$_2$EDTA-pretreated copper powder afforded the expected 4,4'-dimetylbiphenyl with a three times faster reaction.

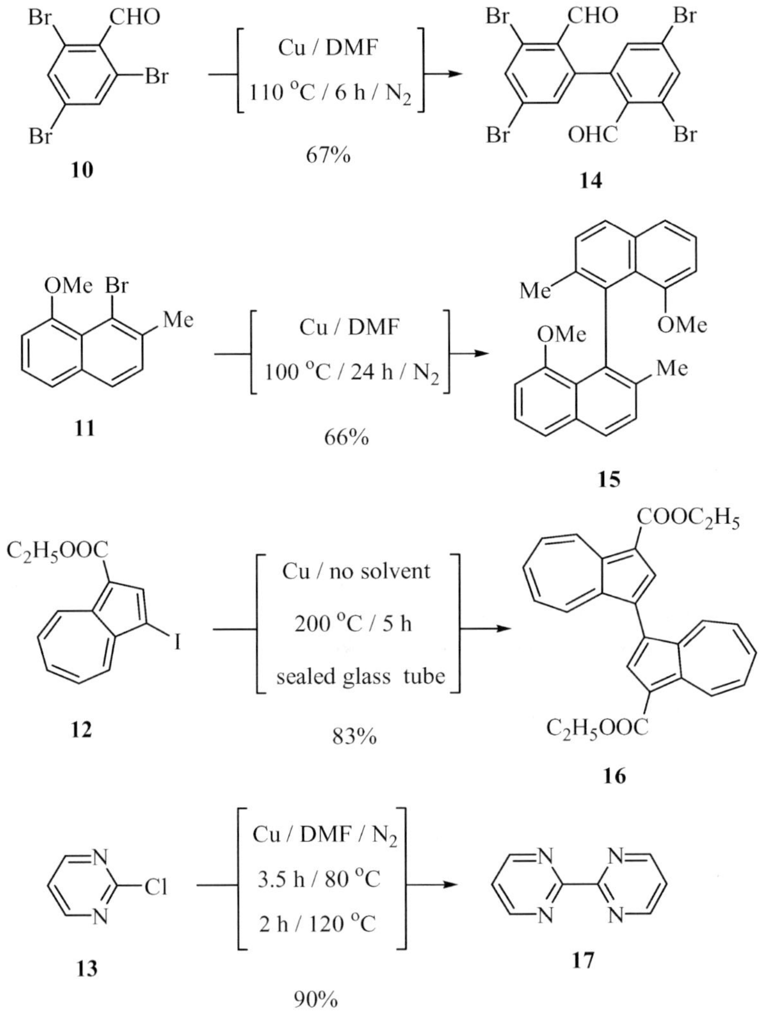

Scheme 2

Moreover, the copper metal obtained by reduction of CuCl, CuI(PEt$_3$) or Cu(SMe$_2$)Cl with the lithium naphthalide, known as Rieke copper, is able to couple a range of aryl halides to biaryls at room temperature in very high yields [27]. For example, simple halobenzenes react with Rieke copper at room temperature to form phenylcopper(I) reagents which, upon refluxing in DME at 80 °C for 24 h, gave 66% of biphenyl. The

2-nitro-iodobenzene (**18**) *via* 2-nitrophenylcopper (**19**) gave the 2,2'-dinitrobiphenyl (**20**) with a 87% yield [27], Scheme 3.

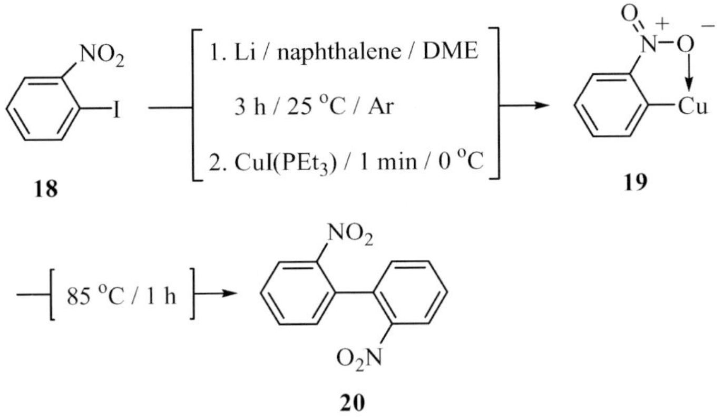

Scheme 3

The homo-coupling Ullmann reaction can be performed at room temperature or bellow by metallation of suitable substrates such as relatively acidic arenes or aryl halides with the *n*-BuLi (or other strong bases) followed by addition of Cu(II) salts [28]. Aryllithium reagent undergoes transmetallation with the Cu(II) salts to form diarylcopper(II) which is prone to rapid reductive elimination of the biaryl with extrusion of a copper mirror. Benincori and coworkers [28] described an efficient procedure for the homo-coupling reaction of bromide **21** *via* diarylcopper(II) **22** under very mild reaction conditions, where the 3,3'-bibenzo[*b*]thiophene (**23**) was obtained in 56% yield, Scheme 4.

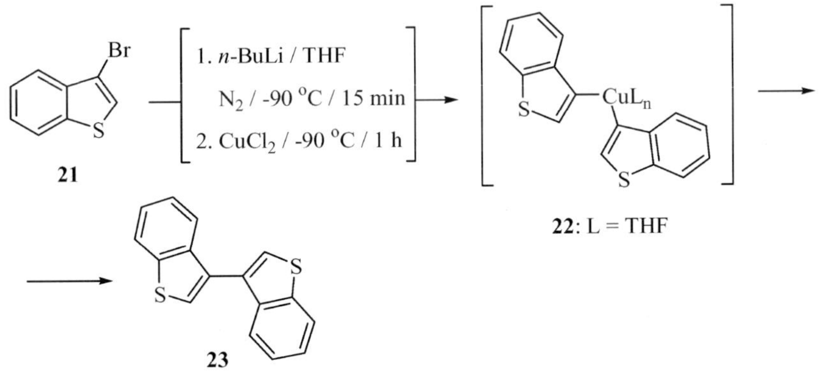

22: L = THF

Scheme 4

Similar variants of Ullmann-type reactions based on intra- as well as intermolecular homo-couplings of various arylmetallic reagents, beside aryllithiums, also aryl Grignard reagents, arylzincs etc. with copper(II) salts give intermediate diarylcopper(II) reagents which subsequently produce respective biaryls, usually in good yields, see Chapter 7.

The mechanism of the Ullmann reaction involves the oxidative addition of the aryl halide to the copper metal with formation of arylcopper(I) compounds [27,29-31]. These organometallics are relatively stable and isolable compounds which are stabilized by forming highly aggregated clusters in non-coordinating solvents [32-35]. The oxidative addition of aryl halide to copper metal is favoured by presence of electronegative substituents in the *ortho*-position to halogen atom. Formation of the intramolecular chelate complex additionally stabilizes arylcopper(I) [35]. Relatively stable arylcopper(I) reacts subsequently with the second molecule of the aryl halide to form species related to diarylcopper(III) halide. This undergoes rapid reductive elimination of the biaryl molecule accompanied with generation of copper(I) halide [27,29], Scheme 5.

$$Ar\!-\!X \; + \; 2\,Cu \; \rightleftharpoons \; Ar\!-\!Cu^{I} \; + \; CuX$$

$$Ar\!-\!Cu^{I} \; + \; Ar\!-\!X \; \longrightarrow \; Ar_2Cu^{III}X$$

$$Ar_2Cu^{III}X \; \longrightarrow \; CuX \; + \; Ar\!-\!Ar$$

Scheme 5

Another possible mechanism is thermal decomposition of arylcopper(I) to biaryl and copper metal, which also has a strong experimental background [32], Scheme 6.

$$Ar\!-\!X \; + \; 2\,Cu \; \rightleftharpoons \; Ar\!-\!Cu^{I} \; + \; CuX$$

$$2\,Ar\!-\!Cu^{I} \; \longrightarrow \; Ar\!-\!Ar \; + \; 2\,Cu$$

Scheme 6

For the highly reactive copper metal forms, e.g. Rieke copper, oxidative addition to aryl halide occurs readily at room temperature or below [27]. At these temperatures oxidative addition of the resultant arylcopper(I) compound to the second molecule of aryl halide, or decomposition to biaryl proceeds very slowly or not at all. The arylcopper(I) compounds can then be homo-coupled or react with the aryl halides to form biaryl as the temperature is raised. In the classical Ullmann reaction employing less reactive forms of copper metal, e.g. bronze, the initial oxidative addition does not take place until the temperature is suitably elevated. At these higher temperatures, either thermal decomposition of the arylcopper(I) compound, or reaction with second molecule of aryl halide occurs to form biaryls at a significant rate, often with no detectable traces of any organometallic intermediate. Each of these reaction steps has its own mechanism [36]. The first step is oxidative addition of aryl halide to metallic copper with formation of an arylcopper(I) compound. This probably includes some type of free-radical mechanism at metal surface, similar to the formation of Grignard reagents at the magnesium surface [4,5,36]. However, other phases in the overall Ullmann and related copper-catalysed reactions do not involve free-aryl radicals. For example, Gomberg-Bachmann-Hey arylation of methyl benzoate with phenyl radicals *in situ* generated from benzenediazonium tetrafluoroborate results with formation of all three isomeric methyl 2-, 3- and 4-biphenylcarboxylates. In contrast, during the Ullmann coupling reaction performed in methyl benzoate as the reaction solvent, no such arylation products were found under classical conditions, indicating the absence of the free-aryl radicals [4,5]. Moreover, arylcopper(I) compounds were evidently detected in the Ullmann reactions [29]. These organometallics react either with the parent aryl halides or by thermal decomposition to give symmetrical biaryls. The aggregation degree of arylcopper(I) apparently plays an important role in the cross-coupling reaction with aryl halides. While reactions in less polar solvents such as ethers or aryl halides give complicated reaction mixtures, strongly coordinating solvents like pyridine or quinoline lead to selective cross-coupling reactions [33,34]. The chemistry of the arylcopper(I) compounds is relatively well known, on the other hand the further process including reductive elimination is far more complicated. Further highlights on this field were introduced by van Koten's group, as presented in the Chapter 3.

From this mechanistic point of view it appears there is a possibility for synthesis of the unsymmetrical biaryls by coupling two different aryl halides. The synthesis of the unsymmetrical biaryl must proceed through two stages: generation of arylcopper(I) compound derived from the first aryl halide followed by cross-coupling reaction with second aryl halide. This can be performed by three different methods:

    a) the classical Ullmann reaction under controlled conditions [4-6],

    b) the Ziegler's method for synthesis of unsymmetrical biaryls [35], and

    c) the Nilsson reaction employing 1,3-dinitrobenzenes [37,38]

The Ullmann reaction can be applied in the synthesis of unsymmetrical biaryls when one aryl halide is relatively reactive (reactant **A**) and the other is relatively unreactive (reactant **B**) [4-6]. Reactive aryl halides type **A** have electronegative groups, e.g. NO_2, $COOCH_3$ or CN, *ortho* to the halogen. The use of bromides and chlorides favours the unsymmetrical coupling product **AB**, while iodides generally give more symmetrical biaryls **AA** (homo-coupling product). Unreactive aryl halides type **B**, lack electronegative groups in the *ortho*-position and are usually iodides or bromides, Scheme 7. When the unsymmetrical Ullmann coupling reaction is attempted with two unreactive (or two reactive) aryl halides, three biaryls **AA**, **AB** and **BB** are usually produced in approximately equal amounts. The optimization of unsymmetrical biaryl coupling between two properly selected aryl halides includes determination of initial reaction temperature for homo-coupling reaction of each reactant under the same reaction conditions.

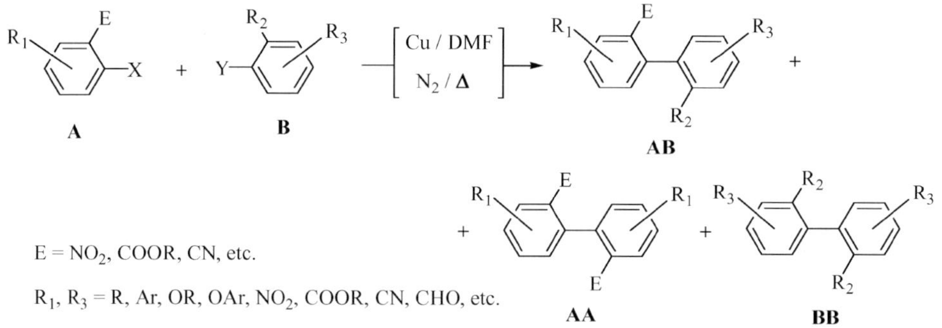

$E = NO_2$, COOR, CN, etc.

R_1, R_3 = R, Ar, OR, OAr, NO_2, COOR, CN, CHO, etc.

R_2 = R, Ar, OR etc, but not NO_2, COOR, etc.

X = I, Br, Cl;   Y = I, Br

Scheme 7

The successful unsymmetrical coupling could be carried out if the difference of the reaction temperatures for homo-coupling reactions is more than approximately 20 °C. Then, the unsymmetrical coupling reaction has to be performed by slow addition of more reactive aryl halide to the solution of less reactive aryl halide and copper powder at the temperature slightly higher than is the homo-coupling reaction temperature for the more reactive aryl halide. Under these reaction conditions more reactive aryl halide forms the arylcopper(I) compound in the fast process, which is subsequently cross-coupled with the less reactive aryl halide present in excess. This conditions suppress the homo-coupling of both aryl halides by:

   a) keeping the reaction temperature below the homo-coupling reaction of less reactive aryl halide, and

   b) slow addition of more reactive aryl halide, providing a low concentration of its arylcopper(I) compound.

The Ullmann synthesis of unsymmetrical biaryls has been employed in a number of cases with remarkable selectivity if the aryl halides were properly selected [2-6,37-41]. Some representative synthesis in the phenyl series are shown in the Table 2.

Table 2. Ullmann synthesis of unsymmetrical biaryls [5]

R_1	R_2	R_3	R_4	X	Y	Temp.($^{\circ}$C)	Yield (%)
NO_2	H	H	H	Br	I	190	60
NO_2	H	Br	H	Br	I	195	55
NO_2	NO_2	H	H	Cl	I	200	42
NO_2	NO_2	H	H	Br	I	175	60
CO_2Me	H	H	H	Br	I	190	45

Further illustrative examples of the unsymmetrical Ullmann coupling reactions where more reactive halides **24-26** were cross-coupled with less reactive halides **27-29** to give the respective biaryls **30-32** are given in the Scheme 8.

An important side-reaction in the unsymmetrical Ullmann reactions is the halogen-exchange reaction [42-43]. Since the copper(I) halide was liberated from the reaction of more reactive aryl halide with the copper metal, it undergoes the halogen-exchange reaction with a less reactive aryl halide. The product of this reaction is an aryl halide which is far less reactive than the parent aryl halide. For example, in the cross-coupling Ullmann reaction of 2,4,6-trinitro-chlorobenzene (type **A**) with 2-bromo toluene (type **B**), the less reactive aryl halide **B** is partially converted to the much less reactive 2-chlorotoluene due to copper(I)-catalysed halogen-exchange reaction between the CuCl (formed from **A** and Cu) and the reactant **B**. To minimize this side-reaction, less reactive aryl halide is usually used in some excess, especially if it is inexpensive and reusable.

The Ziegler's method involves the preparation of the arylcopper(I) compound from the aryl bromides or iodides followed by cross-coupling reaction with second aryl iodide molecule [35]. Both aryl halides bear the ligand atoms, nitrogen or sulphur in the *ortho*-benzylic position, in order to stabilize the transient diarylcopper(III) halide *via* five-membered cyclo-copper complex. *Ortho*-substituents can be *N,N*-dimethylamino methyl, carboxylic esters or acids in an oxazoline masked form, or an aldehyde group in the form of cyclohexylimine, 1,3-dithiane or 1,3-oxathiolane. Arylcopper(I) reagents were prepared either by metallation of corresponding arenes, aryl bromides or iodides with *n*-BuLi. For example, metallation of iodide **33** with *n*-BuLi gave the organolithium reagent which was converted to the organocopper(I) compound **34** with THF-soluble CuI(EtO)$_3$P. The organocopper(I) reagent **34** was treated with iodide **35** to form the biaryl **36** which by deprotection has produced the highly substituted biaryl dialdehyde **37** with a 58% overall yield [35], respectively, Scheme 9.

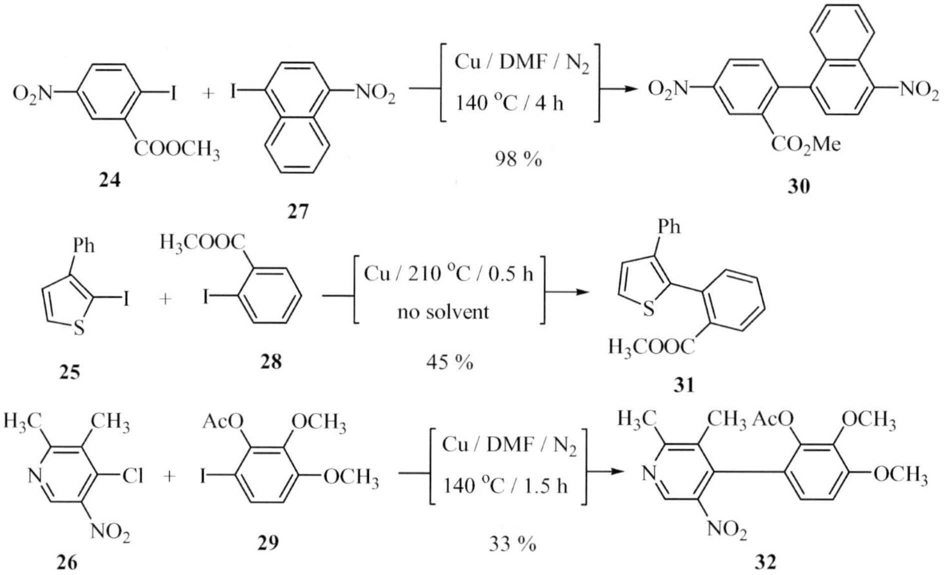

Scheme 8

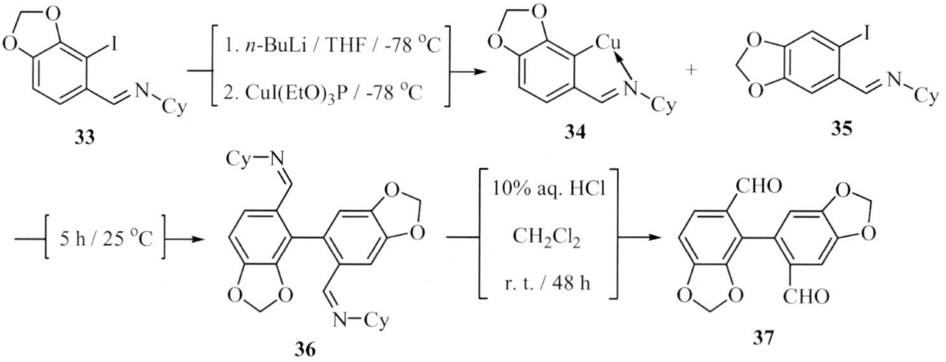

Scheme 9

In some instances, the unsymmetrical coupling reaction of an arylcopper(I) reagent with the aryl halide can be carried out without the *ortho*-coordinating substituents. It has been shown that 2-thienylcopper(I) and 2-furylcopper(I) readily couple with iodobenzene or 4-iodoanisole to give 2-phenylthiophene (50%) and 2-(4-methoxy phenyl)furan (31%), but only in coordinating solvents like quinoline or pyridine, see Chapter 4 [34]. The Nilsson reaction [37] is an additional Ullmann-type reaction which includes the condensation of 1,3-dinitrobenzene(s) with aryl halides with formation of 2,6-dinitrobiphenyls. In this reaction, the arylcopper(I) compound is generated by copper(I) oxide metallation of carbanion formed from the relatively acidic 1,3-dinitrobenzene under an influence of refluxing quinoline. Thus generated 2,6-dinitrophenylcopper(I) reagents readily collapse with aryl iodides under the reaction conditions to afford unsymmetrical biaryl in fair yields. When 1,3-dinitro benzene (**38**) was coupled with 4-iodoanisole (**39**) in refluxing quinoline, 4'-methoxy-2,6-dinitrobiphenyl (**40**) was isolated in 69% yield [37], Scheme 10.

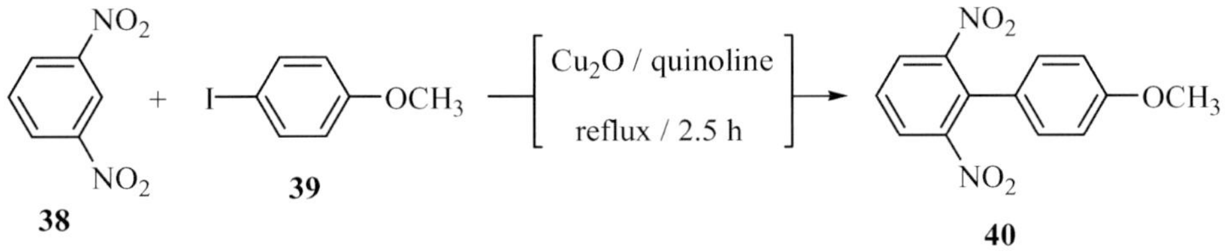

Scheme 10

The reaction of 1,3-dinitrobenzenes with aryl iodides can be accomplished by using copper(I) *tert*-butoxide *in situ* prepared from CuCl and potassium *tert*-butoxide in dry DME. In this case, the reaction can be realized under essentially milder reaction conditions, at 70-110 °C in the presence of pyridine as the base with respective yields. For instance, the biaryl **40** was produced from compounds **38** and **39** after 20 h at 67 °C in DME / pyridine as the reaction solvent in 95% yield, respectively [38]. Dehalogenation products were isolated from many Ullmann reactions performed with the (relatively acidic) nitro-aryl halides, even when exclusion of water or acidic (protic) substances was provided. Furthermore, nitro-aryl halides are partially converted to triarylamines by reduction and subsequent exhaustive *N*-arylation [44,45]. Finnaly, it can be concluded that the Ullmann and related reactions are still valuable tools in the preparative organic chemistry providing a simple and usually efficient approach to simple symmetrical and certain classes of unsymmetrical biaryls.

2.2. The Gomberg-Bachmann-Hey reaction

The arylation of aromatic compounds with aryldiazonium salts (**III**) in the presence of base affording biaryls (**II**) is called the Gomberg-Bachmann-Hey (GBH) reaction [46]. First observations were published by Bamberger [47] and Kühling [48] more than a century ago. However, the first practical synthesis of biaryls was discovered by the former authors, and was further developed by Hey. The original

GBH reaction involves the diazotation of an aromatic amine (**IV**) with sodium nitrite and strong acids, followed by reaction with a liquid aromatic compound to be arylated in the two-phase mixture by adding the 15-40% aqueous sodium hydroxide solution with formation of biaryls (**II**) in low to moderate yields [46,49,50], Scheme 11.

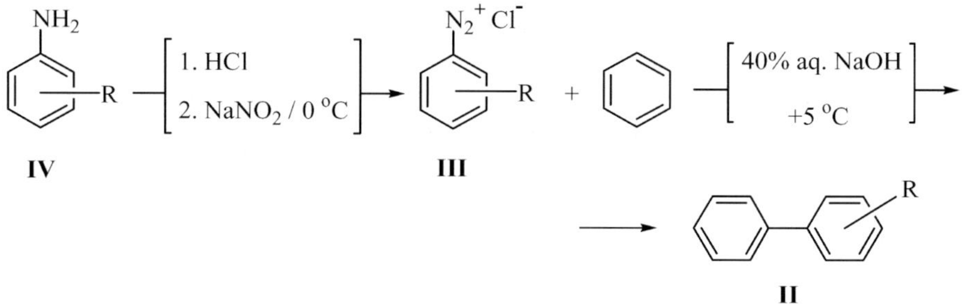

Scheme 11

The base reacts with the aryldiazonium salt to form the covalent aryldiazonium hydroxide which is prone to homolytic decomposition accompanied with generation of the aryl radicals [49-51]. This arylates the liquid aromatic compound (which also serves as solvent) to give the biaryl. Since the radical does not rearrange, a single biaryl is obtained from the symmetric arene, while unsymmetrical ones give a mixture containing all three isomeric biaryls. 4-Chlorodiazonium chloride (**41**) arylates the benzene under the typical GBH conditions to give 40% of the 4-chlorobiphenyl (**42**), whereas benzenediazonium chloride (**43**) reacts with chlorobenzene to produce 2-, 3- and 4-chlorobiphenyls (**44,45,42**) [49], Scheme 12.

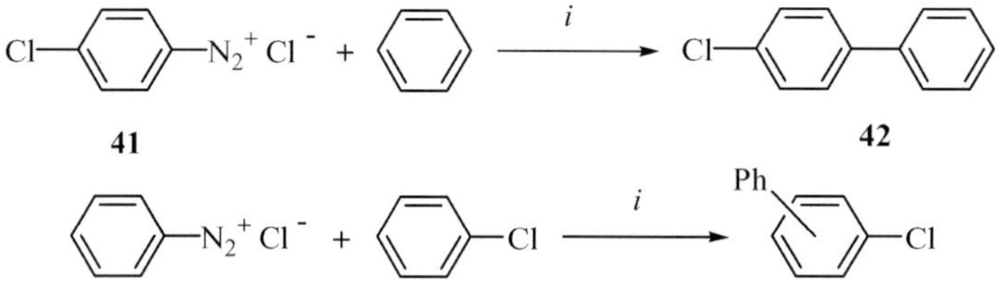

i: 40% aq. NaOH / +5 °C

42: 4-Ph
45: 3-Ph
44: 2-Ph

Scheme 12

When an arenediazonium salt solution was added to a mixture of the liquid arene and sodium hydroxide solution, the excess of base forms the corresponding sodium diazotate, Ar-N=N-O⁻ Na⁺, also an alternative intermediate, which is easily decomposed to form aryl radicals with similar result [50-52]. Several side-reactions occur in the GBH reaction as the reaction conditions are rather unfriendly for highly reactive aryl radicals generated under the influence of the base on the aryldiazonium salts. The common side-reactions are the formation of linear polyaryls, parent arene, and arylazo compounds, which always give deep colours to all GBH reaction mixtures. Polyaryls are obtained by further reaction of the aryl radicals with the biaryl initially produced. The parent arenes are found due to reduction of aryl radicals or other azo-intermediates. The arylazo compounds are formed by the reaction of an aryldiazo-radical rather than by electrophilic aromatic substitution of activated aromatics with the aryldiazonium cation, yet the GBH reactions are carried out under basic conditions. Since the original GBH method rarely gives the biaryls in yields higher than 40-50%, further improvements have been introduced. The use of alternative bases and non-aqueous mediums has brought an apparent increase in yields. The bases such as sodium acetate [53] or pyridine [54-60] react with the arenediazonium salts with formation of different diazo-intermediates which decompose by slower rate to aryl radicals making the overall GBH arylation reaction more chemoselective and clean. These improvements can be realized if the reaction mechanism is clearly introduced. The acetate ion, as the most efficient base, reacts with the aryldiazonium cation to generate the aryldiazonium acetate, ArN=NOAc [53,61]. The latter reacts with an excess of acetate ion to form the corresponding aryldiazotate anion, ArN=N-O$^-$, which is coupled with the excess of the parent diazonium salt to produce an unstable, but isolable, diazo-anhydride, ArN=N-O-N=NAr [52,61]. The diazo-anhydride further decomposes to give the aryldiazotate radicals and aryl-radicals as arylating agent [52,61]. The aryl radicals arylate the arene *via* generation of the corresponding arylcyclohexadienyl radical(s) which is then converted to the biaryl, Scheme 13.

$$Ar-\overset{+}{N}{\equiv}N\ X^- \ + \ CH_3COO^- \ \longrightarrow \ Ar-N{=}N-OOCCH_3 \ + \ X^-$$

$$Ar-N{=}N-OOCCH_3 \ + \ CH_3COO^- \ \rightleftharpoons \ Ar-N{=}N-O^- \ + \ (CH_3CO)_2O$$

$$Ar-N{=}N-O^- \ + \ Ar-\overset{+}{N}{\equiv}N\ X^- \ \rightleftharpoons \ Ar-N{=}N-O-N{=}N-Ar$$

$$Ar-N{=}N-O-N{=}N-Ar \ \longrightarrow \ Ar-N{=}N-O^{\bullet} \ + \ N_2 \ + \ Ar^{\bullet}$$

Scheme 13

Alternative substrates which are capable of generating the aryl radicals are: *N*-nitroso acetanilides [49,62-65], aryltriazenes [66,67], arylazo-triphenylmethanes [68], and arylhydrazines under oxidative conditions. The nitroso-acetanilides (**V**), prepared by reaction of acetanilides with the nitrous fumes or with NOCl [69], are thermally rearranged (presumably *via* **Va**) to the corresponding aryldiazo-acetates (**VI**) which further follow the standard GBH reaction pathway [63,64], Scheme 14.

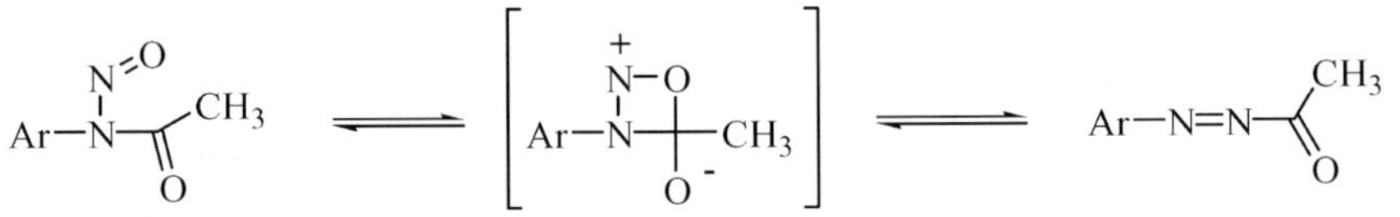

Scheme 14

The aryltriazenes (**VII**), under acidic conditions, are slowly decomposed with generation of aryl radicals [66,67]. The similar reaction pathway follows the GBH reaction accomplished with an equimolar amount of pyridine as the base since it forms the protonated triazenes (**VIIa**) directly [55-60], Scheme 15.

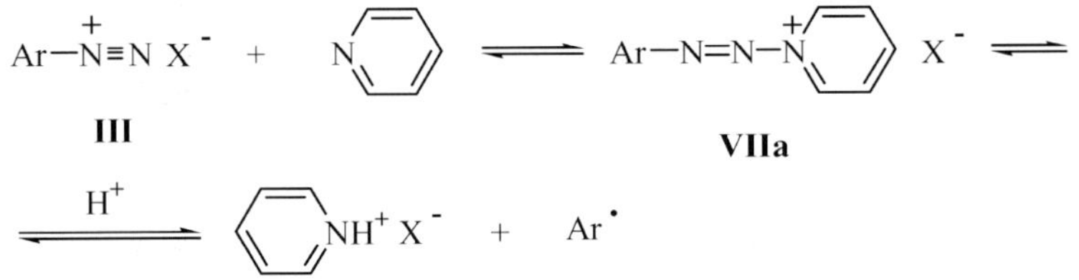

Scheme 15

The use of dry aryldiazonium salts of naphthalene-1-sulfonic [70], naphthalene-1,5-disulfonic [70], $ZnCl_2$ complex [70], hexafluorophosphoric or tetrafluoroboric acid [61] in non-aqueous medium under as mild as possible reaction conditions is substantial to reach higher yields of biaryls. An alternative method for non-aqueous GBH reaction is the aprotic diazotation of aromatic amines with alkyl nitrites such as butyl or pentyl nitrite with subsequent arylation of aromatic compound, as demonstrated by Cadogan [71,72]. This method is realized by simple heating the mixture of aromatic amine, alkyl nitrite and liquid arene at an elevated temperature. When a mixture of 3-aminopyridine (**46**), benzene and pentyl nitrite is heated at reflux, 3-phenylpyridine (**47**) is obtained with a 55% yield [71], Scheme 16.

Today, the GBH reactions are performed by using aryldiazonium tetrafluoroborate salts. An aromatic amine is diazotized at about -10 to 0 °C in the presence of at least 3 equivalents of concentrated hydrochloric acid by adding the solution of sodium nitrite [61]. To the aryldiazonium chloride, 50% aqueous solution of tetrafluoroboric acid is added with additional cooling to precipitate the corresponding aryldiazonium tetrafluoroborate. Alternatively, the diazotation can be carried out in concentrated tetrafluoroboric acid with direct crystallization of the aryldiazonium tetrafluoroborate salt [73,74]. In contrast to chlorides and sulfates which are explosive in dry form, these salts are apparently more stable, and can be safely dried in high vacuum at slightly elevated temperatures, 40 °C, and stored for months at +4 °C without significant decomposition.

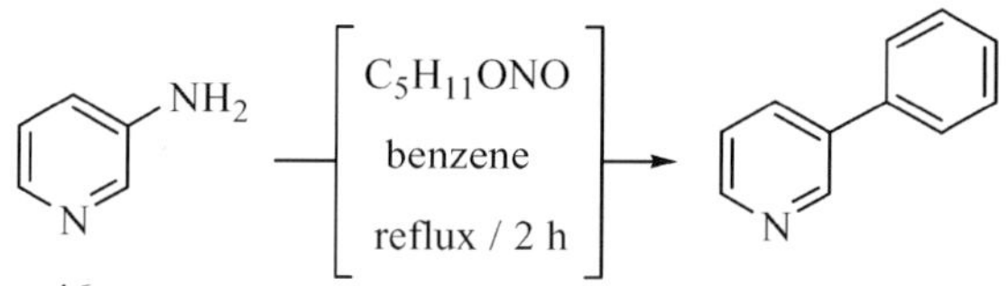

Scheme 16

The modern methods for performing the GBH reaction are based on the thermal decomposition of certain aryltriazenes in the presence of trifluoroacetic acid [66,67], or on the phase transfer catalysed reaction of aryldiazonium tetrafluoroborate in the presence of potassium acetate [61], both methods in an excess of arene as the reactant and reaction solvent. The first method requires previous preparation of the aryltriazene (**VII**) by diazotation of an aromatic amine in hydrochloric acid followed by addition of piperidine [66], other dialkylamine or 5-aminotetrazole [67]. For example, from the diazotized aniline solution **43**, by addition of an equimolar amount of piperidine, the yellow-orange 1-phenyl-3,3-(pentanediyl)triazene (**48**) readily crystallize in 81% overall yield [66], Scheme 17.

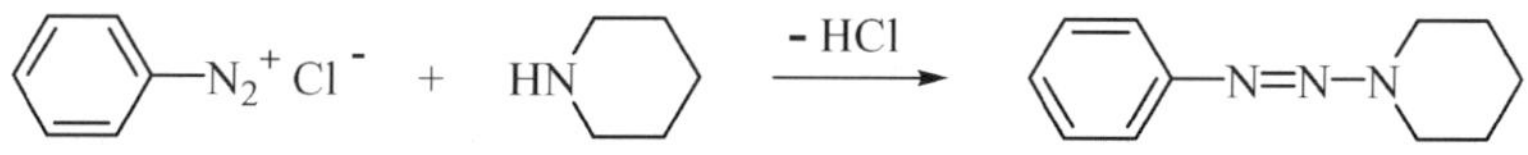

Scheme 17

Patrick and coworkers [66] developed an efficient method for synthesis of biaryls by slow addition of two equivalents of trifluoroacetic acid to the solution of the aryltriazenes (**VII**) in an arene at 65-70 °C. As reaction proceeds, the nitrogen is evolved from the reaction mixture and biaryls (**II**) are formed in fair yields. The efficacy can be judged from the selected examples presented in the Scheme 18. The yields can be further increased for 10-15% by adding a catalytic amount of iodine (5-10 mol%), 5% palladium on charcoal, palladium(II) acetate (5 mol%), or acetonitrile as a cosolvent (ca. 10-25%).

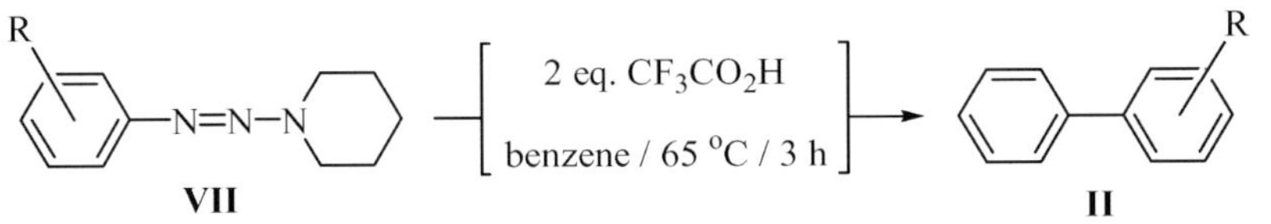

R = H: 40 %   R = 4-Me: 50 %   R = 4-O$_2$N: 66 %

R = 4-Cl: 49 %   R = 4-MeO: 41 %   R = 4-COCH$_3$: 67 %

R = 4-Br: 45 %   R = 4-EtO: 65 %

R = 4-I: 63 %   R = 4-*n*-Bu: 72 %

Scheme 18

The most efficient modern method for performing the GBH reaction is phase transfer catalysed reaction starting from aryldiazonium tetrafluoroborate, introduced by Gokel and his coworkers [61] (ptGBH). This method is accomplished by adding potassium acetate to a well stirred mixture (usually suspension) of aryldiazonium tetrafluoroborate and phase transfer catalyst in an excess of the arene, which serves as the reactant as well as reaction solvent. As the phase transfer catalysts benzyltriethylammonium chloride (5 mol%), (C$_8$H$_{17}$-C$_{10}$H$_{21}$)N(CH$_3$)$_3$Cl (Aliquat 336, 5 mol%), potassium hexanoate or palmitate (200 mol%), or the most efficiently, 18-crown-6 (18-C-6, 5 mol%) have been employed [61,75-77]. Bartsch introduced the polyethylene glycols (e.g. PEG 1000, 30 mol%) as slightly less effective substitutes for toxic and expensive 18-C-6 [78]. However, the addition of a small amount of acetonitrile (10-20% to arene) as a cosolvent is usually as effective as using the crown ethers, and this modification of the GBH reaction is probably the best to date [61,75]. The reactions are usually completed within 1-2 hours at room temperature or below, with evolution of nitrogen, formation of potassium tetrafluoroborate and the biaryls in moderate to good yields. In some ptGBH reactions of aryldiazonium tetrafluoroborates (**III**) with no radical-sensitive substituents, and symmetrical arenes, excellent yields of the (un)symmetrical biaryls were obtained. The yields of biaryls (**II**) produced by the Gokel's method are presented through the following examples of benzene arylation [61], Scheme 19.

As all free-radical arylation reactions, the ptGBH reaction produces a mixture of all three isomers if the arene is unsymmetrical [61,76,79]. The isomer distribution can be seen from the following examples of monosubstituted benzene arylation, Table 3. A number of instances have shown that isomer distribution is indifferent to the catalyst used, this strongly suggests that an aryl radical as an actual arylating agent is formed under either phase transfer or classical GBH conditions.

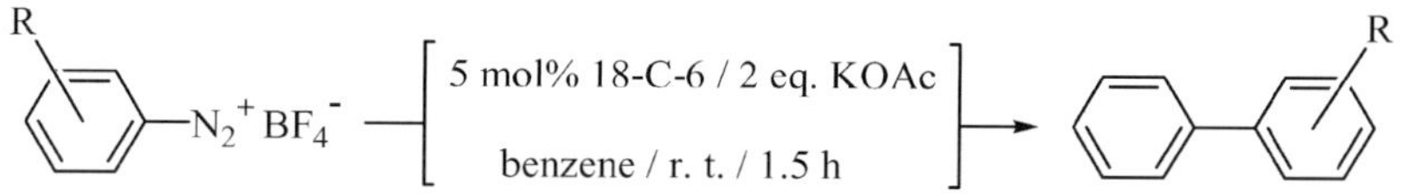

R = H: 62 %   R = 2-Cl: 73 %   R = 2-Br: 81 %   R = 3-Me: 58 %

R = 2-F: 42 %   R = 3-Cl: 79 %   R = 4-Br: 81 %   R = 4-Me: 73 %

R = 3-F: 74 %   R = 4-Cl: 80 %   R = 3,4-Cl$_2$: 53 %   R = 3,4-Me$_2$: 50 %

R = 4-F: 60 %   R = 4-MeO: 80 %   R = 3,5-(MeO)$_2$: 50 %   R = 4-Et: 50 %

Scheme 19

Table 3. Isomer distribution in the ptGBH reaction of monosubstituted benzenes with benzenediazonium tetrafluoroborate [61]

Arene	Total yield (%)	Isomer (%)		
		2-	3-	4-
Fluorobenzene	68	59	31	10
Chlorobenzene	75	85	14	1
Bromobenzene	60	75	25	0
Anisole	55	91	2	7
Benzonitrile	63	60	14	26
Methyl benzoate	29	80	7	13

The GBH arylation reactions of monosubstituted benzene always result in preferential formation of the *ortho*-substituted biaryl. The *ortho*-attack of the aryl radical generates the more stable tertiary (3°) arylcyclohexadienyl radical than at *meta*- or *para*-attack when the transient radicals are secondary (2°) and therefore less stable, Scheme 20.

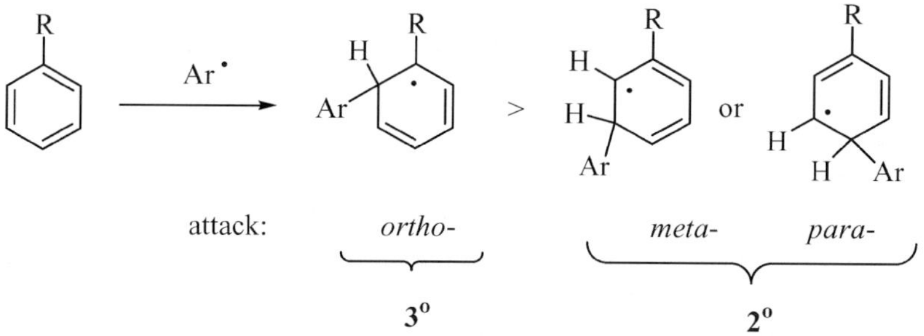

Scheme 20

However, thiophene and furan are arylated selectively at the 2-position as well as symmetrical 1,4-di- and 1,3,5-tri-substituted benzenes giving respective biaryls in good to high yields. Despite to the free-radical nature of the GBH reaction, under certain curcumstances the reaction proceeds, at least partially, by the ionic mechanism [61,80-82]. For instance, 4-ethoxyphenyldiazonium tetrafluoroborate arylates pyridine exclusively at 4-position in dipolar aprotic solvents such as acetonitrile, DMF or DMSO to give 4-(4-ethoxyphenyl)pyridine with 30-48% yields [60,83]. The electron-donating ethoxy-group at *para*-position and polar reaction solvents favour generation of cationic species. In the same conditions 4-nitrophenyldiazonium tetrafluoroborate gives the usual mixture of three isomers, typical for the free-radical GBH reaction.

The major limitation of the all GBH reactions is that the arene has to be in the liquid form to serve as the solvent. All attempts to arylate solid arenes under the GBH reaction conditions were unsuccessful. Furthermore, the presence of radical-sensitive substituents in either diazonium salts or in arene results in lower yields, or the reaction fails completely. Generally, some alkyl, alkoxycarbonyl, formyl, or iodo-substituted arenes or diazonium salts are susceptible for hydrogen- or iodo-abstraction by free-radicals and are not common GBH reactants. For example, 2-tolyldiazonium tetrafluoroborate in the ptGBH reaction with benzene gives indazole in 74% yield, and only 2% of biaryl product [61].

In conclusion, the ptGBH reaction is one of the most efficient methods for synthesis of biaryls where one ring is variously substituted while the second one is phenyl [61], 2-thiophenyl [61,76,84], or 2-furyl [61], Figure 1.

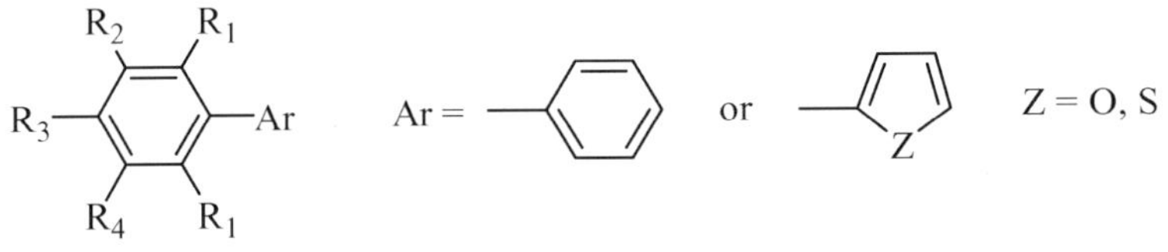

$R_1 \neq$ lower alkyls, CHO

R_{1-4} = F, Cl, Br, NO_2, COOR, CN, Ar, etc.

Fig. 1

This reaction is the most simple method for synthesis of these structures beside several recent methods based on cross-coupling reactions with different organometallic reagents. The starting materials are very inexpensive and easily available, whereas an excess of arene can be regenerated by distillation of the filtrate after the separation of potassium tetrafluoroborate. This can be also reused as a precipitation reagent for preparation of the parent aryldiazonium tetrafluoroborate from the diazonium chloride solutions. In contrast, any method involving organometallics requires relatively more complicated procedures, dry solvents, an inert atmosphere, toxic (nickel-salts) or expensive (Pd-complexes) catalysts. Although the basic problem in the aryl-aryl bond forming reactions, that the organometallic side in any modern cross-coupling reaction can not bring a number of sensitive functional groups, is still at least partially unsolved, the ptGBH reaction plays an important role in the synthesis of some specific biaryl structures.

2.3. The Pschorr cyclization

The intramolecular coupling reaction between an aryldiazonium salt (**III**) with an arene subunit to form five- or six-membered ring by the influence of copper or an acid, is called the Pschorr cyclization reaction [85]. In historical examples the Pschorr cyclization was used in the synthesis of phenanthrene and its derivatives. For example, *cis*-stilbene derivative **49** is diazotized to give **50**, and subsequently treated with the activated copper to obtain phenanthrene (**51**) in 34% yield [86], Scheme 21.

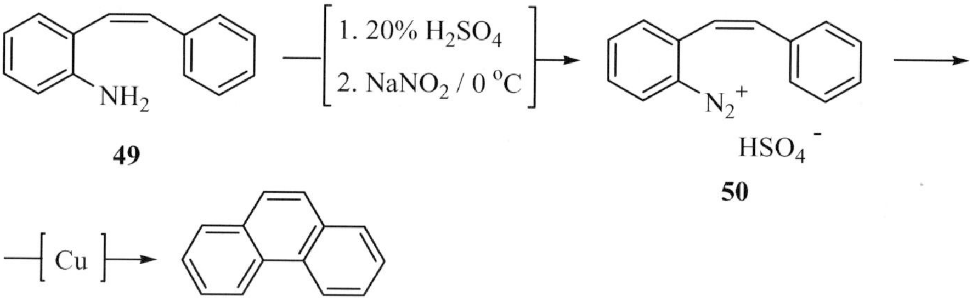

Scheme 21

The copper-induced decomposition of aryldiazonium salt is a well known process where one-electron reduction occurs with the generation of an aryl radical, apparently not free, but held on the copper surface. Generally, the aryldiazonium salt reacts with metals and metalloids such as Hg, Tl, Sn, As, Sb, Ge, Bi in acetone solution to give the corresponding organometall(oid)s [87]. However, when the resulting organometallic reagent is far more reactive, e.g. organocopper(I) reagent, the process finished not in the organometallic stage but in the aryl radicals held on the metal (e.g. copper) surface. Highly reactive arylradical-copper species undergo further intramolecular, and thereby favoured, arylation similarly to the Gomberg-Bachman-Hey reaction pathway. The

Pschorr cyclization reactions have been performed under the following conditions [88-97]:

- a) copper-catalysed decomposition (of Pschorr diazonium salts) in aqueous sulfuric acid what is Pschorr's original method [85,86],
- b) copper-catalysed decomposition in an aqueous alkaline solution [90],
- c) thermal decomposition in an aqueous acid solution in the absence of copper [91],
- d) thermal decomposition of the aryldiazonium tetrafluoroborate in acetone in the presence of copper [92,93], or
- e) the aprotic diazotation with alkyl nitrites in acetone with subsequent reaction with sodium iodide [95].

Only the last method, developed by Rapoport's group [95], is a versatile, chemoselective, and high-yielding. The method is accomplished by adding an alkyl nitrite (2 eq.) to a cold acetone solution of a Pschorr's amine substrate (1 eq.) and sulfuric acid (2 eq.) followed by sodium iodide. Here the iodide serves as a one-electron reductant, which converts an aryldiazonium salt to an aryl radical [95,96]. Then, the fast intramolecular arylation takes place with formation of the phenanthrene structure. While the older Pschorr methods gave very low yields, often under 20%, the Rapoport's method introduced significant improvements as the yields were 45-71% [95]. For example, polymethoxy compound **52** was converted to phenanthrene **53** in 71% yield [95], respectively, Scheme 22.

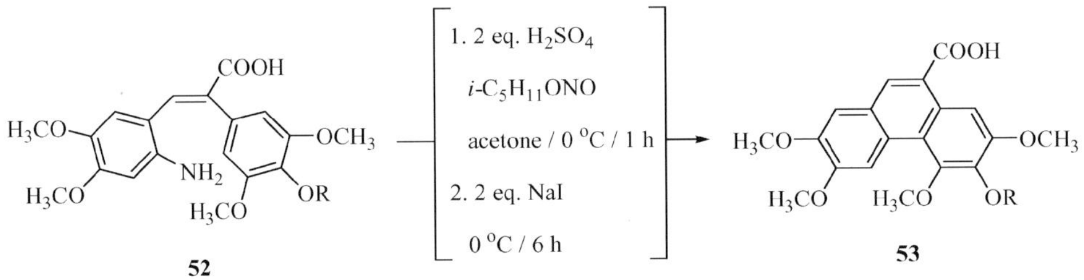

Scheme 22

The only by-product is the 2-iodo-α-arylcinnamic acid as a product of the reaction of the parent diazonium ion with the iodide. This can be also converted to **53** by photoarylation, see Chapter 7.

The Pschorr cyclization reaction was used in the synthesis of a wide variety of phenanthrene and related type of structures [88,89,91-95]. An interesting example is diazotation of **54** to produce **55**, which by intramolecular cyclization gave **56** in 46% yield [92], Scheme 23.

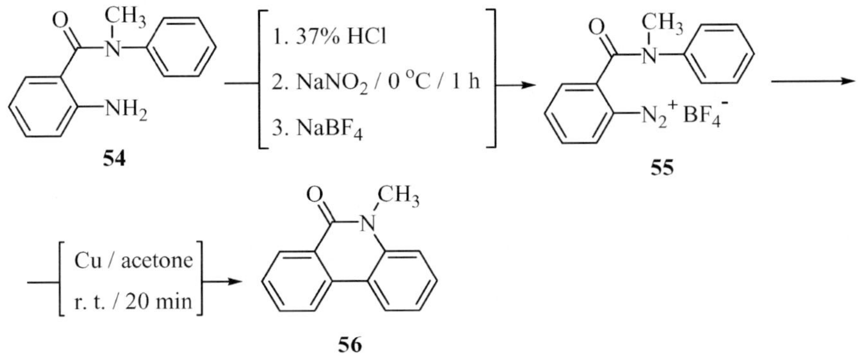

Scheme 23

2.4. The Gatterman synthesis of biaryls

Closely related arylation to Pschorr reaction is the Gatterman synthesis of biaryls (**II**) which involves the coupling reaction of two aryldiazonium salts (**III**) by the influence of copper or copper(I) salts [97,98], Scheme 24.

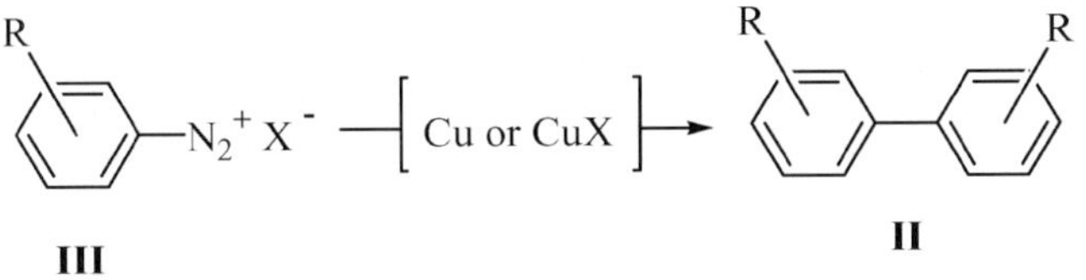

Scheme 24

The reaction apparently involves free-radical mechanism, but arylcopper compounds take a part, at least under certain reaction conditions, as clearly demonstrated through Cohen's concise results [99,100]. The reaction was discovered as a side-reaction during probes of the Gatterman synthesis of aryl halides from diazonium salts and copper(I) halides. Probably the most known example is very practical preparation of diphenic acid (**57**) starting from anthranilic acid (**58**). The reaction is usually conducted by adding an aqueous diazotized anthranilic acid solution (diazonium salt **59**) to the copper(I) reagent, *in situ* obtained by reduction of $CuSO_4$ with an equimolar amount of hydroxylamine in aqueous sodium hydroxide solution, to produce diphenic acid with a 80-90% yield [99,100], Scheme 25.

The yields in the classical variant are rarely as high as that for diphenic acid. However, tetrakis(acetonitrile)copper(I) perchlorate and similar organic solvent soluble copper(I) salts (bearing non-nucleophilic anion) with an excess of $Cu(ClO_4)_2$, are able to couple aryldiazonium tetrafluoroborate salts in acetone solution to furnish the symmetrical biaryls in moderate to high yields [100]. In the absence of Cu(II) salts, a considerable amount of the corresponding azoarene is formed.

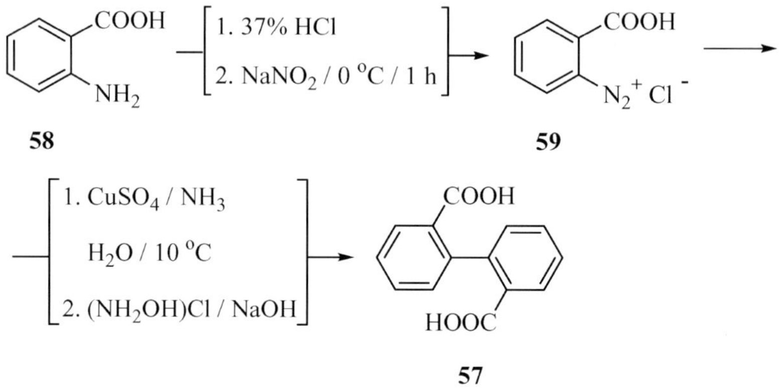

Scheme 25

Cohen has shown that copper(I) salts readily induce the decomposition of the aryldiazonium salt to generate the aryl radical which subsequently reacts with Cu(I) to give an arylcopper(II) species [99,100]. This reacts with another aryl radical to produce the key diarylcopper(III) intermediate. The latter undergoes reductive elimination of the biaryl with regeneration of Cu(I), similar to other copper-based homo-coupling reactions, see Chapter 3, [100] Scheme 26.

$$Ar{-}N_2^+ + Cu^+ \longrightarrow N_2 + Cu^{2+} + Ar^\bullet$$

$$Ar^\bullet + Cu^+ \longrightarrow Ar{-}Cu^{II}$$

$$Ar{-}Cu^{II} + Ar^\bullet \longrightarrow Ar_2Cu^{III}$$

$$Ar_2Cu^{III} \longrightarrow Ar{-}Ar + Cu^+$$

Scheme 26

For example, 4-nitrobenzenediazonium tetrafluoroborate (**60**) was coupled to 4,4'-dinitrobiphenyl (**61**) in 82% yield [100], respectively, Scheme 27. The yield of azoarene side-product **62** is minimized to less than 5% by using Cu(II) salts as an additive.

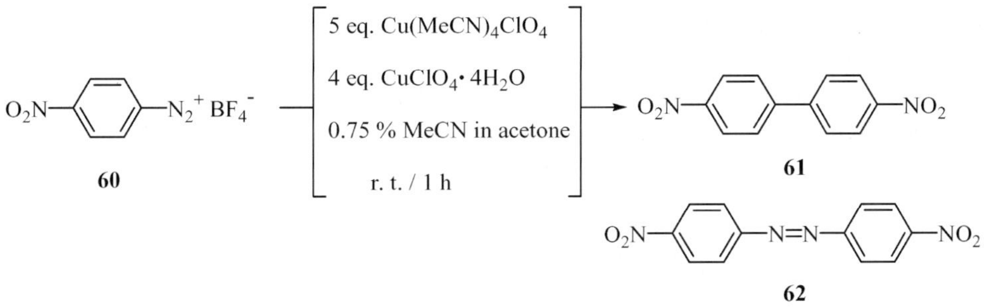

O_2N—⟨⟩—N_2^+ BF_4^-

60

5 eq. Cu(MeCN)$_4$ClO$_4$

4 eq. CuClO$_4 \cdot$ 4H$_2$O

0.75 % MeCN in acetone

r. t. / 1 h

O_2N—⟨⟩—⟨⟩—NO_2

61

O_2N—⟨⟩—N=N—⟨⟩—NO_2

62

Scheme 27

Although the copper-catalysed homo-coupling reaction of aryldiazonium salts is still unemployed, the Cohen's work is a good basis for further achievements of this perspective approach.

2.5. Arylation of arenes with diaroyl peroxides

The thermal decomposition reaction of a diaroylperoxide, (ArCOO)$_2$, in a liquid arene, Ar', furnishes a biaryl, Ar-Ar' [101-107]. At elevated temperatures the diaroyl peroxide dissociates by the homolytic pathway to generate the arylcarboxy-radicals, ArCOO•, which are prone to rapid decarboxylation forming the aryl radicals, Ar•. The latter react with the arene, e.g. benzene, by the free-radical arylation mechanism, closely similar to the classical Gomberg-Bachmann-Hey reaction pathway, to give the biaryls, Ar-Ar', in moderate to good yields [101-103], Scheme 28.

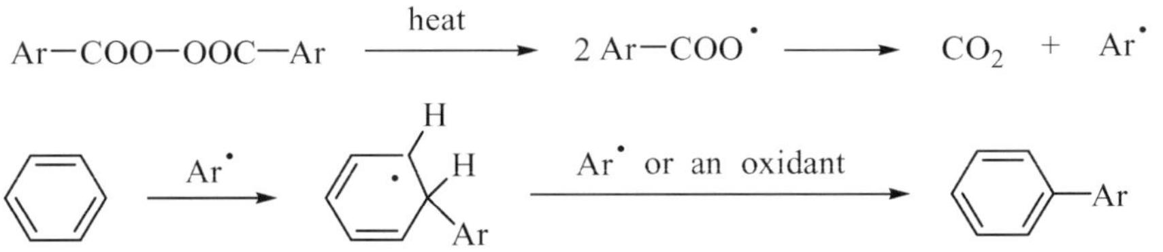

Ar—COO—OOC—Ar $\xrightarrow{\text{heat}}$ 2 Ar—COO$^•$ $\longrightarrow$ CO$_2$ + Ar$^•$

Scheme 28

The reaction is performed by heating the diaroyl peroxide in an excess of arene to be arylated as solvent in an oxygen atmosphere [104], or in the presence of nitrobenzene until the evolution of carbon dioxide ceased [105]. For example, the yield of biphenyl can be increased to almost 80% by the reaction of dibenzoyl peroxide in refluxing benzene, if nitrobenzene or similar oxidant is present. Nitrobenzene efficiently oxidizes transient arylcyclohexadienyl radicals to biaryls, thus preventing several side-reactions such as formation of variously unsaturated cyclohexylaryls and polyarylation products. The latter side-reaction is usually supressed by employing a higher molar ratio of the parent arene versus diaroyl peroxide. For example, 3-chlorobenzoyl peroxide (63) undergoes thermal decomposition in refluxing benzene (high dilution factor) to give 3-chlorobiphenyl (45) in almost 70% yield, Scheme 29.

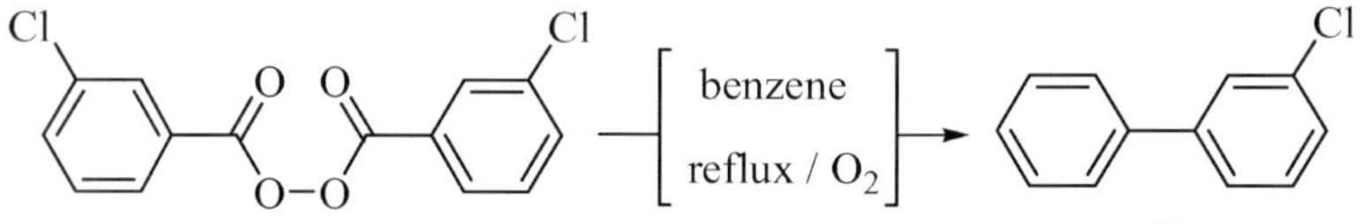

Scheme 29

When a monosubstituted benzene is attacked by aryl radical, a mixture of all three isomeric biphenyls is produced, as well as in the GBH reaction [105-107]. The nature of substituents apparently does not play any important role except those of bulky *ortho*-groups. The *ortho*-substituted biaryl is always the main product followed by *para*- and *meta*-arylation product. The presence of a bulky *ortho-tert*-butyl group significantly decreases the amount of *ortho*-arylation product, indicating the strong *ortho*-steric effect. The typical isomer distribution and the yields of biaryls in the arylation of four mono-substituted benzenes with dibenzoyl peroxide are given in the Table 4.

Table 4. The isomer distribution of biaryls in the reaction of some mono-substituted benzenes with dibenzoyl peroxide in presence of oxygen [105]

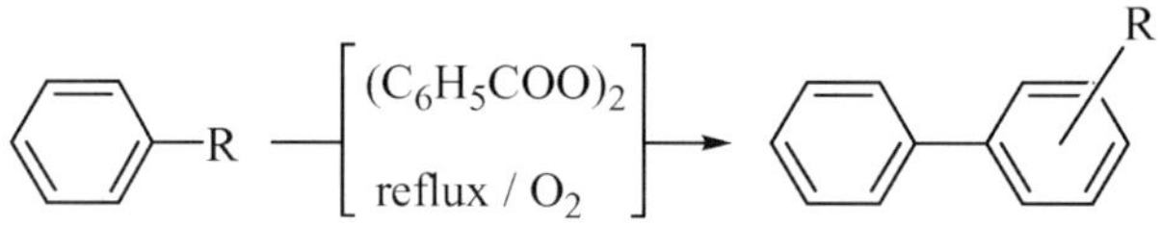

R	Yield (%)	Isomer distribution (%)		
		2-	*3-*	*4-*
CH_3O	68	70	15	16
Br	79	55	29	16
NO_2	34	63	10	28
t-C_4H_9	70	21	50	29

The trichloromethyl group causes the strong steric effect that results in a high chemoselective formation of *meta*-trichloromethylbiphenyls during the arylation of α,α,α-trichlorotoluene with different diaroyl peroxides [50]. Since the trichloromethyl group is a masked carboxylic function, thus obtained biaryls are useful intermediates for the synthesis of, otherwise not easily accessible, biaryl-3-carboxylic acids (by base-catalysed hydrolysis). On the other hand, if the isomers are difficult to separate, the reaction with unsymmetrical arenes obviously does not have any synthetic value. Finally, although the arylation reaction of arenes with diaroyl peroxides does not have wide practical importance today, it can be employed in the synthesis of certain simple biaryls bearing the substituent(s) in only one ring, those originating from the parent diaroyl peroxide.

2.6. Selected synthetic procedures

2.6.1. The Ullmann synthesis of symmetrical biaryls: Preparation of 2,2'-dimethoxy-4,4'-dimethoxycarbonylbiphenyl (64) [15]

A mixture of methyl 3-methoxy-4-iodobenzoate (**65**, 5.00 g, 17 mmol) and activated copper bronze* (15.00 g, 236 mmol, 14 eq.) was heated under argon at 210-220 °C (internal temperature) for 3 h. The cooled mixture was exhaustively extracted with boiling ethyl acetate and the residue left on removal of the solvent was crystallized from methanol. The 2,2'-dimethoxy-4,4'-dimethoxycarbonylbiphenyl (**65**, 2.58 g, 91%) was isolated in the form of blades, m.p. 163-165 °C.

The activation of copper bronze:
The suspension of purchased copper bronze (10 g) was stirred in 100 ml of 0.02 M aqueous solution of Na_2EDTA at room temperature for 3 h. The complexing solution was removed by decantation and copper was successively washed with water (2x100 ml), methanol (2x100 ml), and dried in vacuum dessicator for 2 h. The activated copper can be stored under the water previously degassed with nitrogen, whereas the washing with methanol and drying is necessary prior to the further use. The freshly activated copper is a light salmon coloured and shiny, the copper preserved under water develops a black, carbon-like coating, but with maintained activity.

2.6.2. The Ullmann homo-coupling reaction of polyhalogenated substrates: Preparation of 3,3',5,5'-tetrabromobiphenyl-2,2'-biscarboxaldehyde (14) [14]

To a solution of 2,4,6-tribromobenzaldehyde (**10**, 2.64 g, 7.7 mmol) in dry DMF (10 ml) was added activated copper (0.25 g, 3.9 mmol, 0.5 eq.) and the resulting mixture was heated under nitrogen at 110 °C for 6 h giving a greenish solution and apparent absence of copper bronze. Water (100 ml) was added, and organics were extracted into dichloromethane. The solution was dried (Na_2SO_4), and the solvent removed under vacuum leaving a brown solid (2.50 g). The product was purified by fractional sublimation. First fraction (110 °C / 0.1 mmHg, 0.997 g) was 2,4,6-tribromo benzaldehyde (**10**), second fraction (>150 °C / 0.04 mmHg, 0.846 g) was 3,3',5,5'-tetra bromobiphenyl-2,2'-biscarboxaldehyde (**14**, 42%; 67% based on recovered starting bromide), m.p. 215-217 °C (from toluene / *n*-hexane).

2.6.3. The Ullmann homo-coupling reaction of activated heteroaryl halides: Preparation of 2,2'-bipyrimidine (17) [19]

2-Iodopyrimidine (**66**, 15.00 g, 72.8 mmol) and activated copper powder (17.50 g, 275 mmol, 3.98 eq.) were placed in a 250 ml double-necked flask fitted with a reflux condenser, a N_2 inlet, and a magnetic stirrer. Absolute DMF (60 ml) was added, and the flask was flushed with N_2 for 10 min. The reaction mixture was heated to 80 °C with vigorous stirring at the temperature kept between 80 and 85 °C. After 3.5 h, activated copper powder (4.00 g, 63.0 mmol) was added to the mixture, and the N_2 inlet was replaced with a calcium chloride tube. After another 3.5 h, the temperature was increased to 120-130 °C and the stirring was continued for 2 h. The suspension was then cooled to 0 °C, carefully drowned into a solution of potassium cyanide (23.00 g) in 115 ml of 25% aqueous solution of ammonia, and filtered. The solid residue was extracted with the same amount of cyanide solution and filtered again. The combined filtrates were treated with potassium cyanide (1.20 g), and extracted with chloroform (5x300 ml). The combined organic phases were dried (K$_2$CO$_3$), filtered and evaporated to dryness. Recrystallization of the crude product from ethyl acetate / methanol (19:1) with addition of petroleum ether (b.p. 40-70 °C) gave 5.16 g (90%) of pale tan 2,2'-bipyrimidine (17), m.p. 112-114 °C.

Note: The potassium cyanide solution can be replaced with the dilute aqueous ammonia solution, but somewhat less efficiently.

2.6.4. The Ullmann synthesis of unsymmetrical biaryls: Preparation of 1-(2-methoxycarbonyl-4-nitrophenyl)-4-nitronaphthalene (30) [38]

To a mixture of 4-nitro-1-iodonaphthalene (27, 1.20 g, 4 mmol, 4 eq, less reactive aryl halide) and copper bronze (0.50 g, 8 mmol, 2 eq.) was added a solution of methyl 5-nitro-2-iodobenzoate (24, 0.31 g, 1 mmol, more reactive aryl halide) in dry DMF (5 ml) dropwise over 4 h. After the addition, the mixture was cooled and diluted with water (20 ml). An aqueous ammonia solution (25%, 20 ml) was added to dissolve the copper salt and the resulting mixture was extracted with ethyl acetate (5x20 ml), dried (Na$_2$SO$_4$), and evaporated. The crude product was purified by preparative chromatography over silica gel using n-hexane / ethyl acetate (5:1) to afford 345 mg (98%) of pure biaryl 30.

2.6.5. The Nilsson synthesis of unsymmetrical biaryls derived from 1,3-dinitro benzenes: Preparation of 4'-methoxy-2,6-dinitrobiphenyl (40) [37]

To a solution of 1,3-dinitrobenzene (**38**, 1.68 g, 10 mmol) and 4-iodoanisole (**39**, 2.34 g, 10 mmol) in quinoline (20 ml) was added copper(I) oxide (0.72 g, 5 mmol) and the resulting mixture was refluxed under nitrogen for 2.5 h. After being cooled, a yellow quinoline complex of copper(I) iodide, which crystallizes quantitatively by adding diethyl ether (50 ml), was filtered. The filtrate was washed with 10% aqueous HCl (5x20 ml) to remove the residual quinoline. The organic extract was dried (Na$_2$SO$_4$), filtered, and evaporated to dryness. From the crude product, 1.89 g (69%) of pure 4'-methoxy-2,6-dinitrobiphenyl (**40**) was isolated by preparative chromatography over silica gel (*n*-hexane / dichloromethane).

2.6.6. The Ziegler synthesis of unsymmetrical biaryls: Preparation of 4,5:5',6'-bis(methylenedioxy)-1,1'-biphenyl-2,2'-biscarboxaldehyde (37) [35]

To a solution of *N*-(2-iodo-3,4-methylenedioxybenzylidene)cyclohexylamine [35] (**33**, 0.36 g, 1 mmol) in dry THF (5 ml) maintained under N$_2$ at -78 °C was added *n*-BuLi (0.53 ml, 2.1 M in *n*-hexane, 1.1 mmol, 1.1 eq.) *via* siringe, forming a pale yellow to orange organolithium reagent. After the mixture was stirred for 15 minutes, CuI[P(OEt)$_3$] (0.54 g, 1.5 mmol, 1.5 eq.) was added in one portion and stirred for 15 min. The resulting clear orange to red solution of **34** was treated with *N*-(2-iodo-4,5-methylenedioxybenzylidene)cyclohexylamine [35] (**35**, 0.36 g, 1 mmol) in one portion and allowed to warm to 25 °C. After 5 h the reaction was complete. The reaction mixture was diluted with dichloromethane (50 ml) and washed with 10% aqueous NaCN (20 ml) solution to remove copper(I) salts. The organic solution was dried (Na$_2$SO$_4$), and evaporated to dryness. The resulting crude bisimine **36** was hydrolyzed by stirring for 48 h with 10% aqueous HCl / dichloromethane (20 / 20 ml). The organic layer was separated, washed with saturated aqueous NaHCO$_3$ solution, dried (Na$_2$SO$_4$), filtered, and evaporated to provide the crude dialdehyde. From the crude product, 173 mg (58%) of pure biaryl **37** was obtained by preparative chromatography, m.p. 144-147 °C.

2.6.7. The phase transfer Gomberg-Bachmann-Hey (ptGBH) synthesis: Preparation of 3-chlorobiphenyl (45) [58]

Potassium acetate (1.20 g, 12.2 mmol, 2 eq.) was added at ambient temperature during 30 min to a vigorously stirred mixture of 3-chlorobenzenediazonium tetrafluoroborate* (**67**, 1.36 g, 6 mmol) and 18-crown-6 (80 mg, 0.3 mmol, 5 mol%) in benzene (60 ml). Stirring was continued for 90 min. The red mixture was filtered to remove solid potassium tetrafluoroborate, and the filtrate washed with brine (30 ml) and water (30 ml). The organic layer was dried (Na_2SO_4), evaporated, and the residue was chromatographed over alumina (50 g) with n-hexane / dichloromethane as an eluent to give 0.89 g (79%) of pure 3-chlorobiphenyl (**45**) as the colourless to slightly yellowish oil.

*A convenient method for the preparation of aryldiazonium tetrafluoroborates:

To a concentrated hydrochloric acid (37%, 250 ml, 3 mol) precooled to -5 °C, a liquid arylamine (1 mol) was added dropwise, or in several portions (for solids) during 15 min. The mixture was stirred at -5 °C for an additional 45 min (solid amines has to be grounded to fine powder). A solution of sodium nitrite (69.00 g, 1 mol, 1 eq.) in distilled water (250 ml) was added dropwise at -10 to -5 °C during 1 h. The reaction mixture was stirred at -5 °C for an additional 1 (-2) h, until the TLC analysis shows the disappearance of the starting amine. To the obtained clear yellow solution of aryldiazonium chloride, 48% aqueous solution of tetrafluoroboric acid (150 ml, 100.80 g, 1.15 mol, 1.15 eq.) was added dropwise at -5 °C during 15 minutes. The mixture was stirred at -10 °C for 30 minutes, and the crystals were collected by filtration, washed with precooled (+5 °C) water (3x20 ml), and dried at 30 °C in high vacuum for 20 h. The pure aryldiazonium tetrafluoroborates were obtained in 60-90% yields.

2.6.8. The Gomberg-Bachmann-Hey synthesis of unsymmetrical biaryls from aryltriazenes: Preparation of 4-acetylbiphenyl (68) [62]

4-Aminoacetophenone (**69**, 13.51 g, 0.1 mol) was added to 300 ml of 1 M hydrochloric acid at room temperature. The solution was stirred and cooled to 0 °C. After 15 min, sodium nitrite (6.90 g, 0.1 mol) in water (50 ml) was added slowly to the stirred mixture. After 30 min at 0 °C, piperidine (11 ml, 0.11 mol, 1.1 eq.) was added dropwise. After 30 min at 0 °C, the yellow-orange precipitate was isolated by filtration. The solid was dissolved in diethyl ether, dried (MgSO$_4$), and concentrated to give 15.02 g (65%) of pure 4-acetylphenyl-3,3-(pentadiyl)triazene (**70**), m.p. 65-66 °C. A solution of triazene **70** (6.93 g, 30 mmol) in dry benzene (300 ml) was heated to 65-70 °C, and trifluoroacetic acid (4.62 ml, 60 mmol, 2 eq.) was added dropwise over 10-15 min. The solution darkened and nitrogen gas evolved over 3 h. The cooled solution was added to 5% sodium carbonate solution (300 ml), followed by extraction with dichloromethane (3x50 ml). The extracts were dried (Na$_2$SO$_4$), and the solvent was removed under reduced pressure. The residue was chromatographed over a column of alumina with *n*-hexane / dichloromethane (3:1) as an eluent to afford 3.94 g (67%) of pure 4-acetylbiphenyl (**68**) as colourless crystals, m.p. 119-120 °C.

2.6.9. The Gomberg-Bachmann-Hey synthesis of unsymmetriacl biaryls from aromatic amines and alkylnitrites: Preparation of 2-phenylthiophene (71) [80]

A mixture of aniline (5.30 g, 57 mmol), *n*-pentyl nitrite (11.5 ml), and thiophene (450 ml) was stirred at 30 °C for 72 h. The solution was washed with dilute aqueous hydrochloric acid (3x100 ml) and water (3x100 ml), dried (Na$_2$SO$_4$), and evaporated. The residue was separated by preparative chromatography over silica gel column with light petroleum ether (b.p. 40-60 °C) as eluent to give 4.14 g (45.4%) of pure 2-phenyl thiophene (**71**), m.p. 35-37 °C, followed by mixed fractions and 0.24 g (2.6%) of pure isomeric 3-phenylthiophene.

2.6.10. The high-yielding Rapoport's method for performing the Pschorr reaction: Preparation of 6-((phenylsulfonyl)oxy)-2,3,5,7-tetramethoxy-9-phenanthrenecarboxylic acid (53) [91]

R = OSO$_2$C$_6$H$_5$

52 → **53**

To a solution of (E)-α-(3,5-dimethoxy-4-((phenylsulfonyl)oxy)phenyl)-2-amino-4,5-dimethoxycinnamic acid [91] (**52**, 0.92 g, 1.9 mmol) in acetone (150 ml), H$_2$SO$_4$ (0.21 ml, 3.8 mmol, 96%) was added dropwise at 0 °C followed by isoamyl nitrite (3.8 mmol, 2 eq.). After 1 h, sodium iodide (1.14 g, 4 eq.) was added in four portions over a 5-h period, the solution was stirred for an additional hour, and enough sodium bisulfite was added to turn the reaction mixture yellow, whereupon it was poured into water (400 ml) and extracted with chloroform (4x50 ml). The organic phases were combined, washed with water (3x100 ml), dried (Na$_2$SO$_4$), and evaporated to yield the crude product which was purified by crystallization from ethyl acetate / n-hexane to afford 629 mg (71%) of pure **53**, m.p. 155.5-158.5 °C.

2.6.11. The Cohen's method for performing the Gatterman synthesis of symmetrical biaryls: Preparation of 4,4'-dinitrobiphenyl (61) [94]

60 → **61**

A solution of 4-nitrophenyldiazonium tetrafluoroborate (**60**, 237 mg, 1 mmol) in acetone (300 ml) was added dropwise to the stirred solution of tetrakis(acetonitrile)copper(I) perchlorate (165 mg, 5 mmol, 5 eq.) and copper(II) perchlorate tetrahydrate (135 mg, 4 mmol, 4 eq.) in acetone (200 ml) containing acetonitrile (3.75 ml, 0.75 vol%) and water (16 μl, 0.9 mmol) at 25 °C under nitrogen. The reaction mixture was stirred for 30 minutes, diluted with diethyl ether (20 ml) and 5% aqueous ammonia solution (20 ml). The resulting mixture was stirred for 10 min. The aqueous layer was separated and extracted with diethyl ether (5x20 ml). The combined organic phases were dried (Na$_2$SO$_4$), and evaporated to dryness. The residue was chromatographed over silica gel with *n*-hexane / dichloromethane as an eluent to give 203 mg (83%) of pure 4,4'-dinitrobiphenyl (**61**).

2.7. Conclusion

In this Chapter, several classical aryl-aryl forming reactions, their scope, limitations and synthetic utility are evaluated. The Ullmann reaction is still an important alternative to any modern method for homo- as well as cross-coupling reactions of aryl halides to biaryls due to its simplicity and wide applicability. Since a great number of aromatic amines are very easily available, the Gomberg-Bachmann-Hey reaction in its modern high-yielding phase transfer catalysed version (ptGBH), makes the most economic approach to certain unsymmetrical biaryls, substituted in only one ring. The improved Pschorr reaction, employing iodide as reducing agent for diazonium salts, plays an important role in the synthesis of phenanthrene and related type of structures. The Gatterman homo-coupling reaction of aryldiazonium salts to biaryls, beside obvious potentials, has not found greater applications since. Finally, the homolytic arylation of aromatics with diaroyl peroxides has limited synthetic value allowing simple access to certain simple biaryls variously substitued in one ring.

2.8. References

1. F. Ullmann, Ber. 29 (1896) 1878.
2. F. Ullmann, Ann. 332 (1904) 38.
3. F. Ullmann, J. Bielecki, Ber. 34 (1901) 2174.
4. P. E. Fanta, Chem. Rev. 38 (1946) 139.
5. P. E. Fanta, Chem. Rev., 64 (1964) 613.
6. P. E. Fanta, Synthesis (1974) 9.
7. W. Davey and R. W. Latter, J. Chem. Soc. (1948) 264.
8. J. Forrest, J. Chem. Soc. (1960) 581.
9. J. Forrest, J. Chem. Soc. (1960) 592.
10. J. Forrest, J. Chem. Soc. (1960) 594.
11. S. Gauthier and J. M. J. Frechet, Synthesis (1987) 383.
12. A. R. deLera, R. Suau and L. Castedo, J. Heterocycl. Chem. 24 (1987) 313.
13. A. J. Pearson and M. V. Chelliah, J. Org. Chem. 63 (1998) 3087.
14. J. M. Farrar, M. Sienkowska and P. Kaszynski, Synth. Commun. 30 (2000) 4039.
15. M. A. Rizzacasa and M. V. Sargent, Aust. J. Chem. 41 (1988) 1087.
16. S. V. Kolotuchin and A. I. Meyers, J. Org. Chem. 64 (1999) 7921.
17. T. Morita and K. Takase, Bull. Chem. Soc. Jpn. 55 (1982) 1144.

18. A. G. Mack, H. Suschitzky and B. J. Wakefield, J. Chem. Soc. Perkin Trans. 1 (1980) 1682.
19. G. Vlad and I. T. Horvath, J. Org. Chem. 67 (2002) 6550.
20. M. Pomerantz, A. S. Amarasekara and H. V. R. Dias, J. Org. Chem. 67 (2002) 6931.
21. S. Tasler, H. Endress and G. Bringmann, Synthesis (2001) 1993.
22. M. D. Rausch, J. Org. Chem. 26 (1961) 1802.
23. J. C. Salfeld and E. Baume, Tetrahedron Lett. (1966) 3365.
24. C. F. Carvalho and M. V. Sargent, J. Chem. Soc. Perkin Trans. 1 (1984) 1621.
25. C. F. Carvalho, M. V. Sargent and E. Stanojević, Aust. J. Chem. 37 (1984) 2111.
26. A. H. Lewin, M. J. Zovko, W. H. Rosewater and T. Cohen, J. Chem. Soc., Chem. Commun. (1967) 80.
27. G. W. Ebert and R. D. Rieke, J. Org. Chem. 53 (1988) 4482.
28. T. Benincori, E. Brenna, F. Sannicolo, L. Trimarco, P. Antognazza, E. Cesarotti, F. Demartin and T. Pilati, J. Org. Chem. 61 (1996) 6244.
29. A. H. Lewin and T. Cohen, Tetrahedron Lett. (1965) 4531.
30. A. Cairncross and W. A. Sheppard, J. Am. Chem. Soc. 90 (1968) 2186.
31. D. H. Wiemers and D. J. Burton, J. Am. Chem. Soc. 108 (1986) 832.
32. H. Hashimoto and T. Nakano, J. Org. Chem. 31 (1966) 891.
33. M. Nilsson, Tetrahedron Lett. (1966) 679.
34. C. Ullenius, Acta Chem. Scand. 26 (1972) 3383.
35. F. E. Ziegler, I. Chliwner, K. W. Fowler, S. J. Kanfer, S. J. Kuo and N. D. Sinha, J. Am. Chem. Soc. 102 (1980) 790.
36. F. A. Cotton and G. Wilkinson, Advanced Inorganic Chemistry, John Wiley & Sons, New York, 1996.
37. C. Bjorklund and M. Nilsson, Tetrahedron Lett. (1966) 675.
38. J. Cornforth, A. F. Sierakowski and T. W. Wallace, J. Chem. Soc. Perkin Trans. 1 (1982) 2299.
39. H. Suzuki, T. Enya and Y. Hisamatsu, Synthesis (1997) 1273.
40. A. L. Johnson, J. Org. Chem. 41 (1976) 1320.
41. P. J. Wittek, T. K. Liao and C. C. Cheng, J. Org. Chem. 44 (1979) 870.
42. J. Forrest, J. Chem. Soc. (1960) 589.
43. J. Lindley, Tetrahedron 40 (1984) 1433.
44. J. Forrest, J. Chem. Soc. (1960) 566.
45. J. Forrest, J. Chem. Soc. (1960) 574.
46. M. Gomberg and W. E. Bachmann, J. Am. Chem. Soc. 46 (1924) 2339.
47. E. Bamberger, Ber. 28 (1895) 403.
48. O. Kühling, Ber. 28 (1895) 41.
49. W. E. Bachmann and R. A. Hoffman, Org. React. 2 (1944) 224.
50. O. C. Dermer and M. T. Edmison, Chem. Rev. 57 (1957) 77.

51. D. R. Augood and G. H. Williams, Chem. Rev. 57 (1957) 123.

52. C. Rüchardt and E. Merz, Tetrahedron Lett. (1964) 2431.

53. J. Elks, J. W. Haworth and D. H. Hey, J. Chem. Soc. (1940) 1284.

54. R. A. Abramovitch and J. G. Saha, Can. J. Chem. 43 (1965) 3269.

55. R. A. Abramovitch and J. G. Saha, Tetrahedron 21 (1965) 3297.

56. R. A. Abramovitch and O. A. Koleoso, J. Chem. Soc. B (1968) 1292.

57. R. A. Abramovitch and F. F. Gadallah, J. Chem. Soc. B (1968) 497.

58. R. A. Abramovitch and A. Robson, J. Chem. Soc. C (1967) 1101.

59. D. H. Hey, G. H. Jones and M. J. Perkins, Chem. Commun. (1970) 1438.

60. M. Gurczynski and P. Tomasik, Org. Prep. Proc. Int. 23 (1991) 438.

61. J. R. Beadle, S. H. Korzeniowski, D. E. Rosenberg, B. J. Garcia-Slanga and G. W. Gokel, J. Org. Chem. 49 (1984) 1594.

62. W. S. M. Grieve and D. H. Hey, J. Chem. Soc. (1937) 1797.

63. R. Huisgen and H. Reimlinger, Justus Ann. Chem. 599 (1970) 161.

64. C. Rüchardt and C. C. Tan, Chem. Ber. 103 (1970) 1774.

65. D. F. DeTar and H. J. Scheifele, J. Am. Chem. Soc. 73 (1951) 1442.

66. T. B. Patrick, R. P. Willaredt and D. J. DeGonia, J. Org. Chem. 50 (1985) 2232.

67. R. N. Butler, P. D. O'Shea and D. P. Shelly, J. Chem. Soc. Perkin Trans. 1 (1987) 1039.

68. M. Kobayashi, H. Minato, N. Watanabe and N. Kobori, Bull. Chem. Soc. Jpn. 43 (1970) 258.

69. H. France, I. M. Heilbron and D. H. Hey, J. Chem. Soc. (1940) 369.

70. H. H. Hodgson and E. Marsden, J. Chem. Soc. (1940) 208.

71. J. I. G. Cadogan, J. Chem. Soc. (1962) 4257.

72. L. Friedman and J. F. Chlebowski, J. Org. Chem. 33 (1968) 1633.

73. E. Kalatzis, J. Chem. Soc. B (1967) 277.

74. O. Daněk, D. Šnobl, I. Knížek and S. Nouzová, Coll. Czech. Chem. Commun. 32 (1967) 1642.

75. D. E. Rosenberg, J. R. Beadle, S. H. Korzeniowski and G. W. Gokel, Tetrahedron Lett. (1980) 4141.

76. S. H. Korzeniowski, L. Blum and G. W. Gokel, Tetrahedron Lett. (1977) 1871.

77. S. H. Korzeniowski and G. W. Gokel, Tetrahedron Lett. (1977) 1637.

78. R. A. Bartsch and I. W. Yang, Tetrahedron Lett. (1979) 2503.

79. S. S. Hecht, K. El-Bayoumy, L. Tulley and E. LaVoie, J. Med. Chem. 22 (1979) 981.

80. G. A. Olah and W. S. Tolgyesi, J. Org. Chem. 26 (1961) 2053.

81. B. L. Kaul and H. Zollinger, Helv. Chim. Acta 51 (1968) 2132.

82. M. Kobayashi, H. Minato, E. Yamada and N. Koboti, Bull. Chem. Soc. Jpn. 43 (1970) 215.

83. J. W. Haworth, I. M. Heilbron and D. H. Hey, J. Chem. Soc. (1940) 349.

84. C. M. Camaggi, R. Leardini, M. Tiecco and A. Tundo, J. Chem. Soc. B (1970) 1683.

85. R. Pschorr, Chem. Ber. 29 (1896) 496.

86. P. Ruggli and A. Staub, Helv. Chim. Acta, 20 (1937) 37.

87. W. A. Waters, J. Chem. Soc. (1937) 2007.

88. M. P. Cava, I. Noguchi and K. T. Buck, J. Org. Chem. 38 (1973) 2394.

89. W. Herz and D. R. K. Murty, J. Org. Chem. 26 (1961) 418.

90. A. J. Floyd, S. F. Dyke and S. E. Ward, Chem. Rev. 76 (1976) 509.

91. D. H. Hey, J. A. Leonard, C. W. Rees and A. R. Todd, J. Chem. Soc. C (1967) 1513.

92. D. H. Hey, C. W. Rees and A. R. Todd, J. Chem. Soc. C (1967) 1518.

93. D. M: Collington, D. H. Hey and C. W. Rees, J. Chem. Soc. C (1968) 1017.

94. D. M. Collington, D. H. Hey and C. W. Rees, J. Chem. Soc. C (1968) 1021.

95. R. I. Duclos, J. S. Tung and H. Rapoport, J. Org. Chem. 49 (1984) 5243.

96. A. L. J. Beckwith and G. F. Meijs, J. Org. Chem. 52 (1987) 1922.

97. E. R. Atkinson, H. J. Lawler, J. C. Heath, E. H. Kimball and E. R. Read, J. Am. Chem. Soc. 63 (1941) 730.

98. E. R. Atkinson and H. J. Lawler, Org. Synth. Coll. Vol. I (1941) 222.

99. T. Cohen and A. H. Lewin, J. Am. Chem. Soc. 88 (1966) 4521.

100. T. Cohen, R. J. Lewarchik and J. Z. Tarino, J. Am. Chem. Soc. 96 (1974) 7753.

101. D. F. DeTar, R. A. J. Long, J. Rendleman, J. Bradley and P. Duncan, J. Am. Chem. Soc. 89 (1967) 4051.

102. D. F. DeTar, J. Am. Chem. Soc. 89 (1967) 4058.

103. M. M. Henry, J. M. Dou, G. Vernin and J. Metzger, Bull. Chim. Soc. Fr. (1971) 4593.

104. M. Eberhardt and E. L. Eliel, J. Org. Chem. 27 (1962) 2289.

105. R. T. Morison, J. Cazes, N. Samkoff and C. A. Howe, J. Am. Chem. Soc. 84 (1962) 4152.

106. C. S. Rondestvedt and H. S. Blanchard, J. Am. Chem. Soc. 77 (1955) 1769.

107. E. L. Eliel, S. Meyerson, Z. Welvart and S. H. Wilen, J. Am. Chem. Soc. 82 (1960) 2936.

CHAPTER
3

3. COUPLING REACTIONS OF ARYL HALIDES AND SULFONATES WITH METAL COMPLEXES AND ACTIVE METALS

3.1. Introduction

Aryl halides and sulfonates can be coupled to give biaryls using various metal salts and complexes. A number of methods have been described as a more successful alternative to classical Ullmann coupling reaction for aryl-aryl bond formation [1-8]. These methods are based on oxidative addition of aryl halides to metal complexes. Many low valent metal salts and complexes of nickel(0) [9], palladium(0) [10], copper(I) [11], cobalt(I) [12], rhodium(I), iridium(I) and some other metals [13], have such unique property. Among them, the nickel(0) and palladium(0) complexes, as well as copper(I) salts, are the most popular and widely used catalysts and reagents in the synthesis of symmetrical and unsymmetrical biaryls. In the early phases, investigators used metal complexes as stoichiometric reagents [1]. In contrast, recent methods require only a catalytic amount of, either expensive (Pd-compounds) or toxic (Ni-salts), complexes in the presence of the inexpensive ultimate reductant. Older Ullmann reactions require rather high reaction temperatures (150-280 °C), while these methods include reaction temperatures between room and 100 °C. Among substrates used, not only more reactive aryl iodides and bromides, but also chlorides [14] and even aryl sulfonates [15], can be coupled under relatively mild reaction conditions and in good to high yields. A great deal of progress has been achieved in the coupling reactions of aryl halides mediated by copper(I) salts which are inexpensive, readily availabe and nontoxic. However, these methods require stoichiometric amounts of copper(I) salts and specific substrate structures [6].

3.2. Reaction mechanism

Mechanistic studies from Semmelhack's group showed that in the first step aryl halide undergoes oxidative addition to the low valent metal complex with formation of organometallic intermediate in a higher oxidation state [1,2]. For example, bis(1,5-cyclooctadienyl)nickel(0), $Ni(COD)_2$, reacts with aryl halide (**I**) to give arylnickel(II) halide (**VIII**), which further reacts with another aryl halide molecule to form diarylnickel(IV) halide (**IX**). Each oxidative addition step includes substitution of ligands at metallic centre. 1,5-Cyclooctadiene (COD) dissociates from the nickel to form a coordinatively unsaturated metallic-centre, which does react with aryl halide. Biaryl **II** is formed by reductive elimination step from **IX** with liberation of nickel(II) halide [1,2], Scheme 1.

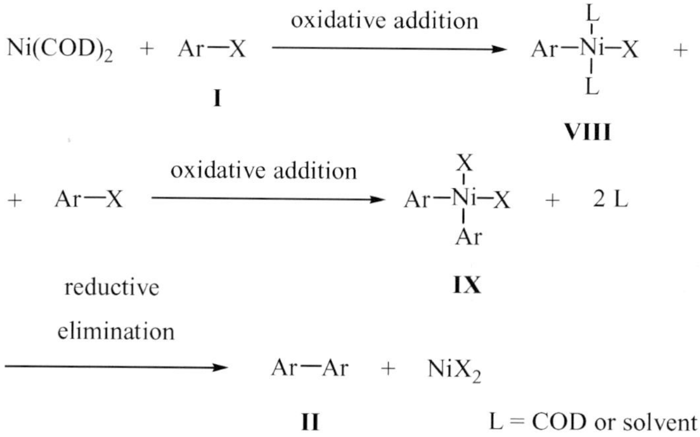

$$Ni(COD)_2 \;+\; Ar-X \xrightarrow{\text{oxidative addition}} Ar-\underset{\underset{L}{|}}{\overset{\overset{L}{|}}{Ni}}-X \;+$$

$$\textbf{I}$$

$$\textbf{VIII}$$

$$+\; Ar-X \xrightarrow{\text{oxidative addition}} Ar-\underset{\underset{Ar}{|}}{\overset{\overset{X}{|}}{Ni}}-X \;+\; 2\,L$$

$$\textbf{IX}$$

$$\xrightarrow[\text{elimination}]{\text{reductive}} Ar-Ar \;+\; NiX_2$$

$$\textbf{II} \qquad\qquad L = COD \text{ or solvent}$$

Scheme 1

Tsou and Kochi [9] concluded that the arylnickel(I) and arylnickel(III) species are the reactive intermediates, rather than arylnickel(IV) intermediate, in the radical chain process, Scheme 2.

$$Ni^I X \;+\; Ar-X \longrightarrow ArNi^{III}X_2$$

$$ArNi^{III}X_2 \;+\; ArNi^{II}X \longrightarrow NiX_2 \;+\; Ar_2Ni^{III}X$$

$$Ar_2Ni^{III}X \longrightarrow Ni^I X \;+\; Ar-Ar$$

Scheme 2

However, these studies were carried out in hydrocarbon solvents and the results, as pointed out by Semmelhack, are not completely consistent with the product studies performed in *N,N*-dimethylformamide [1,2]. Alternatively, arylnickel(II) species are converted to diarylnickel(II) species and nickel(II) halide, Scheme 3. The final step is the formation of biaryl by reductive elimination of diarylnickel(II) intermediate under reaction conditions.

$$Ni^0L_n \ + \ Ar{-}X \ \longrightarrow \ ArNi^{II}XL_n$$

$$2\,ArNi^{II}XL_n \ \longrightarrow \ NiX_2 \ + \ n\,L \ + \ Ar_2Ni^{II}L_n$$

$$Ar_2Ni^{II}L_n \ \longrightarrow \ Ni^0L_n \ + \ Ar{-}Ar$$

Scheme 3

Although proposed mechanisms are different, they all include:
  a) formation of arylnickel(II) intermediate by oxidative addition of aryl halide to nickel(0) complex,
  b) formation of diarylnickel(II, III or IV) intermediate by:
    - disproportionation reaction of two moles of arylnickel(II) halide, Scheme 3
    - reaction of an arylnickel(III) halide with an arylnickel(II) halide, Scheme 2, or,
    - oxidative addition of an aryl halide to the already formed arylnickel(II) halide, Scheme 1, and
  c) reductive elimination of biaryl molecule from the diarylnickel intermediates.

Formation of arylnickel(II) complexes by oxidative addition of aryl halides to nickel(0) complexes is a well documented process [1,2,9]. This kind of intermediate was also studied in the interesting reaction of pentafluorophenyl iodide with activated nickel powder where the bis(pentafluorophenyl)nickel(II) iodide was isolated after addition of triphenylphosphine (2 eq.) to the reaction mixture [16]. Arylnickel(II) halides are relatively stable crystalline substances, soluble in a variety of nonpolar organic solvents. When pure, they are stable at elevated temperatures for prolonged periods, but decompose readily in the presence of aryl halides, or in refluxing toluene, to produce biaryls. Considering the different reaction conditions and observations reported, it is unlikely that one mechanism is operating under all reaction conditions. Oxidative addition of aryl halides to palladium(0) complexes is also well known process with formation of arylpalladium(II) halides [10,17,18]. The formation of various diarylnickel species is less known, obviously because of the lower stability. On the other hand, diarylplatinum complex **72** is stable and decomposes possibly *via* transition state such as **72a** to form biphenyl (**8**) and Pt(PEt₃)₂ [13], Scheme 4.

Similar process takes place during palladium- and nickel-catalysed couplings of aryl halides *via* various diarylnickel(II, III or IV)- or diarylpalladium(IV) intermediates. In the case of the related Heck reaction [19], palladium(IV) intermediate has been isolated and characterized [20].

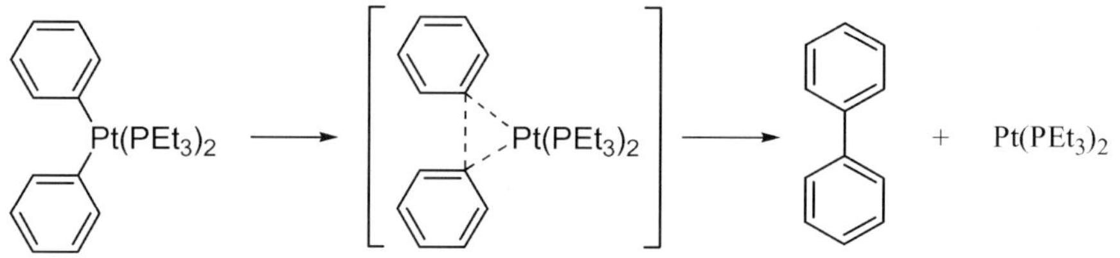

72 **72a** **8**

Scheme 4

In the copper-based processes, the reductive elimination step is far more complicated. First, aryl halides (**I**) react with copper(I) halides to give arylcopper(III) halides, $ArCuX_2$ [6,11,12]. These are reduced with an excess of copper(I) salts to arylcopper(I) species (**X**) [21], which further form stable polynuclear complexes with copper(I) salts, e.g. $Ar_4Cu_6X_2$ [22,23]. These complexes, under prolonged mixing or heating in an inert solvent like benzene, decompose with subsequent formation of biaryls (**II**). Reductive elimination in copper-based processes is likely to involve the arylcopper(II) intermediate (**XI**), as suggested by van Koten's group [22], Scheme 5 (simplified).

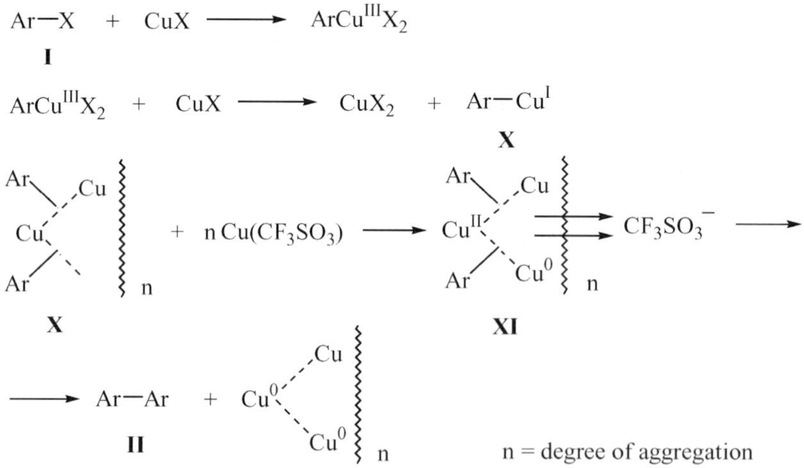

n = degree of aggregation

Scheme 5

3.3. Homo-coupling reactions of aryl halides to biaryls catalysed by nickel complexes

Semmelhack and coworkers [1,2] described homo-coupling of aryl halides to biaryls mediated by stoichiometric amount of Ni(COD)$_2$ in DMF as solvent at 30-60 °C in good to high yields. 4-Bromoacetophenone (73) was converted to 4,4'-diacetylbiphenyl (74) by this method in 93% yield [1], Sheme 6.

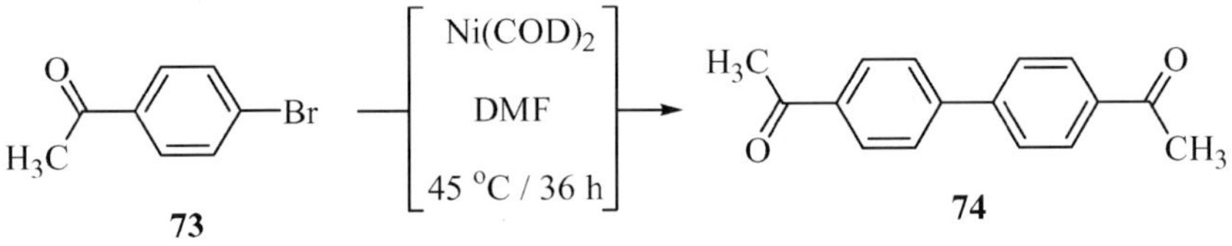

Scheme 6

Two equivalents of aryl halide require one equivalent of Ni(COD)$_2$ for the coupling reaction. This method proved to be efficient in a number of examples, Table 1.

Table 1. Synthesis of biaryls by Semmelhack's Ni(COD)$_2$ method [1]

Aryl halide	Reaction conditions: temp. (°C) / time (h)	Yield (%)
Chlorobenzene	50 / 29	14
Bromobenzene	52 / 25	82
Iodobenzene	40 / 21	71
4-Bromoanisole	40 / 23	83
4-Bromobenzonitrile	36 / 11	81
4-Bromobenzaldehyde	35 / 20	79
4-Bromoaniline	40 / 90	54
Ethyl 4-bromobenzoate	50 / 19	81

However, aryl halides containing acidic functional groups hydroxyl, or carboxylic acid fail to give the coupling reaction. In these cases, dehalogenation products were isolated in considerable amounts. Amino group does not inhibit the reaction, e.g. benzidine was successfully obtained from 4-bromoaniline. Easily avilable Ni(COD)$_2$ was the first nickel(0) complex used in aryl-aryl bond forming reactions [24]. DMF is only reaction solvent which has been found to be satisfactory for the coupling reaction. No reaction of aryl halides and bis(1,5-cyclooctadiene)nickel(0) occurs at moderate temperatures in less polar solvents such as THF or toluene. In these solvents, Ni(COD)$_2$ decomposes to a nickel mirror more rapidly than in DMF. However, the addition of a few molar equivalents of DMF or PPh$_3$ is sufficient to promote an efficient biaryl formation. Still, Semmelhack's method suffers from several disadvantages:

   a) stoichiometric amount of expensive and toxic Ni(COD)$_2$,
   b) Ni(COD)$_2$ is air-sensitive and requires careful handling,
   c) aryl chlorides give poor yields, and
   d) aryl halides containing nitro group fail to give the coupling reactions.

The last disadvantage is typical for all nickel-based processes. The reactivity of the aryl halides is aproximately in the order I > Br > Cl, while tosylates are completely unreactive under these reaction conditions.

Further improvements have been made by introduction of phosphine complexes such as tris- or tetrakis(triphenylphosphine)nickel(0), Ni(PPh$_3$)$_3$ or Ni(PPh$_3$)$_4$, which are far more soluble in less polar solvents and easier to prepare, either by electrochemical method [25], or by simple *in situ* reduction with zinc, magnesium or manganese [3,5,14]. Mori and coworkers [25] used Ni(PPh$_3$)$_4$, *in situ* obtained from Ni(PPh$_3$)$_2$Cl$_2$ and triphenylphosphine (PPh$_3$) by electrochemical reduction, as an efficient stoichiometric reagent for homo-coupling reactions of aryl halides to biphenyls in fairly good yields, Table 2. A major advantage of Mori's method is avoiding the usage of chemical reducing reagent in the reaction medium, but still requires a stoichiometric amount of nickel reagent.

Table 2. Synthesis of biaryls from aryl halides catalysed with Ni(PPh$_3$)$_4$ prepared by electrochemical reduction [25]

R	X	Time (h)	Yield (%)
H	Br	4	79.9
COCH$_3$	Br	14.5	46.9
COOCH$_3$	Cl	21.5	51.1
CH$_3$	Br	20	74.5
OCH$_3$	Br	17	35.9

However, the homo-coupling of aryl halides can be accomplished at catalytic nickel loading with Ni(bpy)Br$_2$, e.g. 15-30 mol% [26]. The reaction of 2-bromopyridines in DMF or acetonitrile was affected with zinc as the sacrificial anode at ambient temperature affording 2,2'-bipyridines in moderate to good yields [26]. Moreover, the reactions of variuos bromobenzenes have been conducted in water as the solvent in the presence of sodium chloride, a nonionic surfactant, Brij 35^R (1.5 mol%), and Ni(bpy)Br$_2$ (7.5 mol%) furnishing the appropriate biaryls in fair yields [27].

Kende used zinc powder to reduce Ni(PPh$_3$)$_2$Cl$_2$ in the presence of triphenylphosphine in oxygen-free DMF to obtain a red-brown slurry of Ni(PPh$_3$)$_4$ [3]:

$$Ni(PPh_3)_2Cl_2 \ + \ 2 \ PPh_3 \ -\!\!\left[\ Zn \ / \ DMF \ / \ 50\ ^oC \ / \ 1\ h \ \right]\!\!\longrightarrow \ Ni(PPh_3)_4$$

Using such prepared nickel(0) complex as the stoichiometric reagent, several methyl-, methoxycarbonyl-, methoxy-, and formyl-substituted aryl iodides and bromides were coupled affording the respective biaryls in DMF as the reaction solvent at 50 °C for 24 h in 42-85% yields.

Nickel(II) complexes formed during nickel-catalysed biaryl synthesis can be *in situ* reduced with an ultimate reductant. This is the basic principle of several catalytic methods which use: NiX$_2$ (X = Cl, Br, I) [14,28], Ni(PPh$_3$)$_2$Cl$_2$ with or without the presence of various amounts of triphenylphosphine [14], Ni(PEt$_3$)$_2$Cl$_2$ [29], Ni(PBu$_3$)$_2$Cl$_2$ [29], Ni(PCy$_3$)$_2$Cl$_2$ [29], and Ni(bpy)Cl$_2$ as precatalysts [30,31]. Coupling reactions were carried out in the presence of a catalytic amount (2.5-10 mol%) of given nickel(II) salt or phosphine complex and stoichiometric amount of zinc [5,14,28,29,32], aluminum [31], magnesium [14], and manganese [14,30] in HMPA [28,29], NMP [28,29], 1,1,3,3-tetramethylurea (TMU) [29], DMAc [14,28], DMF [5,28,29], THF [32], acetonitrile [32] and acetone [32] as reaction solvents in an inert atmosphere at temperatures between room and max. 80 °C. It has been found that iodides accelerate the coupling reactions, acting as bridging ligands between nickel and ultimate heterogeneous reductant (Zn) in the electron-transfer process [5,9,29,32]. Another assumption is that the coordination of the iodide ion to the low valent nickel complexes facilitates the oxidative addition of aryl halides by increasing the electron density at the nickel centre [9]. For this reason iodides were used as additives; potassium iodide in more polar solvents like HMPA or NMP [28,29], and Et$_4$NI (TEAI) for use in less polar THF [32]. Efficiency of these methods, as well as some related but specific ones, can be seen from the result in the model coupling reactions of brombenzene (*methods A-D*), chlorobenzene (*method E*) and iodobenzene (*method F*) to biphenyl, Table 3.

Reducing metals (*methods A-D*) act as ultimate reductants which reduce Ni^{II}(PPh$_3$)$_2$X$_2$ in the presence of triphenylphosphine to the catalytically active Ni^0(PPh$_3$)$_3$ or Ni^0(PPh$_3$)$_4$ complexes. Aryl halides (**I**) react with, *in situ* generated, Ni^0(PPh$_3$)$_4$ to form biaryls (**II**) and nickel(II) halide *via* arylnickel(II) (**VIII**) and diarylnickel(IV) (**IX**) intermediates. Thus obtained, Ni^{II}(PPh$_3$)$_2$X$_2$ is *in situ* reduced back to the catalytically active Ni^0(PPh$_3$)$_4$. The catalytic cycle is presented in the Scheme 7.

Table 3. Synthesis of biphenyl (**8**) by catalytic nickel-based methods from bromobenzene (*A,B,C,D*), chlorobenzene (*E*) and iodobenzene (*F*)

Method	Pre-catalyst (mol%)	Yield (%)	Ultimate reductant (eq.)	Additives (eq.)	Reaction condition [a] (°C) / (h)	Lit. ref.
A	Ni(PPh₃)₂Cl₂ (5) PPh₃ (40)	89 [b]	Zn (1)	-	DMF 50 / 20	5
B	NiBr₂ (2.5)	98	Zn (1.25)	KI (2.5)	HMPA or NMP 50 / 7	28
C	Ni(PEt₃)₂Cl₂ (4)	74	Zn (1)	KI (2.0)	NMP 30 / 2	29
D	Ni(PPh₃)₂Br₂ (10)	99	Zn (1.5)	Et₄NI (0.1-1)	THF 50 / 2	32
E	NiCl₂ (5) PPh₃ (50)	98 [c]	Zn, Mg, Mn (2)	NaBr (0.15)	DMAc or DMF 80 / 0.1	14
F	NiCl₂ (10) bpy (10)	68 [d]	Al (1.5)	PbBr₂ (0.1) KI (1.5)	MeOH 25 / 22	31

[a] All reactions were carried out in an inert atmosphere (N₂ or Ar)
[b] At room temperature, 81% of biphenyl was obtained after 24 h
[c] Starting from chlorobenzene
[d] Starting from iodobenzene

Method E, developed by Colon and Kelsey [14], is efficient for coupling reactions of much less reactive aryl chlorides under the similar reaction conditions. The role of triphenylphosphine in all these methods is only to stabilize catalytically active nickel(0) complexes, arylnickel(II) and diarylnickel intermediates, providing high selectivity and turnover number.

Some described methods use triphenylphosphine in amounts larger than it is necessary for formation of either Ni(PPh₃)₃ or Ni(PPh₃)₄. In more polar reaction solvents, e.g. DMF or DMAc, triphenylphosphine reacts with zinc halides, formed during the coupling reaction, to give zinc complexes. To avoid this problem, triphenylphosphine is, in some instances, used in increased quantities, e.g. 40-50 mol%, in order to achieve high conversion of aryl halides. However, this causes the difficulty of separating excess of triphenylphosphine from the reaction mixture. Beside the standard techniques, it can also be removed from the crude biaryl product by stirring in xylene or toluene with excess of methyl iodide. Thus formed, insoluble Wittig salt is easily removed by filtration. The pure product is isolated after the solvent and methyl iodide were evaporated.

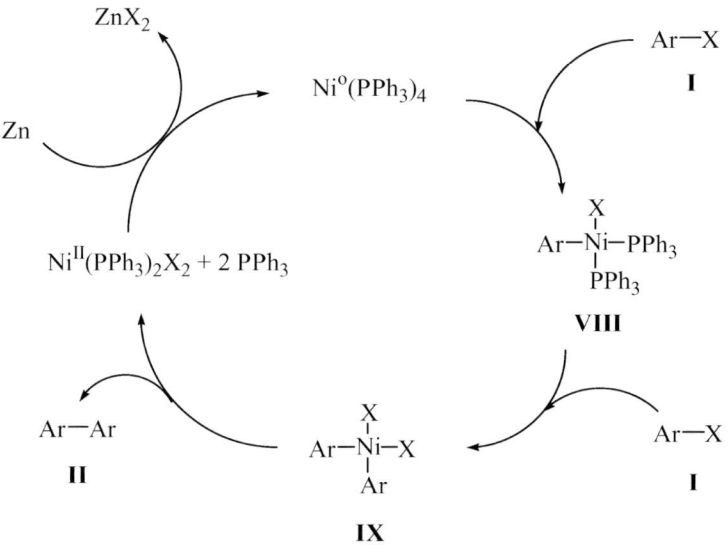

Scheme 7

The last method, *F*, which uses NiCl$_2$-2,2'-bipyridine complex, works well only for the most reactive aryl iodides [31]. It is mechanistically more complicated because here lead(II) bromide acts as the source of lead(0) which actually reduces Ni^{II}(bpy)X$_2$ back to the catalytically active Ni^0(bpy). Since the 2,2'-bipyridine **75** is somewhat more electron-rich ligand than simple 2,2'-bipyridine (**5**), the resulting nickel(II) complex employed in the related Chen's method is slightly more reactive with electron-poor aryl halides, e.g. bromides, but generally the method is applicable only with aryl iodides [30]. For instance, both iodo- and bromobenzene gave biphenyl (**8**) in 98 and 88% yields, respectively, but 2-bromothiophene was coupled in only 22% yield [30], Scheme 8. Chromium(II) chloride in a catalytic amount was used as an intermediate reductant as it provides an effective redox system, Cr(II) / Cr(III) between Ni(0) / Ni(II) and metallic manganese as the ultimate reductant.

Selected preparations of various biphenyls through presented methods are shown in the Table 4. *Para-* and *meta-* substituted aryl bromides and iodides were successfully coupled to the corresponding biphenyls in good to high yields using all methods. Coupling reactions of *ortho*-substituted aryl bromides and iodides were performed by using *methods B* [28], *C* [29], and *D* [32]. Other methods are not as effective because of low conversion (steric hindrance), even at elevated temperatures where dehalogenation occurs [5,28,31]. *Ortho*-substituted aryl chlorides are the least reactive and, because of steric encumbrances, react very slowly. *Method E* is the most efficient for *para*- and *meta*-substituted aryl chlorides. As previously mentioned, nitro-substituted aryl halides cannot be coupled to biaryls by any nickel-catalysed synthesis. The reason for this is probably the deactivation of nickel by the nitrosonickel(0) complexes formation [14,33].

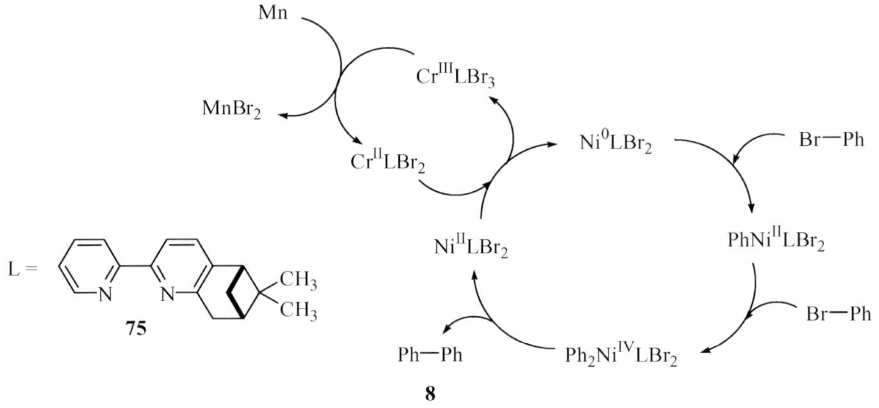

Scheme 8

During nickel-catalysed coupling reactions of aryl halides, following side-reactions were observed:

a) reduction or dehalogenation, and
b) formation of monosubstituted biphenyls by aryl-transfer from triphenylphosphine.

In the presence of the acidic (protic) substituents: phenols, amines, carboxylic acids, or contaminates such as water or lower alcohols, the substantial reduction occurs. The mechanism of dehalogenation is either by reaction of starting nickel(0) complex with the active hydrogen compounds to produce nickel hydride which reduces the starting aryl halide, or the arylnickel intermediates react with the protic sources to yield arene [1,2,9,14]. Reduction was minimized when the active hydrogen source was removed. However, in the coupling reactions of some aryl iodides substituted with electron-withdrawing groups in *ortho*-position by *method B*, the major reaction was dehalogenation [28]. Moreover, Semmelhack developed a versatile method for mild dehalogenation of aryl bromides and iodides using $Ni(COD)_2$ in the presence of acetic or trifluoroacetic acid [2,34].

Formation of monosubstituted biphenyl as side-reaction product is favoured by strong electron-donating groups in aryl halides. It is well known that nickel-metal can insert into aromatic C-O or C-S bonds [35,36]. Under reaction conditions used in all presented methods, the insertion reaction can take place and then the subsequent reaction with solvent or trace of protic source (e.g. water) could cause the loss of the heteroatom.

Table 4. Synthesis of various biphenyls using different nickel-based catalytic methods

Method	R	X	Yield (%)	Lit. ref.
A	4-CH$_3$	Br	73	5
A	4-OCH$_3$	Br	73	5
A	4-NHCOCH$_3$	Br	37	5
A	2-COOCH$_3$	Br	33	5
B	2-COOCH$_3$	I	22 [a]	28
B	3-COOCH$_3$	I	96	28
B	2-CH$_3$	I	83	28
C	3-COOCH$_3$	Br	91	29
C	2-COC$_6$H$_5$	I	67	29
D	4-COCH$_3$	Br	71	32
D	4-CH$_3$	Br	89	32
D	4-CH$_3$	Cl	81	32
D	4-CHO	Br	75	32
D	4-CHO	Cl	70	32
D	2-COOCH$_3$	Br	90 [b]	32
E	4-COCH$_3$	Cl	100	14
F	4-OCH$_3$ [c]	I	68	31

[a] Considerable amount of dehalogenation product, methyl benzoate was isolated
[b] For *ortho*-substituted aryl bromides and iodides, 20-50 mol% of Ni(PPh$_3$)$_2$Br$_2$ was used
[c] 2-Iodoanisole gave anisole in almost quantitative yield.

Another real possibility for the formation of monosubstituted biphenyls is phenyl-migration from triphenylphosphine. For example, 4-chloroanisole (**76**) coupled under reaction conditions of *method E*, beside homo-coupling product, 4,4'-dimethoxy biphenyl (**77**), gave the reduction product, anisole, and a significant amount of 4-metoxybiphenyl (**78**) as side-product, Scheme 9. In the presence of 2,2'-bipyridine (bpy) as an additional ligand, formation of 4-methoxybiphenyl was completely inhibited. More stable and with enough steric hindrances, Ni(0)(bpy) supresses oxidative addition to aromatic C-O bond. Side-product is formed from triphenylphosphine through the *ipso*-migration, *via* **79**, of nickel for phosphorus on the phenyl group to generate **80** [35,36], Scheme 10. Phenyl-migration does not involve *ortho*-metallation process and this is clearly shown in the example where 4-substituted phosphines did not form the isomeric 3-substituted biphenyl side-products [14].

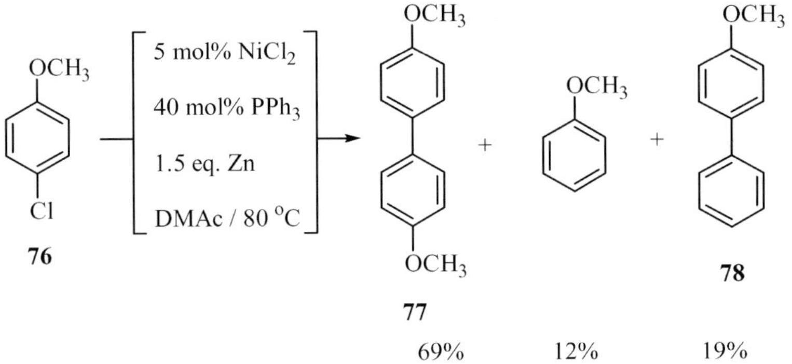

Scheme 9

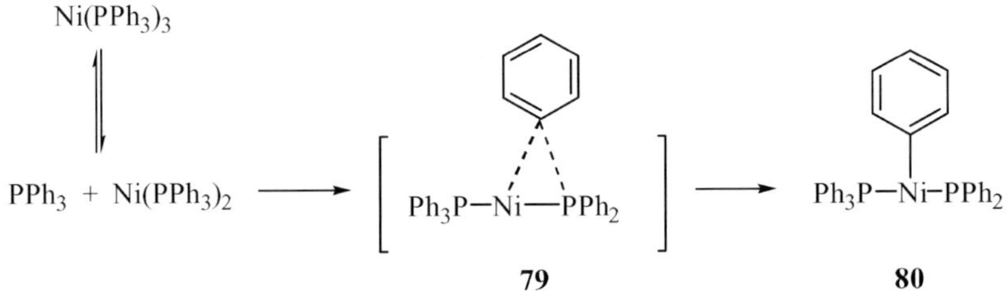

Scheme 10

Once formed arylnickel compound **80** reacts with, for example, 4-chloroanisole (**76**) to form unsymmetical diarylnickel complex **81** which undergoes reductive elimination to give unsymmetical biaryl, 4-methoxybiphenyl (**78**), Scheme 11.

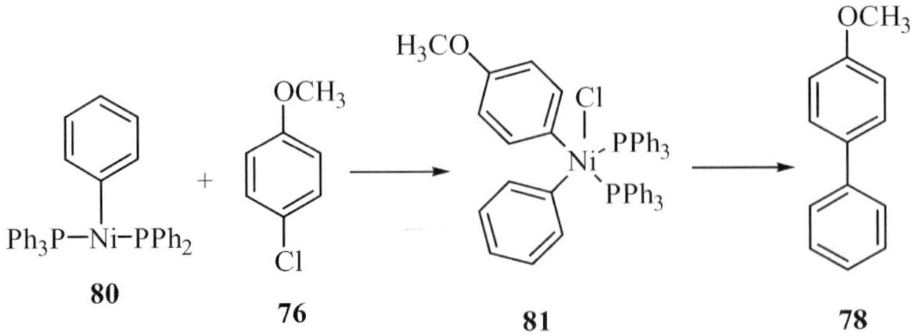

Scheme 11

There were no attempts to use this observation in order to develop any practical synthesis of unsymmetical biaryls from triarylphosphines and aryl halides.

Selected examples of homo-coupling reactions involving heteroaromatic halides using described methods are shown in the Table 5.

Table 5. Nickel-catalysed synthesis of biaryls derived from heteroaromatic halides using *methods A, C, D* and *E*

$$Ar\!-\!X \longrightarrow Ar\!-\!Ar$$

Method	Heteroaryl halide	Yield (%)	Lit. ref.
A	2-bromothiophene	41	5
C [a]	2-bromothiophene	7	28
C	2-iodothiophene	87	28
C	3-iodothiophene	83	28
C	3-bromofuran	80	28
D [b]	2-bromopyridine	72	29
D	2-chloropyridine	60	29
D	3-bromopyridine	73	29
D	3-methoxycarbonyl-2-chloropyridine	53	29
D	6-methoxy-2-chloropyridine	90	29
D	4-chloroquinoline	77	29
D	1-chloroisoquinoline	37	29
E	2-chlorothiophene	98	14
E	2-chloropyridine	70	14

[a] 2-and 3-Iodopyridine failed to give coupling reaction by *method C*
[b] Et$_4$NI has to be used in stoichiometric amount

There exists a few nickel-based methods for homo-coupling of aryl halides to biaryls, but with no basic improvements to that already presented. For example, Tiecco and coworkers [37] described the efficient method for synthesis of various bipyridines and biquinolines, by using a stoichiometric amount of Ni(PPh$_3$)$_4$. 2-Methoxy-3-chloropyridine (**82**) was converted by this method to bipyridine derivative **83** in 51% yield [37], Scheme 12.

Exactly the same method was successfully employed in the synthesis of 2,2'-bipyridines bearing a free amino group. Thus 4,4'-diamino-2,2'-bipyridine was prepared from 5-amino-2-chloropyridine in 60% yield, respectively [38]. Nickel-catalysed synthesis of biaryls and related vinyl halides were successfully used in a great number of examples [39,40], including the strained aromatic ether-sulfone oligomers [41]. For instance, compound **84** was cyclized to a very strained cyclic structure **85** with a 25% yield using a high dilution technique [41], respectively, Scheme 13.

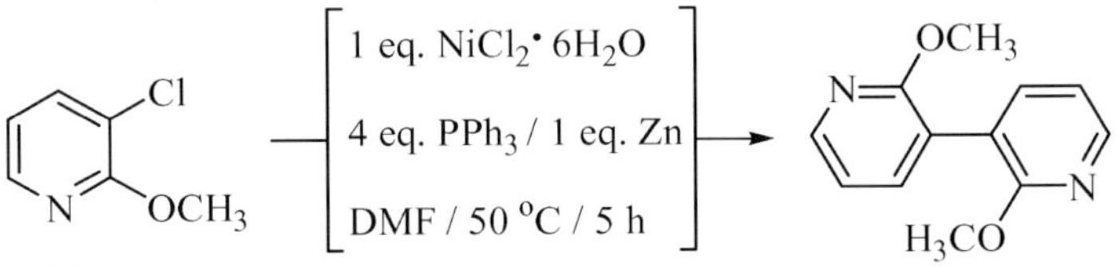

Scheme 12

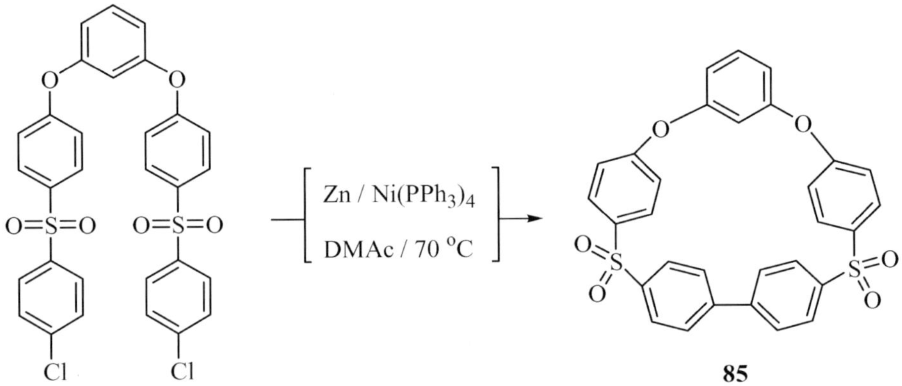

Scheme 13

Apart from aryl halides, other very easily available substrates for nickel-catalysed biaryl (**II**) synthesis are aryl sulfonates (**XII**). *Method D* is very efficient in the homo-coupling reactions of substituted aryl sulfonates in good to excellent yields [15], Table 6. Substituted aryl sulfonates are readily obtained from phenols and trifluoromethanesulfonic anhydride, benzenesulfonyl-, tosyl- or methanesulfonyl chloride in pyridine, or in a suitable inert solvent such as dichloromethane in the presence of triethylamine or Hünig's base. Among other nickel complexes, Ni(dppe)Cl$_2$ and Ni(dppf)Cl$_2$ have been used (10 mol%) as slightly less versatile catalysts for the homo-couplings of naphthyl sulfonates in refluxing THF, DMF or their mixtures [42].

Caubère and coworkers [43] developed another specific method, based on the unique stoichiometric nickel reagent. Reagent NiCRA (*complex reducing agent*) was prepared by reaction of sodium hydride (6 eq.) and *tert*-amyl alcohol (2 eq.) followed by addition of anhydrous nickel(II) acetate (1 eq.) and 2,2'-bipyridine (bpy, 2 eq.; or other nickel(0)-stabilizing ligand) in refluxing THF or mixture of benzene and THF [44], Scheme 14.

Table 6. Homo-coupling of aryl sulfonates (**XII**) to biaryls (**II**) using *method D* [15]

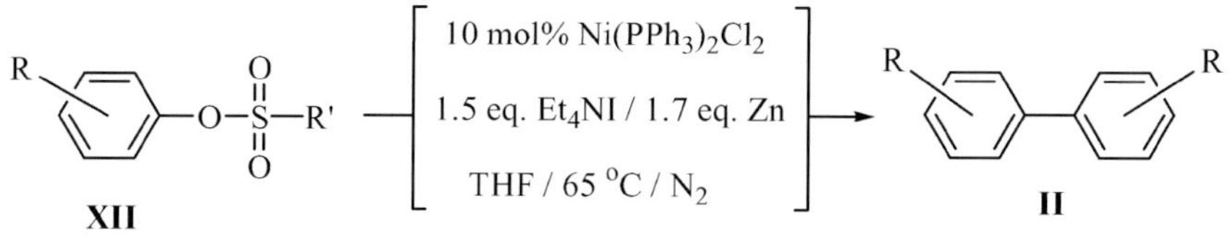

R	R'	Time (h)	Yield (%)
4-COOCH$_3$	C$_6$H$_5$	5	83
4-CH$_3$	C$_6$H$_5$	10	90
4-OCH$_3$	C$_6$H$_5$	10	94
2-COOCH$_3$	CH$_3$	24	99
4-COOCH$_3$	4-CH$_3$C$_6$H$_4$	10	99
4-COOCH$_3$	CF$_3$	5	99

$$2 \text{ NaH} + 2 \text{ } t\text{-AmOH} \xrightarrow{\text{THF}} 2 \text{ } t\text{-AmONa}$$

$$\text{Ni(OAc)}_2 + 4 \text{ NaH} + 2 \text{ } t\text{-AmONa} \xrightarrow{2 \text{ bpy}} \left[(t\text{-AmO}^-)_2\text{Ni}^0(\text{H}^-)_2(\text{bpy})_2(\text{Na}^+)_4 \right]$$

NiCRA-bpy

Scheme 14

The reagent is a black solid and consists of finely divided, very reactive and complex zero-valent nickel-species with bounded sodium *tert*-amyloxide as well as hydride and 2,2-bipyridine at the nickel surface. Therefore it is not the simple physical mixture of sodium hydride, sodium *tert*-amylate and finely divided nickel [45]. NiCRA proved to be an efficient stoichiometric reagent for the homo-coupling of various aryl and heteroaryl bromides and chlorides containing methyl, methoxy, fluoro, cyano, trifluoromethyl, ethyleneglycol-acetal, *N,N*-dimethylamino, and also phenolic hydroxy substituents to the respective biaryls with 57-84% yields [43,46,47]. Addition of potassium iodide accelerates the coupling reactions, similar to other nickel-catalysed processes. However, the NiCRA method has the following drawbacks:

    a) it requires a stoichiometric amount of nickel reagent, 2 eq., to aryl halide.

    c) The reagent must be prepared before the coupling reaction. It is *in situ* preparation, but still needs a few hours.

c) Once again, nickel salts are toxic, allergenic and cancer suspect agents. Therefore use of any stoichiometric nickel method must have a good reason.

d) Nitro-substituted aryl halides cannot be coupled with NiCRA because of the reduction of the nitro group. For that reason, an aldehyde group must be protected.

The only basic advantage of using the NiCRA reagent over other catalytic nickel methods is its lower sensitivity to *ortho*-steric hindrance. Moreover, the NiCRA method allows the coupling of the least reactive aryl chlorides even with significant *ortho*-steric encumbrances. For example, 2-chlorobenzaldehyde ethyleneglycol-acetal (**86**) was coupled to the sterically hindered biphenyl **87** in 64% yield [43], Scheme 15.

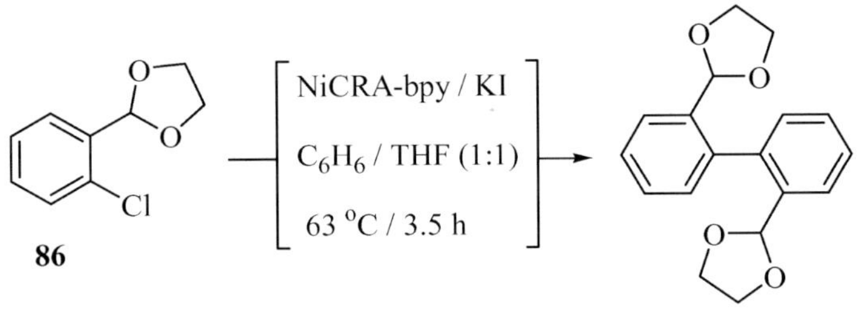

Scheme 15

In contrast to a few older reports, Lin and coworkers [48] described a versatile method for synthesis of tetra-*ortho*-substituted biaryls *via* the Ni(PPh₃)₂Cl₂ / 2PPh₃ catalysed homo-coupling reaction in the presence of Bu₄NI, which, like other iodides, strongly facilitates the reaction, but in the presence of sodium hydride as an ultimate reductant. Although the method is far from clear mechanistically, it appears to be related to Caubère's NiCRA method. In this manner, Lin's method is not sensitive to the *ortho*-steric hindrances and gives a significant amount of reductive dehalogenation side-products, as well as NiCRA's do this also.

A somewhat specific application of nickel-catalysed coupling of aryl halides is the production of polyaryls. The latter materials have been widely utilized in the electronic industry. Yamamoto's group [49] studied catalytic polymerization of *p*-dibromo- and *p*-dichlorobenzene, as well as several di- and tri-chlorobenzenes and pyridines which, upon Ni(bpy)Cl₂, Ni(PPh₃)₂Br₂, or Ni(COD)(PPh₃)₂-catalysed reactions in the presence of magnesium as an ultimate reductant, gave the respective polyaryls in high to excellent yields. In this manner, poly-*p*-phenylene (**PPP**) was obtained from *p*-dibromobenzene (**88**) in 95% yield [49], respectively, Scheme 16. Although this reaction proceeds *via* formation of the respective Grignard reagent followed by the Kharasch cross-coupling reaction with *p*-dibromobenzene, the overall process is closely related to all reactions presented in this Chapter, where a terminal reductant serves only as reducing agent for nickel(II) to the catalytically active nickel(0) complexes.

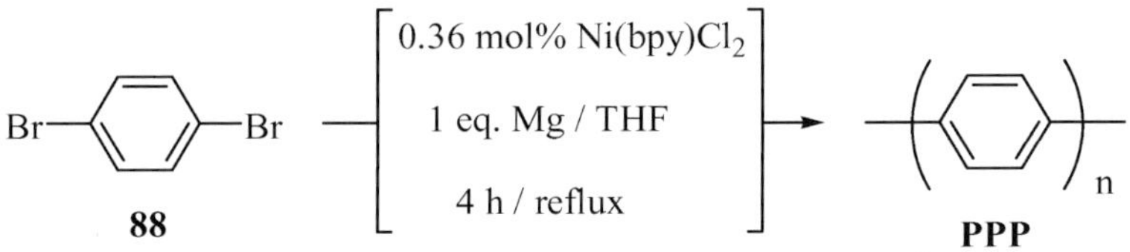

$$Br\!-\!\!\bigcirc\!\!-\!\!Br \quad \left[\begin{array}{c} 0.36\ mol\%\ Ni(bpy)Cl_2 \\ 1\ eq.\ Mg\ /\ THF \\ 4\ h\ /\ reflux \end{array}\right]\!\longrightarrow\ \!-\!\!\left(\!\!\bigcirc\!\!\right)_{\!n} $$

88   **PPP**

Scheme 16

Coupling reactions involving magnesium as an ultimate reductant often can not be clearly distinguished from the former, one pot, two step, process: tandem Grignard plus Kharasch reactions. Beside the above mentioned nickel complexes, simple metal salts such as nickel(II) chloride, iron(II) chloride, and cobalt(II) chloride, analogously to the Kharasch reaction (see Chapter 4) have also shown remarkable catalytic activity in the same reaction [49].

3.4. Homo-coupling reactions of aryl halides to biaryls catalysed by palladium complexes

Palladium salts and complexes are capable of coupling aryl halides under certain reaction conditions to yield biaryls. It is well known that palladium salts (as pre-catalysts) can be easily reduced to metallic palladium or palladium(0) complexes if the reduction was carried out in the presence of coordinating molecules (ligands, solvents). Once again, oxidative addition of aryl halide (**I**) to palladium(0) species occurs to form arylpalladium(II) halide (**XIII**). The latter subsequently reacts with another molecule of aryl halide to produce transient diarylpalladium(IV) intermediate (**XIV**). It undergoes rapid reductive elimination of a biaryl molecule (**II**) and forms the starting palladium(II) complex which is reduced by stoichiometric reductant - closing the catalytic cycle [17], Scheme 17.

Alternatively, under strong reductive conditions, e.g. electrochemical reduction, arylpalladium(II) halide (**XIII**) is reduced to arylpalladium(0) which, upon the second oxidative addition, gives diarylpalladium(II) complex, also capable of undergoing reductive elimination of biaryl similarly to **XIV** [10]. Isopropanol [50-52], amines [4,17,52-54], hydrazine [55], tetrakis(dimethylamino)ethylene [18], hydroquinone [56], tetrabutylammonium fluoride [57], formates [58-60], zinc [61-64], and molecular hydrogen [65] have been used as stoichiometric reductants. Electro-reductive homo-coupling of aryl halides was also reported [10]. The most important methods for coupling of aryl halides are based on a catalytic amount of palladium complexes. Since palladium compounds are very expensive, the stoichiometric palladium-based approach does not have a practical value.

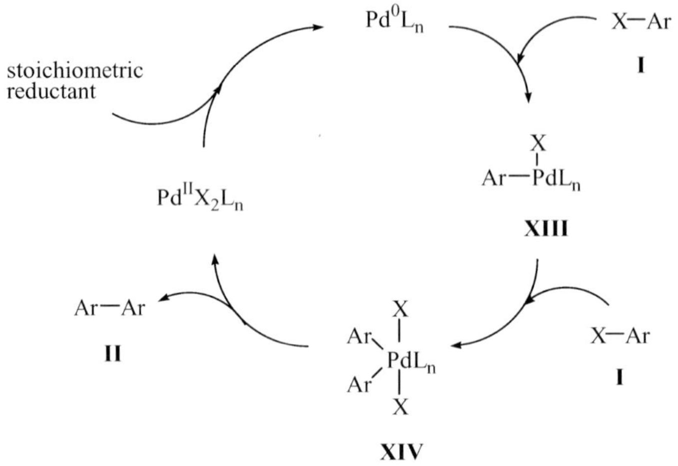

Scheme 17

In early 1970's Norman and coworkers [4] reported the first coupling reactions of aryl halides using palladium(II) acetate under reaction conditions typical for the Heck reaction. They found that aryl iodides readily couple when they are treated with the catalytic quantity of palladium(II) acetate in refluxing triethylamine or tri-*n*-butyl amine to produce biaryls in moderate to good yields, Table 7.

Table 7. Homo-coupling reaction of aryl iodides to biaryls using Norman's $Pd(OAc)_2$ / Et_3N method [4]

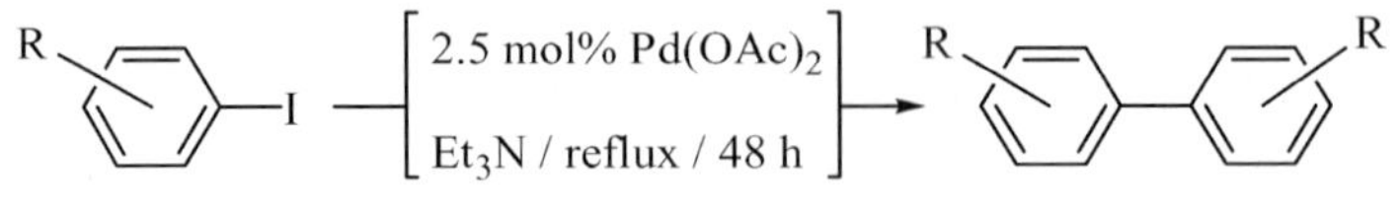

R	Yield (%)
H	54
4-OCH$_3$	39
4-NO$_2$	54
2-CH$_3$	10
2,4,6-(CH$_3$)$_3$	0

The coupling reaction is very sensitive (more than in the case of nickel-catalysed methods) to the nature of the substituent and its position in the aromatic ring. Aryl halides bearing an electron-withdrawing substituent (e.g. NO_2) easily undergo oxidative addition to the active palladium(0) catalyst, while those with electron-donating groups (e.g. OCH_3) lead to this process less readily. Aryl halides with significant steric hindrances, e.g. iodomesitylene, as well as less reactive halides, bromides and chlorides, do not couple under these reaction conditions. However, Norman's method has been the basis for further achievements with the palladium-based catalytic methods.

Recent methods for coupling reactions of aryl halides by palladium-based catalytic methods and their relative efficacy are presented in the Table 8.

Methods A and *B* have a significant difference in stoichiometric reductant. In *method A* for this purpose serves isopropanol while in *method B* diisopropylethylamine (Hünig's base). Among all tertiary amines used, diisopropylethylamine proved to be the most effective in the palladium-catalysed systems [17,52,53]. Some other tertiary amines, e.g. tri-*n*-butylamine, lead to the formation of butyrophenone side-products due to partial acylation of electron-rich aromatic rings with its oxidation products, *n*-butylidene-dibutylamine [4]. The palladium acetate reacts with Hünig's base at elevated temperature to form catalytically active nano-sized palladium particles (clusters), active palladium clusters. Polar solvent such as DMF, as well as soluble bromide source, e.g. *n*-Bu$_4$NBr, stabilize these nano-sized catalytically active particles from agglomeration to the catalytically inactive agglomerated palladium black. This and several related ways of palladium clusters stabilization have been successfully employed in a number of palladium-catalysed reactions: Heck, Stille, Suzuki-Miyaura, see Chapter 5. *Methods A* and *B* work well for coupling reactions of aryl iodides and bromides while only activated aryl chlorides, e.g. 4-nitro-chlorobenzene, give a significant amount of coupling product [52]. Despite good to high yields of biaryls, these methods require high reaction temperatures (>100 °C) and, in a number of cases, long reaction times (>24 h). Another practical problem may arise during extraction work-up procedures due to the surfactant properties of *n*-Bu$_4$NBr, which is used as an additive. The required amount of Pd(OAc)$_2$ is the major disadvantage of these methods. However, this can be successfully solved by using palladacycle catalyst **89** which is efficient at only 0.5 mol%-catalyst loading [53]:

89

Table 8. Homo-coupling reactions of aryl halides to biaryls using different palladium-based catalytic methods

Method	Aryl halide	Reaction conditions	Yield, % (time, h)	Lit. ref.
A	4-iodoanisole	5 mol% Pd(OAc)$_2$ / 115 °C 50 mol% n-Bu$_4$NBr / 1 eq. K$_2$CO$_3$ DMF / H$_2$O / i-PrOH (3:1:4)	81 (8)	50, 52
B	4-iodoanisole	5 mol% Pd(OAc)$_2$ 50 mol % n-Bu$_4$NBr 1.2 eq. (i-Pr)$_2$NEt / DMF 115 °C	80 (8)	52, 53
C	iodobenzene	2 mol% PdCl$_2$ / 2 mol% HgCl$_2$ 50 mol% N$_2$H$_4$ · H$_2$O / MeOH 65 °C	90 (6)	55
D	4-iodoanisole	2 mol% Pd(OAc)$_2$ 4 mol% PTol$_3$ 50 mol% hydroquinone 1 eq. Cs$_2$CO$_3$ / DMAc / 75 °C	95 (3)	56
E	4-iodoanisole 4-bromoanisole 4-chloroanisole	5 mol% Pd(PhCN)$_2$Cl$_2$ 200 mol% TDAE DMF / 50 °C	70 (5) 98 (4) trace (24)	18
F	iodobenzene	2-3 mol% [Pd(π-C$_3$H$_5$)Cl]$_2$ 2 eq. n-Bu$_4$NF DMSO / 120 °C	82 (4.5)	57
G	4-bromoanisole	3% Pd-C / HCOONa 32% NaOH (aq.) / CTAB / 95 °C	49 (24)	58-60
H	4-iodoanisole	1-20 mol% Pd-C / 3 eq. Zn water / acetone (1:1) / r. t.	92 (24)	61-63
I	bromobenzene	18 mol% PdCl$_2$/ 24 mol% PPh$_3$ 2.4 eq. Zn / PTC / DMF / 70 °C	82 (10-16)	64
J	bromobenzene	0.5 mol% Pd-C / 7 mol% PEG-400 / 2.8 eq. NaOH (aq.) water / 110 °C / 4 atm. H$_2$	73 (2.5)	65

TDAE = tetrakis(dimethylamino)ethylene
CTAB = cetyltrimethylammonium bromide, usually 10-20 mol% to aryl halide

Alternatively, the Hünig's base mediated coupling reactions of aryl iodides have been effected in supercritical carbon dioxide ($scCO_2$) as the reaction medium with complexes of $Pd(OAc)_2$ or $Pd(CF_3COO)_2$ and tri(2-furyl)phosphine (PFu_3) at 2 mol% catalyst loading to afford the respective biaryls in excellent yields [54].

Method C is based on hydrazine as the stoichiometric reductant. Active palladium(0) reagent is palladium-amalgam *in situ* formed by the reduction of $PdCl_2$ and $HgCl_2$ in the presence of hydrazine. Using this method, *meta-* and *para-*substituted aryl iodides were smoothly converted to the respective biaryls in moderate to excellent yields, while less reactive substrates, bromides and chlorides and sterically hindered *ortho-*disubstituted aryl iodides fail to give coupling reactions or give low conversions [55].

Method D is conducted with hydroquinone as the homogeneous stoichiometric reductant with complexes of $Pd(OAc)_2$ and tri(2-tolyl)phosphine ($PTol_3$) or tri(2-tolyl) arsine ($AsTol_3$) providing a very convenient and high-yielding approach to biaryls from a wide variety of aryl iodides and bromides, including mono *ortho-*substituted substrates [56].

Method E is based on the tetrakis(dimethylamino)ethylene (TDAE) as a soluble and very mild organic reductant with $Pd(PhCN)_2Cl_2$ as the catalyst in the coupling reactions of aryl iodides and bromides [18]. Excellent yields of biaryls were obtained under mild reaction conditions, at 50 $^{\circ}$C in DMF as the solvent. The latter method is the most efficient homogeneous palladium-based catalytic system for the coupling of aryl iodides and bromides to biaryls.

Tetra-*n*-butylammonium fluoride in the *method F* acts as the ultimate reductant with subsequent oxidation to tri-*n*-butylamine, and a fluoride ion, as strong base, is apparently involved in this conversion. This method was found to be applicable in the homo-couplings of aryl iodides and bromides. Since the use of quaternary ammonium salts as oxidants has not been widely explored, this reaction requires further mechanistic investigations.

Method G uses sodium formate as ultimate reductant with palladium on charcoal as the heterogeneous catalyst [58]. In this case, very similar reaction mechanism is operating with fine palladium particles like with palladium(0) complexes. The coupling reactions of aryl bromides and chlorides are carried out in concentrated aqueous sodium hydroxide solution [60] in the presence of surfactant like cetyltrimethylammonium bromide [58], benzyltriethylammonium chloride [59], or tetra-*n*-butylammonium bromide [60] which apparently play an important role. Xylene is also used as cosolvent [58]. This method is effective in the coupling reactions of aryl iodides and bromides affording the corresponding symmetrical biaryls in good yields. Although the major side-reaction, the reductive dehalogenation, is sometimes extended to approx. 40%, this method is simple and cheap because the expensive Pd-C catalyst can be easily removed from the reaction mixture by simple filtration and reused several times [58-60]. Once again, presence of the surfactant may cause difficulties during the work-up procedure. However, under these reaction conditions, many sensitive functional groups like nitro, *O*-benzyl, aldehyde, ester and some other can be reduced, hydrogenolysed or saponified.

Method H is closely related to the previous method since it is catalysed by Pd-C, however, with zinc as terminal reductant. The reactions have been performed in water / acetone (1:1) mixture [61], or water [62,63] as the only reaction solvent. In the latter case, the addition of a nonionic surfactant, polyethylene glycol (PEG-400) [62] or 18-crown-6 [63] was essential to obtain high conversions and high yields of homo-coupled products. The PEG and crown ethers probably provide water-soluble forms of aryl halides, otherwise almost water-insoluble materials. The method is efficient in the homo-couplings of all types of aryl halides, including chlorides at elevated temperatures [62].

The catalytically active nano-sized palladium particles can be also stabilized by using triphenylphosphine and a polar solvent like in *method I*. Similar to *methods A, B* and *F*, the palladium(0) catalyst, *in situ* prepared from palladium(II) chloride and triphenylphosphine (only 1.3 eq. to Pd) and additionally stabilized with tetra-*n*-butyl ammonium bromide (as PTC) and polar reaction solvent, DMF, effectively accelerates the homo-couplings of aryl bromides and chlorides with zinc as the ultimate reductant to give the respective biaryls in good yields. Once again, the reductive dehalogenation as the only side-reaction occurs in considerable amount (max. 20%) [64].

Palladium on charcoal, as the most simple Pd(0)-species, is capable of catalysing the coupling of aryl halides to produce biaryls. The molecular hydrogen is also a suitable stoichiometric reductant, compatible with Pd-C, which converts the Pd(II) back to the catalytically active Pd(0)-species as illustrated in the *method J*. The reactions have been conducted in water in the presence of nonionic surfactant, PEG-400, to provide the water soluble forms, e.g. micelles, of hydrophobic substrates such as chlorobenzene. Due to the nature of reductant, molecular hydrogen, the reactions were performed in an autoclave under pressure of hydrogen (4 atm.). This reaction is apparently an important basis for environmental friendly, inexpensive and economic process for scale-up production of symmetrical biaryls, however, the formation of dehalogenated products (up to 45%) is still an unsolved side-reaction.

Synthesis of biaryls derived form heteroaromatic halides have been carried out by four of above described methods, Table 9.

In comparison to the nickel-catalysed aryl halide coupling reactions, the palladium-catalysed analogues are not sensitive to the steric hindrances and to the presence of nitro group. The nickel(0) complexes are more reactive than the appropriate palladiums, and thus more effective in the reactions with the least reactive aryl chlorides. However, the palladium complexes are more convenient and selective, even in the cross-coupling reaction of two electronically different aryl halides.

Table 9. Selected coupling reactions of heteroaromatic halides using palladium-based catalytic methods

Method [a]	Aryl halide	Time (h)	Yield (%)	Lit. ref.
A	2-bromopyridine	45	92	50
B	3-bromoquinoline	22	79	52
	2-chloroquinoline	96	62	
E	2-bromopyridine	3	75	18
	4-bromopyridine	6	52	
G [b]	2-chloropyridine	24	52	58
	4-chloropyridine	24	46	

[a] In *methods A, B* and *C*, 5 mol% of Pd(OAc)$_2$ was used as precatalyst.
[b] Unoptimized yields

3.5. Cross-coupling reactions of aryl halides to biaryls catalysed by nickel- and palladium-complexes

Synthesis of unsymmetrical biaryls *via* nickel- or palladium-catalysed coupling reaction of aryl halides can be achieved using certain methods [10,17,66-71]. Generally, more reactive nickel(0) complexes provide lower selectivity in the cross-coupling reaction of two different aryl halides than analogous palladium complexes [2,9]. Kochi and coworkers [9] studied reaction of bis(triethylphosphine)-2-methoxyphenylnickel bromide (**90**) with bromobenzene (in sealed tube) where increasing selectivity for cross-coupling was obtained. Cross-coupling product, 2-methoxybiphenyl (**91**) was the major product together with smaller amounts of homo-coupling products, 2,2'-dimethoxybiphenyl (**92**), and biphenyl (**8**), Scheme 18.

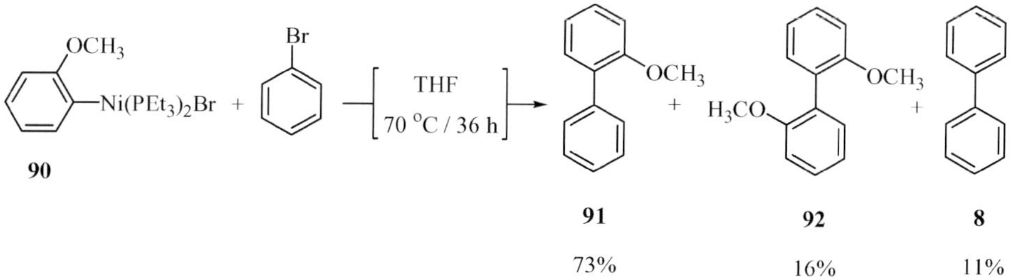

Scheme 18

In the cases of *ortho*-substituted aryl bromides or iodides, selectivities are rarely higher than is the statistic ratio from coupling of two different aryl halides [17,71]. Practical methods for the synthesis of unsymmetrical biaryls using nickel- and palladium-catalysed couplings of two different aryl halides **Ia** and **Ib** and their relative efficiency are presented in the Table 10. The most important fact is that there is no method which is completely chemoselective. Beside the unsymmetrical biaryl **IIa**, the symmetrical ones **IIb** and **IIc** as products of homo-couplings are always produced.

Method A is based on the nickel-catalysed coupling reaction where an increasing selectivity in cross- vs. homo-coupling reactions has been observed in pyridine as the reaction medium [66,67]. The method has some practical importance in the synthesis of unsymmetrical biaryls from chlorobenzenes with electron-withdrawing substituents in the *ortho*-position, e.g. CN group, and bromobenzenes with electron-donating substituents in different positions, e.g. methyl group. Presence of an electron-withdrawing group in the *ortho*-position to the halogen activates them to the oxidative addition reaction. Increased reactivity of one aryl halide, e.g. 2-chlorobenzonitrile, over another one, e.g. 4-bromotoluene, causes the preferential formation of the cross-coupling product. At the same time, homo-coupling reactions are much slower because of either lower reactivity of 4-bromotoluene or significant sterical hindrances of 2-chlorobenzonitrile. However, intramolecular synthesis of unsymmetrical biaryls can be readily affected under ordinary nickel-catalysed conditions, by employing Ni(PPh$_3$)$_4$ *in situ* prepared from Ni(PPh$_3$)$_2$Cl$_2$, PPh$_3$ (2 eq. to Ni) and a reducing agent, such as *n*-BuLi [68].

Method B involving the NiCRA stoichiometric reagent stabilized with 2,2'-bipyridine (bpy), is a very powerful and general method for the smal-scale synthesis of unsymmetrical biaryls [69,70]. Relatively low cross- vs. homo-coupling selectivity is usually observed, thus chromatographic separations are almost always necessary. The method offers some advantages like possibility to couple the least reactive aryl chlorides and halophenols. However, previous preparation of NiCRA reagent is reqiured, with toxic and hazardous nickel waste being produced. Reduction of aryl iodides is another serious side-reaction.

Method C involves the palladium-catalysed coupling of two different aryl halides with Hünig's base as an ultimate reductant. Selectivity for cross- vs. homo-coupling reactions has been achieved in the following instances [17,71]:

  a) one reactant must be a more reactive aryl iodide while another is a less reactive aryl bromide, and

  b) the less reactive aryl bromide must be in four-fold excess to achieve better selectivity.

This method is the most selective, but still suffers from several disadvantages:

  a) it requires relatively drastic reaction conditions, reflux of *p*-xylene, 130 °C for 24-96 h,

b) separation of excess of the less reactive aryl bromides is often difficult,

c) the method requires presence of a stoichiometric amount of n-Bu$_4$NBr which, as a surfactant, may cause problems during work-up of reaction mixtures, and

d) often a large amount of the expensive Pd(OAc)$_2$, 5 mol%, is required.

Table 10. Synthesis of unsymmetrical biaryls using nickel- and palladium-based methods

Method	R$_1$	X$_1$	R$_2$	X$_2$	Yield (%)			Lit.
					IIa	IIb	IIc	ref.
A [a]	4-Me	Br	2-CN	Cl	69	11	12	66, 67
A	4-Me	I	2-CN	Cl	45	21	18	66, 67
A	4-OMe	Br	2-CN	Cl	52	11	10	66, 67
B [b]	4-Me	Br	4-CF$_3$	Cl	58	20	22	69, 70
B	4-OMe	Br	4-CH(OMe)$_2$	Cl	76	12	13	69, 70
C [c]	H	Br	3-NO$_2$	Br	100	-	-	17, 71
C	H	Br	3-CN	Br	79	21	-	17, 71
C	4-OMe	Br	3-CN	Br	71	29	-	17, 71
C	4-Me	Br	2-CN	Br	86		-	17, 71

[a] Method A: 10 mol% NiCl$_2$, 40 mol% PPh$_3$, 4 eq. Zn dust, pyridine, 80 °C, 6 h.

[b] Method B: NiCRA-bpy (2 eq.), KI (1 eq.), C$_6$H$_6$ / THF (1:1), 63 °C , 3 h. NiCRA-bpy-reagent was prepared by dropwise addition of the sodium *tert*-amylate solution in THF to the suspension of NaH and Ni(OAc)$_2$ in refluxing THF containing 2,2'-bipyridine. After 2 h of stirring at 63 °C, the NiCRA-bpy reagent was ready for use.

[c] Method C: 5 mol% Pd(OAc)$_2$, (*i*-Pr)$_2$NEt (1 eq.), n-Bu$_4$NBr (1 eq.), *p*-xylene, 130 °C, 24-112 h. Four-fold excess of less reactive aryl halide was used

Palladium-based methods for coupling of aryl halides always require more drastic reaction conditions than nickel based methods because of lower reactivity of palladium-complexes in the same reactions. However, Pd(PPh$_3$)$_4$, *in situ* formed by electrochemical reduction of mixture of Pd(PPh$_3$)$_2$Cl$_2$ and PPh$_3$, catalyses the cross-coupling of two different aryl iodides with similar selectivity, but at room temperature [10]. Unsymmetrical biaryls prepared by any of these methods were usually purified from homo-coupling side-products using preparative chromatography, preparative TLC [17,71] or fractional distillation [66,67].

A somewhat specific method for synthesis of unsymmetrical phenylpyridines from iodopyridines and various 2-nitro-bromobenzene derivatives was developed by Shimizu's group [72]. It has been known for a long time that the reactive 2-nitro-bromobenzenes readily undergo an oxidative addition to the copper bronze under relatively mild reaction conditions, e.g. 100-120 °C (see Chapter 2), to give the respective arylcoppers. Shimizu's method is based on this reaction with a subsequent transmetallation reaction of thus formed arylcoppers to the arylpalladium reagents *in situ* generated from the less reactive aryl halide, iodopyridines. The unsymmetrical biaryl is actually formed *via* reductive elimination from the corresponding diarylpalladium(II) complexes, and not from diarylcoppers. In this manner, Shimizu's method is really an original alternative to the above presented palladium-catalysed methods. It also has some similarities to the cross-coupling reactions of aryl halides with separately prepared arylmetallic reagents. Thus 2-nitro-bromobenzene derivatives readily react with iodopyridines in the presence of copper bronze (4 eq.) and a catalytic amount (2-5 mol%) of $Pd(PPh_3)_4$ in dimethylsulfoxide at elevated temperatures to produce the unsymmetrical 2-nitrophenylpyridine derivatives in good to excellent yields. In this fashion, the reaction of 3-iodopyridine (**93**) with 2-nitro-bromobenzene (**94**) afforded the cross-coupling product **95** with a 90% yield, respectively, as well as symmetrical biaryls side-products, 3,3'-bipyridine (**96**) in 6% and 2,2'-dinitrobiphenyl (**20**) with a 54% yield [72], Scheme 19.

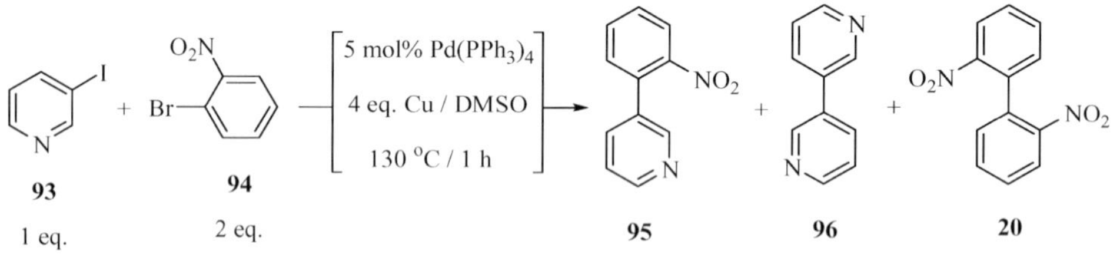

Scheme 19

Unfortunately, this interesting reaction has not been studied in the synthesis of other classes of biaryls.

3.5. Homo-coupling reactions of aryl halides to biaryls mediated by copper(I) salts

Copper(I) salts like CuCl, CuBr, Cu(I) trifluoromethanesulfonate (CuOTf) and above all, copper(I) thiophene-2-carboxylate (CuTC), promote the coupling reactions of aryl, heteroaryl, and vinyl iodides and bromides to biaryls at room temperature [6,11,21-23,73]. With simple copper(I) salts, e.q. CuCl or CuBr, the coupling of activated methyl 2-iodobenzoate in *N*-methylpyrrolidinone (NMP) at room temperature is very slow [73]. Moreover, the halide-exchange reaction occurs between the substrate and CuX. However, aryl iodide or bromide must possess a coordinating *ortho*-substituent, otherwise the copper(I) salts fail to give the coupling reaction. The oxidative addition product, arylcopper(III) intermediate, is stabilized by *ortho*-coordinating substituent [11,73]. Although Cohen [6,11] was successfully used CuOTf in the homo-couplings of 2-nitro-bromobenzene and a few other substrates, Libeskind's CuTC-based method is very efficient and has a huge practical value [73]. Thiophene-2-carboxylic acid salt was chosen because of additional stabilization of oxidative addition product, arylcopper(III) intermediate **XV**, as well as because of its stability to air and rate of reaction with aryl halides.

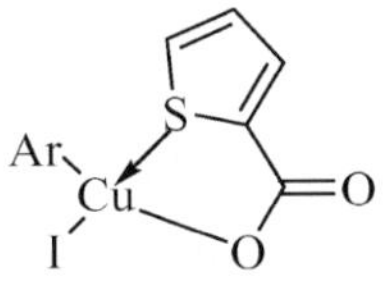

XV

Cu-TC promoted homo-couplings of aryl halides are presented in the Table 11. Among heteroaryl halides, 2-iodothiophene (**97**) was coupled to 2,2'-bithiophene (**98**) with a 77% yield [73], respectively, Scheme 20.

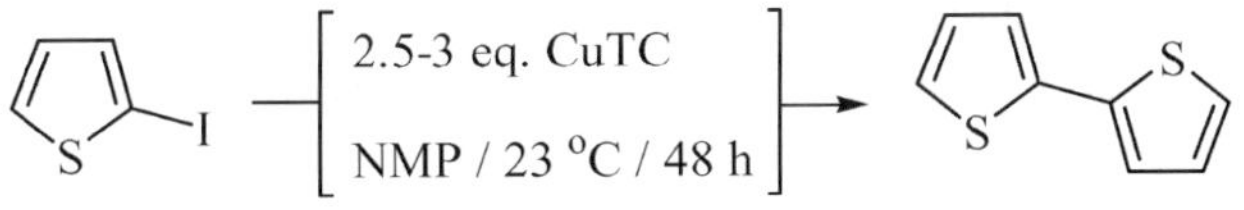

97   **98**

Scheme 20

Whiteside's group has described the oxidative coupling of lower order cuprate reagents, Ar$_2$CuLi, in the presence of oxygen at -78 °C, to give the symmetrical biaryls in good yields [74]. On this basis, Lipshutz and coworkers [75] have found that arylcyanocuprates, ArCu(CN)Li, with an equimolar amount of an aryllithium, Ar'Li, at -125 °C give higher order reagents, Ar(Ar')Cu(CN)Li$_2$. The latter, upon molecular oxygen-oxidation at this temperature lead to the formation of unsymmetrical biaryls in good yields, whereas the amounts of homo-coupling side-products are minimized.

Table 11. Copper(I) thiophene-2-carboxylate (CuTC) induced reductive coupling of aryl halides [73]

R	X	Reaction conditions	Yield (%)
COOMe	I	23 °C / 1 h	97
NO$_2$	I	23 °C / 0.5 h	92
NO$_2$	Br	70 °C / 2 h	86
CH$_2$NMe$_2$	I	23 °C / 1 h	97
CH$_2$CONHMe	I	23 °C / 12 h	51
CH$_2$OH	I	70 °C / 24 h	48
NMeCOMe	I	70 °C / 24 h	83

In this manner, 2-tolyllithium (**99**) was reacted with CuCN to give pale yellow arylcyanocopper reagent **100**. This was colled to -125 °C using a pentane / liquid nitrogen bath to rich the conditions for the reaction with the second aryllithium, 2-metoxyphenyllithium (**101**). At such low temperature the ligand-exchange reaction is too slow to produce the statistic mixture of all three diarylcuprates from the originally generated unsymmetrical diarylcuprate **102** type Ar(Ar')Cu(CN)Li$_2$. The latter was converted with oxygen at this temperature to the respective biaryl **103** in 81% yield [75], Scheme 21. The Lipshutz method is successful in the synthesis of variously substituted biaryls, however, more reactive substituents are precluded and cryogenic reaction conditions are the most serious limit of this interesting reaction.

The mechanism of copper(I) induced coupling of aryl halides has some similarities with nickel(0) catalysed reaction. It is proved that aryl halides undergo oxidative addition to copper(I) reagent through the reversible process to from arylcopper(III) intermediate [11,21,73]. The nature of further process to biaryl is not quite clear [21]. A number of evidences show that reduction of arylcopper(III) intermediate with excess of Cu(I) salt occurs to form organometallic of lower oxidation state, obviously arylcopper(I) compounds [22]. Arylcopper(I) triflate compounds having *ortho*-coordinating substituents, e.g. dimethylamino group, form complicated polynuclear arylcopper triflate complexes, for example $Ar_4Cu_6OTf_2$ [22]. These complexes smoothly decompose in benzene solution at room temperature to form biaryls in high yields. This method does not have a practical value for synthesis of symmetrical biaryls. However, it is very fruitful contribution to understanding the reductive elimination of biaryl molecule from the polynuclear arylcopper intermediate [22,23].

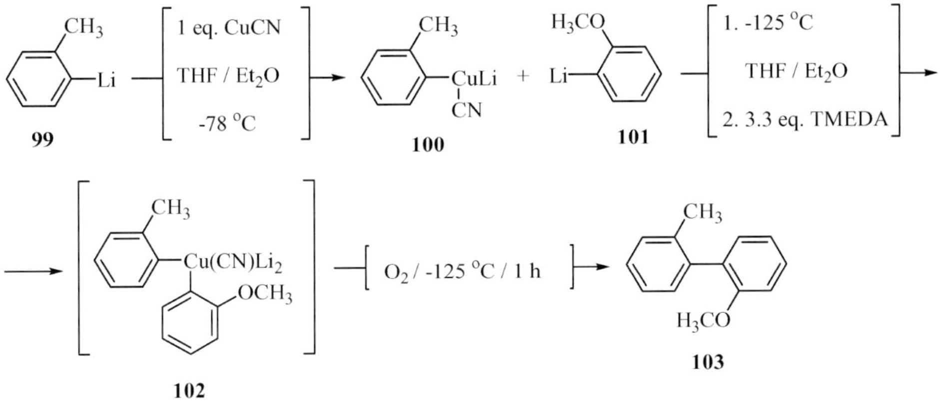

Scheme 21

Copper-based methods for synthesis of biaryls from aryl halides may become dominant in the future research.

3.6. Homo-coupling reactions of aryl halides to biaryls promoted by active metals

Closely related to the coupling reactions with zero-valent nickel or palladium complexes, the active forms of nickel [16,76,77], cobalt [77], as well as indium [78] are effective reagents for the homo-coupling of aryl halides. Thus Rieke nickel, obtained by the well known reduction of nickel(II) iodide with lithium (2.3 eq.) in the presence of naphthalene (10 mol%) as an electron carrier in diglyme as solvent at room temperature, readily affects the homo-coupling reactions of *para-* and *meta-* substituted aryl iodides at elevated temperatures. However, the reactions with aryl chlorides failed, whereas *ortho-*substituted aryl bromides and iodides gave predominant reductive dehalogenation products [16,76]. Moreover, nitro-substituted aryl iodides failed to undergo the coupling reaction with Rieke nickel, analogously to all nickel-catalysed processes, due to deactivation of nickel, see above [16]. Activated nickel, prepared by electrochemical reduction according to the Cheng's procedure [77], is apparently very reactive and slightly more selective, providing an effective homo-coupling reaction of *para-* and *meta-*substituted aryl iodides and bromides. Cheng's activated nickel readily couples even aryl chlorides in the presence of a stoichiometric amount of potassium iodide, which facilitates the oxidative addition of these, the least reactive substrates. The efficiency of homo-coupling reactions promoted by activated nickel is shown in the Table 12. Cheng's activated nickel is obtained by electrolysis of aqueous nickel(II) sulfate solution with mercury as the cathode, followed by high-vacuum distillation of mercury from the nickel-amalgam, leaving a finely divided and very reactive black nickel powder.

Table 12. Synthesis of symmetrical biaryls by homo-coupling reactions with activated nickel powder [77]

Aryl halide	Time (h)	Yield (%)
Iodobenzene	19	73
Bromobenzene	21.5	71
Chlorobenzene	60	90 [a]
p-Bromotoluene	23	71
m-Bromotoluene	22	68
o-Bromotoluene	21.5	27
p-Chlorotoluene	58	83 [a]
p-Chloroacetophenone	22	90 [a]

[a] In the presence of KI (1 eq.)

In comparison, bromobenzene, when heated with commercial nickel powder under the same reaction conditions for 80 h gave only 2.9% biphenyl (**8**) [77]. Activated cobalt powder, prepared by the Cheng's amalgam-method, behaves similarly to nickel powder and affected the homo-coupling of bromobenzene to biphenyl in 71% yield [77].

Furthermore, besides activated nickel, cobalt, and copper (see Chapter 2), indium has been proved as an effective active metal for the coupling of aryl iodides. Thus by heating an equimolar amount of indium and aryl iodides in refluxing DMF, symmetrical biaryls were furnished with high to excellent yields [78]. The indium-promoted homo-coupling reaction tolerates the presence of free hydroxy and carboxylic acid groups well. For example, 2-iodobenzoic acid (**104**) was converted to diphenic acid (**57**) in 75% yield, Scheme 22.

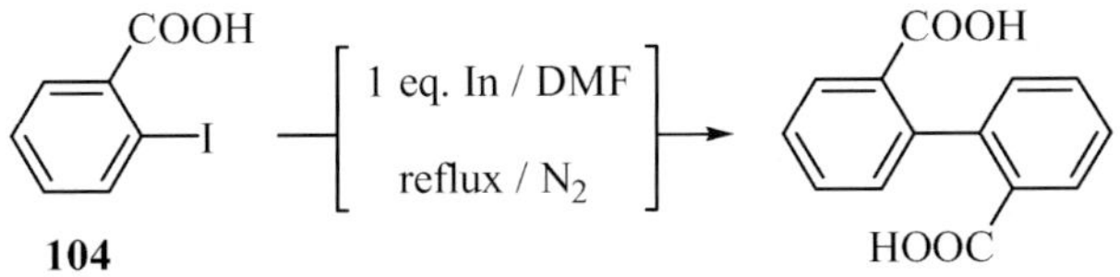

Scheme 22

3.7. Selected synthetic procedures

3.7.1. Preparation of 4,4'-diformylbiphenyl (105) from 4-bromobenzaldehyde (106) using Semmelhack's Ni(COD)$_2$ method [1]

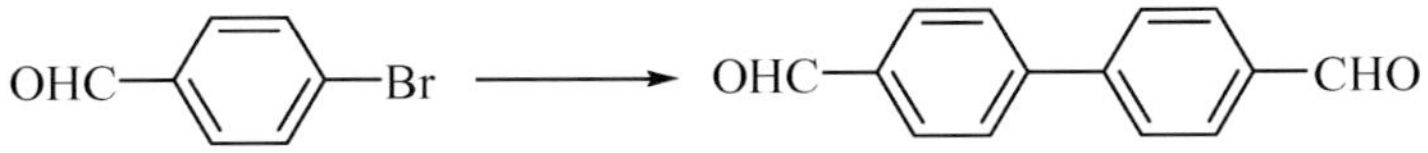

A solution of 4-bromobenzaldehyde (**106**, 1.85 g, 10 mmol) in dry *N,N*-dimethyl formamide (10 ml) was added dropwise to a suspension of bis(1,5-cyclo octadiene)nickel(0) (Ni(COD)$_2$, 1.53 g, 5.6 mmol) in *N,N*-dimethylformamide (10 ml) under nitrogen at 25 °C. The reaction mixture was stirred at 35 °C for 20 h, then partitioned between 3% aqueous hydrochloric acid (20 ml) and dichloromethane (20 ml). After being filtered to remove a finely divided nickel, the organic layer was washed with water, dried (Na$_2$SO$_4$), and concentrated to dryness. The residue was purified with diethyl ether (2-3 ml) to obtain 0.83 g (79%) of pure 4,4'-diformyl biphenyl (**105**) as colourless crystals.

3.7.2. Preparation of 4,4'-diformylbiphenyl (105) from 4-chlorobenzaldehyde (107) catalysed by Ni(PPh$_3$)$_4$ in the presence of Zn-dust [14]

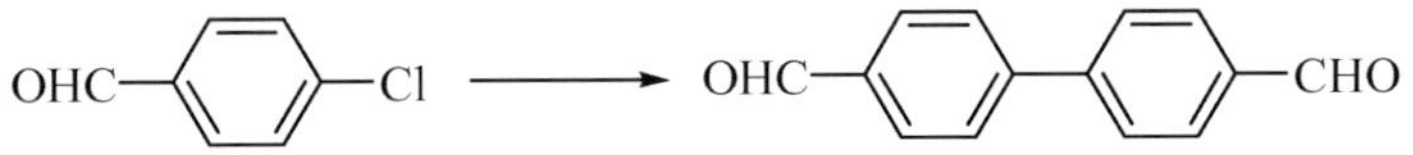

A 2-l, four necked, dried round bottomed flask equiped with a 500 ml addition funnel, mechanical stirrer, thermometer, and stopcock adapter was charged with Zn-powder (131.00 g, 2.0 mol), sodium bromide (15.00 g, 0.15 mol), triphenylphosphine (PPh$_3$, 131.00 g, 0.5 mol) and anhydrous nickel(II) chloride (6.50 g, 0.05 mol). The reaction flask was evacuated and filled with nitrogen several times. Nitrogen was then swept through the system by opening the stopcock to the addition funnel for 15 min. Nitrogen-purged, dry DMF (600 ml) was added through the addition funnel and the mixture was heated by an electric mantle. The solution quickly became red-brown (from Ni(PPh₃)₄). A solution of 4-chlorobenzaldehide (**107**, 145.00 g, 1.0 mol) in dry DMF (100 ml) was placed in the addition funnel. When the catalyst mixture reached 60 °C the heating mantle was removed and slow addition of the 4-chlorobenzaldehyde solution was begun. The reactant was added dropwise during 60 minutes. The reaction mixture was then maintained at 70 °C for 3 hours. The mixture was cooled, filtered to remove excess Zn and the filtrate diluted with dichloromethane (2000 ml). This solution was extracted with distilled water (4x500 ml) and with saturated NaCl solution (200 ml). The organic layer was dried (Na₂SO₄) for 1 h, filtered and evaporated. The residue was extracted with *n*-heptane / toluene (1000 ml, 1:1, v/v) solvent mixture. The product was filtered and dried to give 65 g (62%) of pure (> 99%, HPLC) 4,4'-diformylbiphenyl (**105**).

3.7.3. Preparation of 2,2'-biquinoline (108) from 2-chloroquinoline (109) catalysed by Ni(PPh₃)₂Br₂ in the presence of Zn and Et₄NI [32]

A 50-ml, round-bottomed, two necked flask containing a magnetic stirring bar was charged with bis(triphenylphosphino)nickel(II) bromide (743 mg, 1 mmol, 10 mol%), zinc-dust (981 mg, 15 mmol, 1.5 eq.) and tetraethylammonium iodide (2.57 g, 10 mmol, 1 eq.). A rubber septum was placed over one neck of the flask and a 3-way stopcock adapter attached and filled with argon several times. Dry *N,N*-dimethyl formamide (20 ml) was added *via* syringe through the septum. The reaction mixture was stirred at room temperature. After the dark brown catalyst had formed (30 min), an argon-purged solution of 2-chloroquinoline (**109**, 1.64 g, 10 mmol) in THF (10 ml) was added *via* syringe to the reaction mixture. The resulting mixture was heated at 50 °C for 20 h and then cooled and filtered. The solid mass was washed with dichloromethane (3x20 ml), and the filtrate and washings were evaporated to dryness. The residue was separated by column chromatography on silica gel (benzene / ethyl acetate or dichloromethane / *n*-hexane) to give 1.08 g (84%) of pure 2,2'-biquinoline (**108**).

3.7.4. Preparation of 4,4'-dimethoxybiphenyl (77) from 4-chloroanisole (76) by NiCRA-bpy method [43]

H$_3$CO—⟨ ⟩—Cl ⟶ H$_3$CO—⟨ ⟩—⟨ ⟩—OCH$_3$

76   **77**

A mixture of *tert*-amyl alcohol (2.2 ml, 1.77 g, 20 mmol) and anhydrous THF (10 ml) was added dropwise to a suspension of degreased sodium hydride (1.44 g, 60 mmol) and dried nickel(II) acetate (1.77 g, 10 mmol) in refluxing THF (20 ml) containing 2,2'-bipyridine (**5**, 3.12 g, 20 mmol). The reaction mixture was refluxed for 2 h to obtain a solution of NiCRA-bpy reagent. Then, the solution of 4-chloroanisole (**76**, 1.43 g, 10 mmol) in THF (10 ml) was added dropwise. Reaction mixture was heated under reflux for 10 h. After completion of the reaction, the excess hydride was carefully destroyed by dropwise addition of ethanol at 25 °C until hydrogen evolution ceased. The mixture was then acidified and organic phase was extracted into diethyl ether and dried (MgSO$_4$). After removal of the solvents, the product was separated by flash chromatography to obtain 0.93 g (87%) of pure 4,4'-dimethoxybiphenyl (**77**), m.p. 172-173 °C.

3.7.5. Preparation of 4,4'-dimethoxybiphenyl (77) from 4-bromoanisole (110) catalysed by Pd(OAc)$_2$ in the presence of Hünig's base [52]

H$_3$CO—⟨ ⟩—Br ⟶ H$_3$CO—⟨ ⟩—⟨ ⟩—OCH$_3$

110   **77**

A mixture of diisopropylethylamine (Hünig's base, 1.74 ml, 10 mmol), palladium acetate (112 mg, 0.5 mmol, 5 mol%), tetra-*n*-butylammonium bromide (1.61 g, 5 mmol) and 4-bromoanisole (**110**, 1.87 g, 10 mmol) in DMF (1.25 ml) was stirred under nitrogen at 115 °C for 96 h. After cooling to room temperature, water (50 ml) and diethyl ether (50 ml) were added. The organic phase was separated, washed with water and dried (MgSO$_4$). The solvent was evaporated *in vacuo*. Pure 4,4'-dimethoxy biphenyl (**77**, 514 mg, 48%) was obtained after purification with preperative chromatography, m.p. 172-173 °C.

3.7.6. Preparation of 4,4'-dimethoxybiphenyl (77) from 4-bromoanisole (110) catalysed by Pd(PhCN)Cl₂ in the presence of TDAE [18]

A mixture of 4-bromoanisole (**110**, 1.87 g, 10 mmol), tetrakis(dimethylamino)ethylene (TDAE, 1.4 ml, 1.21 g, 6 mmol, 60 mol%), and Pd(PhCN)$_2$Cl$_2$ (192 mg, 0.5 mmol, 5 mol%) in DMF (5 ml) was heated at 50 °C for 4 h. After cooling to room temperature, water (50 ml) and dichloromethane (50 ml) were added. The organic layer was separated, washed with water, dried (Na$_2$SO$_4$), filtered and evaporated. From the residue, 986 mg (92%) of pure 4,4'-dimethoxybiphenyl (**77**) was obtained by flash chromatography, m.p. 172-173 °C.

3.7.7. Preparation of 4,4'-dimetoxybiphenyl (77) by Pd-C catalysed coupling of 4-bromoanisole (110) in the presence of sodium formate [58]

Sodium formate (6.80 g), cetyltrimethylammonium bromide (4.00 g), 3% palladium on charcoal (2 g, 50% paste), sodium hydroxide (64.00 g), and 4-bromoanisole (**110**, 1.87 g, 10 mmol) were mixed in distilled water (260 ml) and then stirred under reflux for 4 h. A further quantity of sodium formate (6.80 g) was added and stirring at the boiling point under reflux continued for 20 h. After cooling to room temperature, water (50 ml) and dichloromethane (50 ml) were added. The organic layer was separated, washed with water, dried (Na$_2$SO$_4$), filtered and evaporated. From the residue, 525 mg (49%) of pure 4,4'-dimethoxybiphenyl (**77**) was obtained by crystallization from chloroform, m.p. 172-173 °C.

3.7.8. Preparation of 2,2'-dimethoxycarbonylbiphenyl (111) from methyl 2-iodobenzoate (25) by copper(I) thiophene-2-carboxylate method [73]

25

111

Copper(I) thiophene-2-carboxylate (CuTC, 5.70 g, 30 mmol, 3 eq.) was added in one portion to a solution of methyl 2-iodobenzoate (**25**, 2.62 g, 10 mmol) in *N*-methyl pyrrolidinone (30 ml, NMP) under a nitrogen atmosphere. After being stirred at room temperature for 1 h, the mixture was diluted with ethyl acetate (100 ml), and the resulting slurry was passed through a plug of silica gel using ethyl acetate as eluent (500 ml). Solvents were removed by evaporation and then vacuum distillation (NMP).

From the residue, 2.62 g (97%) of pure 2,2'-dimethoxycarbonylbiphenyl (**111**) was obtained by preparative chromatography (dichloromethane / *n*-hexane).

3.7.9. Preparation of 4'-methyl-2-cyanobiphenyl (112) by cross-coupling of 4-bromotoluene (113) with 2-chlorobenzonitrile (114) applying Ni(PPh₃)₂Cl₂ / Zn method [66,67]

Into a 1-l four necked flask, dry pyridine (500 ml), 4-bromotoluene (**113**, 44.30 g, 0.259 mol), 2-chlorobenzonitrile (**114**, 35.60 g, 0.259 mol), anhydrous nickel(II) chloride (3.24 g, 0.025 mol, 10 mol%), triphenylphosphine (PPh₃, 13.10 g, 0.05 mol, 2 eq. to Ni) and zinc-powder (67.60 g, 1.034 mol, 4 eq.) were charged under a nitrogen atmosphere and stirred at room temperature for 30 minutes (formation of Ni(0) catalyst). The temperature was raised to 80 °C and the mixture was stirred for additional 6 hours. Then, toluene (400 ml) was added, and the solid filtered off while it was still hot. The solvents were recovered under reduced pressure. Then, toluene (500 ml) was added to the residue, and the solid was filtered off while it was still hot and washed with hot toluene (100 ml). The warm toluene layer was washed with 5% aqueous hydrochloric acid (300 ml), then with 5% aqueous sodium hydrogencarbonate solution (300 ml) and then with warm water (2x300 ml). The toluene layer was concentrated to obtain 28.4 g (57%) of pure (ca. 97%, HPLC) 4'-methyl-2-cyanobiphenyl (**112**) as a white crystalline substance, m.p. 49-50 °C (cyclohexane).

3.7.10. Preparation of 4'-methyl-2-cyanobiphenyl (112) by cross-coupling of 4-iodotoluene (115) with 2-bromobenzonitrile (116) using Pd(OAc)₂ - Hünig's base method [17]

A mixture of diisopropylethylamine (Hünig's base, 1.74 ml, 10 mmol), palladium(II) acetate (112 mg, 0.5 mmol, 5 mol%), tetra-*n*-butylammonium bromide (1.61 g, 5 mmol), 4-iodotoluene (**115**, 2.18 g, 10 mmol) and 2-bromobenzonitrile (**116**, 7.28 g,

40 mmol, 4 eq.) in *p*-xylene (10 ml) was stirred under a nitrogen atmosphere at 130 °C for 63 h. After cooling to room temperature, water (50 ml) and diethyl ether (50 ml) were added. The organic phase was separated, washed with water and dried ($MgSO_4$). The solvent was evaporated and residue distilled under reduced pressure to eliminate excess of 2-bromobenzonitrile. Then, the 4'-methyl-2-cyanobiphenyl (**112**) was purified by preparative chromatography to obtain 638 mg (33%) of pure product, m.p. 48-49 °C.

3.8. Conclusion

This chapter has attempted to show recent achievements in the methods for the synthesis of biaryls by coupling reactions of aryl halides and related aryl sulfonates. Any kind of substrate can be homo-coupled to form symmetrical biaryls by using the correct method, while cross-coupling reactions of two different aryl halides have more limitations. The most important practical methods use nickel-phosphine complexes in catalytic amounts with a suitable ultimate reductant. Nickel complexes, $Ni(PPh_3)_2X_2$ (X = Cl, Br, I) in the presence of a varying amount of triphenylphosphine have proved to be the most efficient pre-catalysts for the coupling reaction of aryl halides or aryl sulfonates with excess of active metals (Zn, Mg, Al, Mn) as ultimate reductants. Methods based on far more expensive palladium salts as catalysts have given better results in the cross-coupling reactions. Lower reactivity of palladium(0) complexes versus nickel(0) complexes with aryl halides is crucial for better selectivity in the synthesis of unsymmetrical biaryls. Among copper(I) salts, copper(I) thiophene-2-carboxylate (CuTC), is the most versatile stoichiometric reagent for synthesis of symmetrical biaryls derived from aryl iodides and bromides having *ortho*-coordinating substituents. Beside nickel, palladium, and copper(I) complexes, activated nickel-, cobalt, and indium-powders are effective reagents for the homo-couplings of *para*- and *meta*-substituted aryl iododes and bromides to symmetrical biaryls.

3.9. References

1.   M. F. Semmelhack, P. M. Helquist and L. D. Jones, J. Am. Chem. Soc. 93 (1971) 5908.
2.   M. F. Semmelhack, P. Helquist, L. D. Jones, L. Keller, L. Mendelson, L. S. Royono, J. G. Smith and R. D. Stauffer, J. Am. Chem. Soc. 103 (1981) 6460.
3.   A. S. Kende, L. S. Libeskind and D. M. Braitsch, Tetrahedron Lett. (1975) 3375.
4.   F. R. S. Clark, R. O. C. Norman and C. B. Thomas, J. Chem. Soc., Perkin Trans 1 (1975) 121.

5. M. Zemabayashi, K. Tamao, J. Yoshida and M. Kumada, Tetrahedron Lett. (1977) 4089.
6. T. Cohen and I. Cristea, J. Org. Chem. 40 (1975) 3649.
7. M. Sainsbury, Tetrahedron 36 (1980) 3327.
8. V. N. Kalinin, Synthesis (1992) 413.
9. T. T. Tsou and J. K. Kochi, J. Am. Chem. Soc. 101 (1979) 7547.
10. C. Amatore, E. Carre, A. Jutand, H. Tanaka, Q. Ren and S. Torii, Chem. Eur. J. 2 (1996) 957.
11. T. Cohen and I. Cristea, J. Am. Chem. Soc. 98 (1976) 748.
12. S. Iyer, J. Organomet. Chem. 490 (1995) C27.
13. F. A. Cotton and G. Wilkinson, Advanced Inorganic Chemistry, John Wiley & Sons, New York, 1988.
14. I. Colon and D. R. Kelsey, J. Org. Chem. 51 (1986) 2627.
15. V. Percec, J. Y. Bae, M. Zhao and D. H. Hill, J. Org. Chem. 60 (1995) 176.
16. H. Matsumoto, S. Inaba and R. D. Rieke, J. Org. Chem. 48 (1983) 840.
17. J. Hassan, C. Hathroubi, C. Gozzi and M. Lemaire, Tetrahedron 57 (2001) 7845.
18. M. Kuroboshi, Y. Waki and H. Tanaka, Synlett (2002) 637.
19. M. Ohff, A. Ohff and D. Milstein, Chem. Commun. (1999) 357.
20. J. M. Brunel, M. H. Hirlemann, A. Henmann and G. Buono, Chem. Commun. (2000) 1869.
21. T. Cohen, J. Wood and A. G. Dietz, Tetrahedron Lett. (1974) 3555.
22. G. van Koten, J. T. B. H. Jastrzebski and J. G. Noltes, J. Org. Chem. 42 (1977) 204.
23. G. van Koten, J. T. B. H. Jastrzebski and J. G. Noltes, J. Chem. Soc., Chem. Commun. (1977) 203.
24. B. Bogdanović, M. Kroner and G. Wilke, Justus Liebigs Ann. Chem. 699 (1966) 1.
25. M. Mori, Y. Hashimoto and Y. Ban, Tetrahedron Lett. 21 (1980) 631.
26. K. W. R. de Franca, M. Navarro, É. Léonel, M. Durandetti and J.-Y. Nédélec, J. Org. Chem. 67 (2002) 1838.
27. F. Raynal, R. Barhdadi, J. Périchon, A. Savall and M. Troupel, Adv. Synth. Catal. 344 (2002) 45.
28. K. Takagi, N. Hayama and S. Inokawa, Bull. Chem. Soc. Jpn. 53 (1980) 3691.
29. K. Takagi, N. Hayama and K. Sasaki, Bull. Chem. Soc. Jpn. 57 (1984) 1887.
30. C. Chen, Synlett (2000) 1491.
31. H. Tanaka, S. Sumida, N. Kobayashi, N. Komatsu and S. Torii, Inorg. Chim. Acta 222 (1994) 323.
32. M. Iyoda, H. Otsuka, K. Sato, N. Nisato and M. Oda, Bull. Chem. Soc. Jpn. 63 (1990) 80.
33. R. S. Berman and J. K. Kochi, Inorg. Chem. 19 (1980) 248.

34. I. Colon, J. Org. Chem. 47 (1982) 2622.

35. T. Nakamura, Tetrahedron Lett. (1974) 463.

36. A. Kikukawa, Bull. Chem. Soc. Jpn. 52 (1979) 1493.

37. M. Tiecco, L. Testaferri, M. Tingoli, D. Chianelli and M. Montanucci, Synthesis (1984) 736.

38. C. Janiak, S. Deblon and H.-P. Wu, Synth. Commun. 29 (1999) 3341.

39. B. Krische, J. Hellberg and C. Lilja, J. Chem. Soc., Chem. Commun. (1987) 1476.

40. M. Iyoda, K. Sato and M. Oda, Tetrahedron Lett. 26 (1985) 3829.

41. H. M. Colquhoun, Z. Zhu and D. J. Williams, Org. Lett. 3 (2001) 4031.

42. A. Jutand and A. Mosleh, J. Org. Chem. 62 (1997) 261.

43. M. Lourak, R. Vanderesse, Y. Fort and P. Caubère, J. Org. Chem. 54 (1989) 4840.

44. P. Caubère, Angew. Chem., Int. Ed. Engl. 22 (1983) 597.

45. J.-J. Brunet, D. Besozzi, A. Courtois and P. Caubère, J. Am. Chem. Soc. 104 (1982) 7130.

46. R. Vanderesse, J. J. Brunet and P. Caubère, J. Organomet. Chem. 264 (1984) 263.

47. R. Vanderesse, M. Lourak, Y. Fort and P. Caubère, Tetrahedron Lett. 27 (1986) 5483.

48. R. Hong, R. Hoen, J. Zhang and G.-q. Lin, Synlett (2001) 1527.

49. T. Yamamoto, Y. Hayashi and A. Yamamoto, Bull. Chem. Soc. Jpn. 51 (1978) 2091.

50. V. Penalva, J. Hassan, L. Lavenot, C. Gozzi and M. Lemaire, Tetrahedron Lett. 39 (1998) 2559.

51. D. L. Boger, J. Goldberg and C.-M. Andersson, J. Org. Chem. 64 (1999) 2422.

52. J. Hassan, V. Penalva, L. Lavenot, C. Gozzi and M. Lemaire, Tetrahedron 54 (1998) 13793.

53. F. T. Luo, A. Jeevanandam and M. K. Basu, Tetrahedron Lett. 39 (1998) 7939.

54. N. Shezad, A. A. Clifford and C. M. Rayner, Green Chem. 4 (2002) 64.

55. R. Nakajima, Y. Shintani and T. Hara, Bull. Chem. Soc. Jpn. 53 (1980) 1767.

56. D. D. Hennings, T. Iwama and V. H. Rawal, Org. Lett. 1 (1999) 1205.

57. D. Albanese, D. Landini, M. Penso and S. Petricci, Synlett (1999) 199.

58. P. Bamfield and P. Quan, Synthesis (1978) 537.

59. G. R. Newkome, W. E. Puckett, G. E. Kiefer, V. K. Gupta, Y. Xia, M. Coreil and M. A. Hackney, J. Org. Chem. 47 (1982) 4116.

60. S. Mukhopadhyay, S. Ratner, A. Spernat, N. Qafisheh and Y. Sasson, Org. Proc. Res. & Dev. 6 (2002) 297.

61. S. Venkatraman and C.-J. Li, Org. Lett. 1 (1999) 1133.

62. S. Mukhopadhyay, G. Rothenberg, D. Gitis and Y. Sasson, Org. Lett. 2 (2000) 211.
63. S. Venkatraman, T. Huang and C.-J. Li, Adv. Synth. Catal. 344 (2002) 399.
64. N. Qafisheh, S. Mukhopadhyay and Y. Sasson, Adv. Synth. Catal. 344 (2002) 1079.
65. S. Mukhopadhyay, G. Rothenberg, H. Wiener and Y. Sasson, Tetrahedron 55 (1999) 14763.
66. H. Kageyama, Synlett (1994) 371.
67. H. Kageyama, U.S. Pat. 5,380,910 (1995).
68. G. Bringmann, J. Hinrichs, P. Henschel, K. Peters and E.-M. Peters, Synlett (2000) 1822.
69. M. Lourak, R. Vanderesse, Y. Fort and P. Caubère, Tetrahedron Lett. 29 (1988) 545.
70. M. Lourak, R. Vanderesse, Y. Fort and P. Caubère, J. Org. Chem. 54 (1989) 4844.
71. J. Hassan, C. Hathroubi, C. Gozzi and M. Lemaire, Tetrahedron Lett. 41 (2000) 8791.
72. N. Shimizu, T. Kitamura, K. Watanabe, T. Yamaguchi, H. Shigyo and T. Ohta, Tetrahedron Lett. 34 (1993) 3421.
73. S. Zhang, D. Zhang and L. S. Liebeskind, J. Org. Chem. 62 (1997) 2312.
74. G. M. Whitesides, J. San Filippo, C. P. Casey and E. J. Panek, J. Am. Chem. Soc. 89 (1967) 5302.
75. B. H. Lipshutz, K. Siegmann, E. Garcia and F. Kayser, J. Am. Chem. Soc. 115 (1993) 9276.
76. S.-i. Inaba, H. Matsumoto and R. D. Rieke, Tetrahedron Lett. 23 (1982) 4215.
77. C. S. Chao, C. H. Cheng and C. T. Chang, J. Org. Chem. 48 (1983) 4904.
78. B. C. Ranu, P. Dutta and A. Sarkar, Tetrahedron Lett. 39 (1998) 9557.

CHAPTER
4

4. CROSS-COUPLING REACTIONS OF ARYLMETALLIC REAGENTS WITH ARYL HALIDES AND SULFONATES

4.1. Introduction

Arylmetallic reagents, Ar-M, react with aryl halides or triflates in nickel- or palladium-catalysed processes to give biaryls [1]. The aryl-M bond possesses more or less ionic character depending on the electronegativity of metallic counterpart (M). For example, in organolithiums, as the most reactive organometallics of practical value, the nature of aryl-lithium bond is apparently ionic. The latter, as typical nucleophilic reagents as well as very strong bases, react with all common electrophilic species such as H^+, CO_2, aldehydes, ketones, esters, amides, nitriles and even carboxylate anions to furnish the corresponding addition products: parent arenes, arylcarboxylic acids, alcohols, etc. In contrast, the organometallics of the less electronegative metall(oid)s, e.g. Hg, Sn, B, Si, generally do not react in common Grignard-type reactions nor with water or other protic sources (at least not so readily). However in nickel- and palladium-catalysed cross-coupling reactions, the organometallics of both electronegative and electropositive metals act as formal carbanion-donors, whereas the aryl halides are electrophilic reactants. The general reactivity of aryl halides in these reactions follows the usual order: I > Br > Cl. Aryl sulfonates, nitriles and thioethers have been successfully used as alternative reactants, making this approach to biaryls wide applicable. Catalytically active examples are various nickel(0), palladium(0), and sometimes copper(I) complexes. Thus different nickel(II) or palladium(II) pre-catalysts actually used, were *in situ* reduced to the catalytically active zero-valent complexes with an excess of organometallics, prior to the cross-coupling reactions, Scheme 1.

The oldest cross-coupling reactions of aryl halides were accomplished with Grignard and organolithium reagents. Since these do not tolerate several common functional groups, less reactive organometallics were introduced. Herein, beside former reagents, the cross-coupling reactions involving arylmetallic reagents of zinc, tin, silicon, mercury, copper, manganese as well as the slightly less known titanium, indium and germanium reagents are described.

$$Ar-M \quad + \quad X-Ar' \quad \xrightarrow{\text{Pd, Ni (or Cu) - catalysis}} \quad Ar-Ar'$$

M = Li, Mg, Zn, Sn, Si, Hg, Cu, Mn, Ti, In, Ge

X = I, Br, Cl, OTf, etc.

Scheme 1

The cross-coupling reaction of Grignard reagents with aryl halides is well known as the Kharasch reaction. The arylzinc based analogue is called the Negishi reaction, while the organostannanes are employed in the Stille reaction [1]. In general, the reaction is more applicable when the parent organometallic is less reactive and thus more selective, tolerating various functionalities. For instance, the number of successful Negishi cross-coupling reactions is far greater than those involving organolithiums and Grignard reagents all together. However, less reactive and more selective organotins are, after the Suzuki-Miyaura reaction based on arylboronic acids, the most widely used organometallics for the aryl-aryl bond formation. Although certain previously described reactions actually involve organometallics, e.g. arylcopper reagents as the transient Ullmann intermediates, in this Chapter the reactions of well defined and separately prepared organometallics with aryl halides and sulfonates are focused through the following topics:

1. reactions of the organolithium and Grignard reagents with aryl halides and sulfonates
2. the Negishi reaction
3. the Stille reaction
4. reactions of arylsilane reagents with aryl halides and sulfonates
5. reaction of diarylmercurials with aryl halides
6. reaction of arylcopper reagents with aryl halides
7. reactions of miscellaneous organometallics: Mn, Ti, In, and Ge with aryl halides

These reactions are of a great significance in the aryl-alkyl [2-7], aryl-alkenyl [1-3,8], alkenyl-alkenyl [9,10], alkyl-alkenyl [9,10], as well as alkyl-alkyl bond forming reactions, but herein only the aryl-aryl bond formation aspects are presented.

4.2. Reaction mechanism

The nickel- and palladium-catalysed cross-coupling reactions of arylmetallics with aryl halides proceed by the following reaction mechanism and include the following three fundamental processes:

1. oxidative addition of aryl halide (or sulfonate) to the catalytically active nickel(0) or palladium(0) complexes affording arylnickel(II) or arylpalladium(II) intermediates, e.g. $ArPd(PPh_3)_2X$,
2. transmetallation of the aryl-group from the parent organometallic to the arylnickel(II) or arylpalladium(II) species occurs to give the diarylnickel(II) or diarylpalladium(II) complexes, and finally,
3. reductive elimination of a biaryl from the latter complex with regeneration of the catalytically active nickel(0) or palladium(0) complex, thus closing the catalytic cycle.

The reaction mechanism is shown with $Pd(PPh_3)_4$ as the most commonly employed catalyst, Scheme 2.

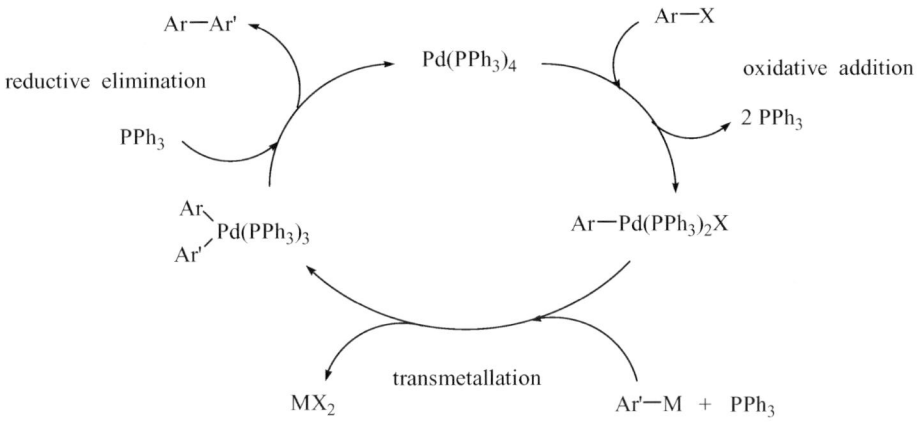

X = I, Br, Cl, OTf, etc.

M = Li, MgX, ZnX, SnR$_3$, Si(OR)$_3$, MnX, Ti(OR)$_3$, InAr'$_2$

Scheme 2

The oxidative addition step probably proceeds through the charge-transfer complex, Ar-X···Pd(PPh$_3$)$_4$, to produce the relatively stable arylpalladium(II) halides (or pseudohalides), see Chapter 3. The formation of a stable ionic metal-salt, e.g. MgCl$_2$, is the driving force for the further reaction step, transmetallation of an aryl-group from the arylmetallics to arylnickel(II) or arylpalladium(II) complex. Generally, all arylmetallics derived from metals which are more electronegative than nickel or palladium can affect the transmetallation reaction. Practically, aryl-Li, Mg, Zn, Hg, B, Al, Si, Sn, Bi, as well as Cu have been found to undergo the transmetallation reaction to nickel or palladium. Finally, the reductive elimination step proceeds *via* the *cis*-diarylnickel(II) or diarylpalladium(II) intermediate. Strong evidences show that the primary generated *trans*-diarylpalladium(II) complex reacts with a coordinating solvent, e.g. THF, DMF or suitable ligand such as PPh₃, to generate an unstable five-coordinative intermediate where the pseudorotation is possible by forming the *cis*-diarylpalladium(II) complex. The dissociation of the ligand occurs with liberation of the *cis*-diarylpalladium(II) complex which is prone to the rapid 1,1-reductive elimination of a biaryl from palladium, see Chapter 3 [11].

4.3. Cross-coupling reactions of aryl halides and sulfonates with organolithium and Grignard reagents, the Kharasch reaction

Aryllithium reagents derived from simple arenes react with aryl halides in the presence of different nickel or palladium catalysts to give biaryls in moderate to good yields. Thus the reaction of 2-lithiated 1-methylindole **117** and 4-iodoanisole (**39**) resulted in the respective biaryl **118** in 68% yield [12], Scheme 3.

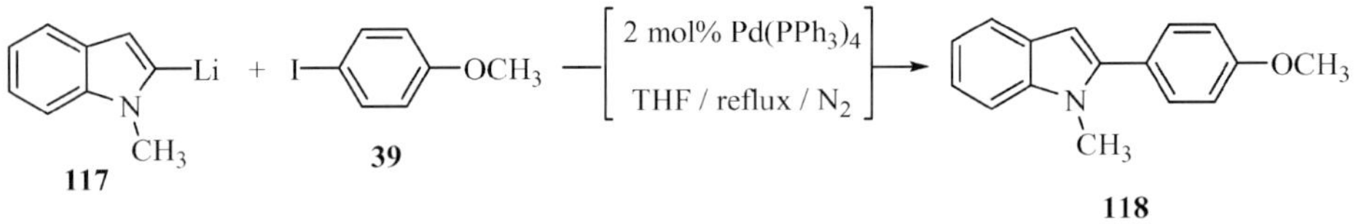

Scheme 3

Since the aryllithiums are highly polar and very strong bases, several side-reactions are possible. Certain aryl halides can undergo the elimination of HX to generate the aryne intermediates that may produce an isomeric biaryl *via* nucleophilic addition to the adjacent carbon atom, Scheme 4.

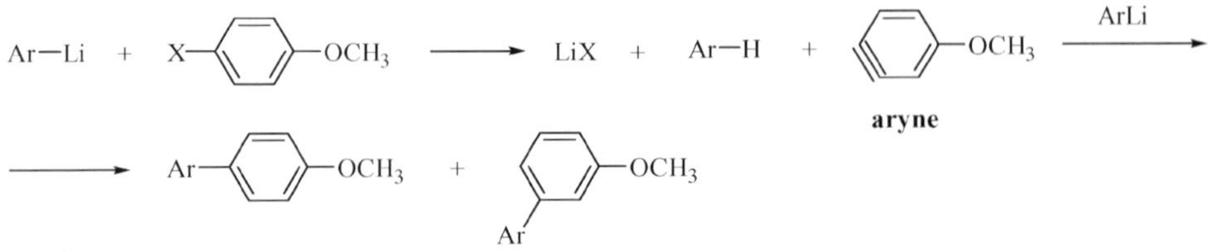

Scheme 4

The latter is known as the elimination-addition mechanism of the nucleophilic aromatic substitution. The high reactivity of aryllithiums precludes the use of several types of functional groups such as aldehydes, ketones, esters, amides, amines, alcohols, phenols, nitro-compounds, carboxylic acids etc. However, properly protected alcohols, phenols or amines can effect the reaction. Various ether protecting groups, e.g. Tr, Me, *i*-Pr, Bn, 4-MeOBn, are employed in the former instances, whereas *N,N*-dibenzyl group is convenient for the amine protection.

Nevertheless, the cross-coupling reactions of simple aryl- or heteroaryl-lithiums have been efficiently accomplished with aryl iodides and bromides bearing no above-mentioned sensitive groups. The aryllithiums are prepared under an inert atmosphere of argon (nitrogen is not completely inert to ArLi) by the following alternative reactions [13]:

- metallation of aryl halides with lithium metal,
- metallation of aryl halides with strong bases: *n-*, *s-*, *t*-BuLi, LDA, etc., well known as halogen-metal exchange reaction [14],
- metallation of an arene with strong bases: *n-*, *s-*, *t*-BuLi, LDA, LIC-KOR (*n*-BuLi / KO*t*-Bu) [15].

Thus obtained aryllithiums are employed in the synthesis of biaryls either directly or *via* transmetallation with $MgBr_2$, $ZnCl_2$, $ClSnR_3$, or $B(OR)_3$, into the ArMgBr, ArZnCl, $ArSnR_3$, $ArB(OR)_2$, as common key-intermediates of the far more convenient Kharasch, Negishi, Stille or Suzuki-Miyaura reactions. The organolithium-based cross-coupling reactions can be catalysed by nickel(0) or palladium(0) complexes of triphenylphosphine (PPh_3), dppe, dppp, and related phosphines. Apart from the catalytic processes, the aryllithiums or aryl Grignard reagents can arylate cyclopalladium complexes to produce the respective *ortho*-functionalized biaryls in moderate yields. The latter complexes are readily obtained in high yields by reaction of *N*-benzylideneanilines with palladium(II) acetate. In this manner, *N*-benzilidene aniline (**119**) was converted to the cyclopalladium complex **120**, which, upon treatment with NaCl, underwent the ligand-exchange reaction affording the complex **120a**. Aryl bromide **121** was lithiated to give **122**. The latter was reacted with complex **120a** to obtain the respective vancomycin precursor, biaryl **123** in 38% overall yield [16], Scheme 5.

Despite the stoichiometric requirements of expensive palladium salts, this reaction can be useful in the synthesis of *ortho*-formyl-, *N,N*-dialkylaminomethyl-, and amino-substituted biaryls since variuos nitrogen containing substrates at benzylic position readily form palladacycles, e.g. *N,N*-dimethylbenzylamine, *N*-benzylidene-*N',N'*-dimethylhydrazone, *O*-methylbenzaldoximes, and *N*-acylanilines [17,18].

The Grignard reagents are far more applicable counterparts in the nickel- and palladium-catalysed Kharasch cross-coupling reactions with aryl halides (I, Br, Cl), as well as aryl triflates, to give biaryls in moderate to excellent yields [19]. In early phase, the reaction was catalysed by simple transitional metal salts (5-10 mol%), whose effectiveness is arranged as: $FeCl_3 < NiCl_2 < CoCl_2$. The Kharasch reaction has been greatly developed by the Kumada's, and many other research groups latter. For example, 2-methoxyphenylmagnesium bromide (**124**) was reacted with iodobenzene (or bromobenzene) to furnish 2-methoxybiphenyl (**91**) in almost quantitative yield [12], Scheme 6.

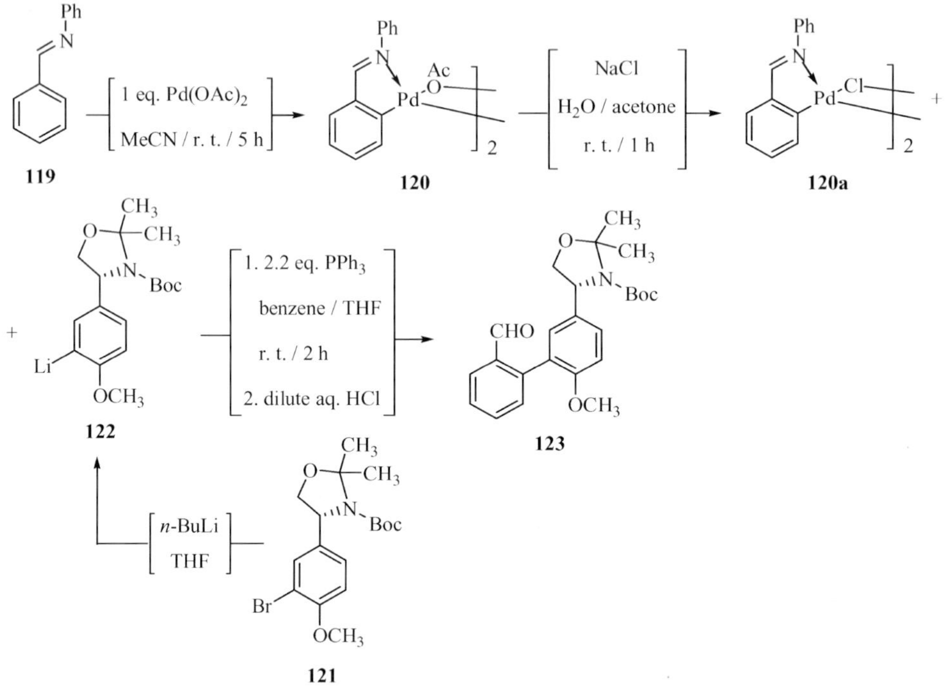

Scheme 5

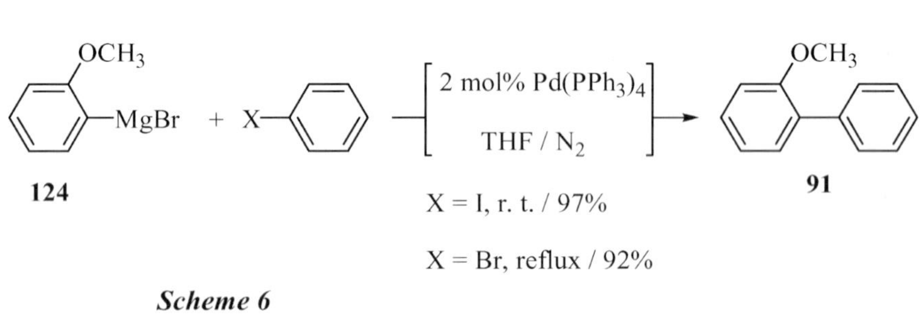

X = I, r. t. / 97%

X = Br, reflux / 92%

Scheme 6

Apart from the original Kharasch reaction-promoters, the following palladium or nickel (pre)catalysts appeared to be efficient: $Pd(PPh_3)_4$ [12], $Pd(PPh_3)_2Cl_2$ [20,21],

Pd(dppb)Cl$_2$ [22], PdCl$_2$ [23], Ni(acac)$_2$ [24-26], Ni(dmpf)Cl$_2$ [3], Ni(dppe)Cl$_2$ [3,24], Ni(dppp)Cl$_2$ [3,27], Ni(PPh$_3$)$_2$Cl$_2$ [3,28], and Ni(bpy)Cl$_2$ [29].

Beside aryl iodides and bromides, less reactive aryl chlorides, triflates, tosylates, and also aryl-methyl thioethers, aryl thiols, diaryl disulfides, diaryl sulfides, diaryl sulfoxides, sulfones, phosphates [30] and diaryliodonium salts [31] were reacted with Grignard reagents under the Kharasch reaction conditions [1]. Recently, closely related Ni(PMe$_3$)$_2$Cl$_2$-catalysed reaction of aryl Grignard reagents with arylnitriles has been reported [32]. The Kharasch reaction is usually performed by adding the desired aryl halide (or pseudohalide), dissolved in dry solvent, to the previously prepared aryl Grignard reagent (in an excess of 20-50 mol%) followed by the corresponding nickel or palladium (pre)catalyst (1-10 mol%), while the cross-coupling reaction is carried out under an inert atmosphere within the temperature rang between room temperature to approximately 100 °C (refluxing 1,4-dioxane). As the reaction medium, all common Grignard solvents are used: THF, DME, 1,4-dioxane, Et$_2$O, various dimethyl glymes dried over LiAlH$_4$, or similar reagents. The main practical difficulty may arise during the preparation of the aryl Grignard reagent. A number of relatively unreactive aryl chlorides and bromides, specially sterically hindered ones react sluggishly with magnesium turnings. Generally, the standard precautions connected with the Grignard reagent preparation:

- dried solvents (a few hours refluxing over CaH$_2$, LiAlH$_4$, other LAH-derivatives, etc., followed by distillation),
- proper activation of magnesium turnings (treatment with iodine alone, or followed by ethyl bromide, 1,2-dibromoethane, tetrachloromethane, methyl iodide, etc.) and / or
- preheating the magnesium metal (to 80-100 °C), even with the least reactive aryl chlorides [33], usually give acceptable results.

However, examples of sluggishly reactive aryl halides have been reported since. Problematic Grignard reactions can be started quite readily under ultrasonic irradiation [28]. In addition, a rather convenient method for activation of magnesium turnings includes the addition of diisobutylaluminum hydride (DIBAlH, 1 mol%) to the starting solution of approximately 5-10% of aryl halide in commonly dried solvent (containing 0.01-0.05% water) providing an effective induction of the Grignard reaction, even at room temperature [34]. Alternatively, certain aryl Grignard reagents can be efficiently prepared by transmetallation reaction with readily available alkyl Grignard reagents, e.g. 3-pyridylmagnesium chloride is easily obtained by reaction of *i*-PrMgCl and 3-chloro- or 3-bromopyridine in THF at room temperature [35]. The latter process is proceeding due to the favourable formation of Grignard reagent derived from a more acidic hydrocarbon (basicity of carbanions: sp^3 > sp^2, thus the acidity of C-H is: sp^2 > sp^3). An additional drawback of the Kharasch reaction appears during the purification of the crude products when the homo-coupling side-product contaminates (10-50%

range) the major, unsymmerical biaryl. This side-product is generated from an excess of aryl Grignard reagent.

The Kharasch cross-coupling reaction is considerably sensitive to the steric hindrances. *Ortho*-monosubstituted biaryls have been prepared in good yields, whereas the 2,2- and 2,2'-disubstituted biaryls furnished in poor to moderate yields. For instance, 2-methoxyphenylmagnesium bromide and 2-bromotoluene gave the appropriate 2,2'-disubstituted biaryl in only a 8% yield. However, this can be improved by using HMPA (10%) as a cosolvent, with a stoichiometric amount of palladium catalyst (1 eq.), or by employing the diphosphine-based catalysts to give the 2,2'-disubstituted biaryl in moderate yield [12]. Additionaly, highly *ortho*-occupied biaryls cannot be prepared by this approach, at least not in synthetically useful yields. Generally, the presence of bulky groups in the Grignard reagents has a significantly lower impact on the yield of the cross-coupling reaction, than the sterically demanded aryl halides counterpart. Selected examples of the Kharasch cross-coupling reactions where aryl Grignard reagents **125-128** were reacted with aryl halides **129**, bromobenzene, **109** and **130** to give the respective biaryls **131-134** are shown in the Scheme 7 [12,24,35,36].

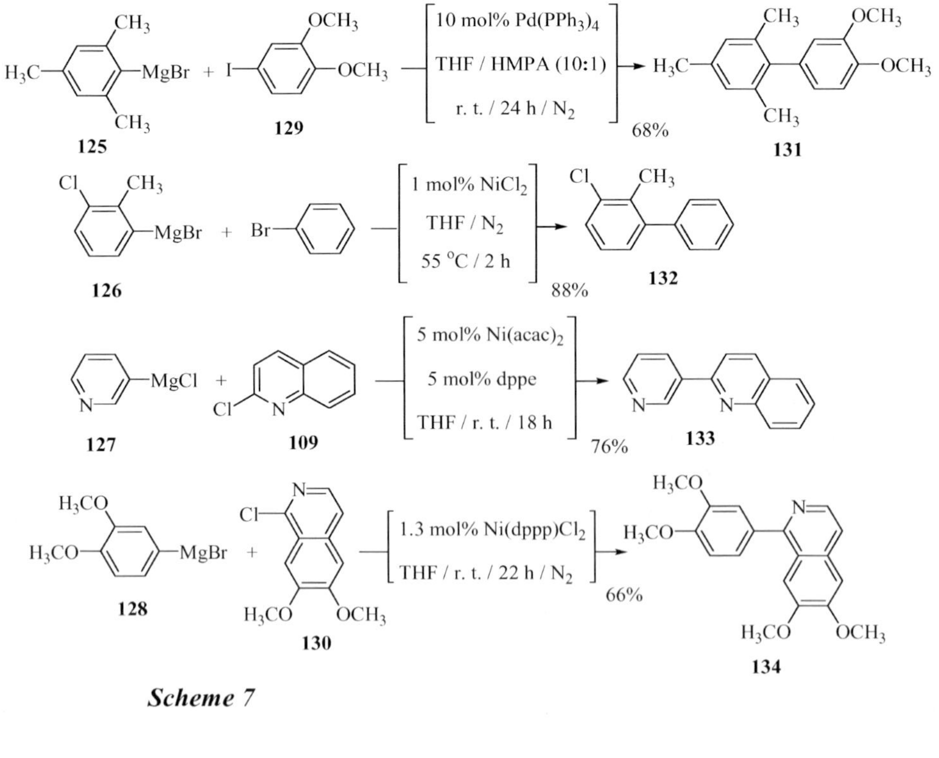

Scheme 7

The Kharasch reaction appeared to be efficient in the synthesis of a wide variety of substituted biphenyls [1,12,19-25], diphenylnaphthalenes [26], binaphthyls [28], terphenyls [1], phenylpyridines [3,35,36], phenylfurans [1], phenylthiophenes [22,36,37], phenylquinolines [1,35], phenylisoquinolines [1], bithiophenes [36], thiophenylpyridines [3,35], thiophenylquinolines [36], thiophenylisoquinolines [36], bipyridines [35], pyridylquinolines [35], as well as polythiophenes [1,36,38], polythiophenylpyridines [1], polyphenylpyridines [1], and various oligo-phenyl structures [39].

Once again, the nickel-catalysed reactions are far more sensitive to the steric encumbrances, than the palladium-catalysed analogues. Aryl triflates and related sulfonates can serve as the electrophilic reactants in the Kharasch reactions in the presence of an equimolar amount of lithium bromide. The latter additive is a reagent for triflate-bromide exchange, which became a very fast reaction under the Kharasch reaction conditions, since the nickel and palladium-catalysed nucleophilic substitutions of aryl halides and sulfonates are well established reactions, see Chapter 3. In this manner, 2-biphenyl triflate (**135**) was reacted with 4-tolylmagnesium bromide (**136**) to give 2-(4-tolyl)biphenyl (**137**) with a 93% yield [40], respectively, Scheme 8.

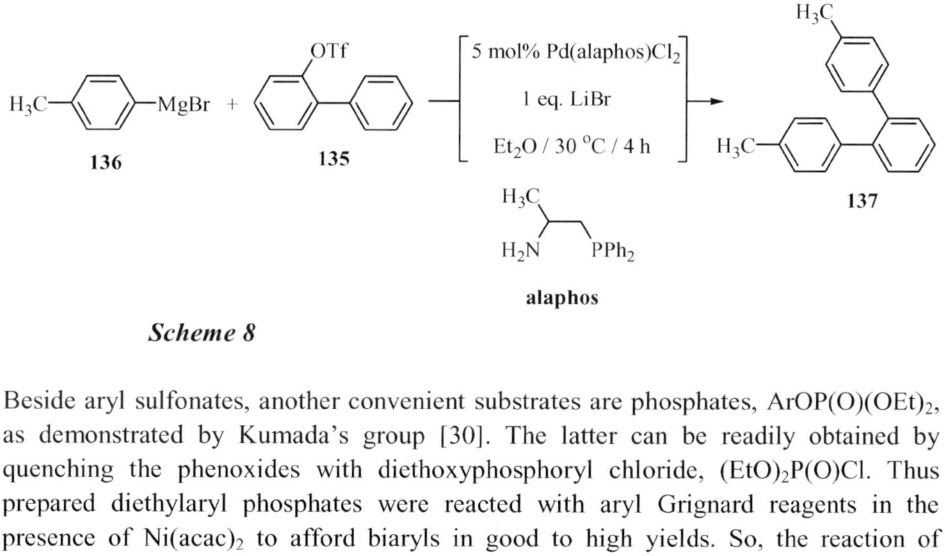

Scheme 8

Beside aryl sulfonates, another convenient substrates are phosphates, $ArOP(O)(OEt)_2$, as demonstrated by Kumada's group [30]. The latter can be readily obtained by quenching the phenoxides with diethoxyphosphoryl chloride, $(EtO)_2P(O)Cl$. Thus prepared diethylaryl phosphates were reacted with aryl Grignard reagents in the presence of $Ni(acac)_2$ to afford biaryls in good to high yields. So, the reaction of diethylphenyl phosphate (**138**) with 4-tolylmagnesium bromide (**136**) yielded 91% of 4-methylbiphenyl (**139**) [30], respectively, Scheme 9.

Further valuable electrophilic reactants in the Kharasch reaction are diaryliodonium salts. They smoothly react with aryl Grignard reagents in the $Pd(PPh_3)_4$-catalysed reactions in the presence of zinc chloride to give biaryls in excellent yields [31]. Although access to some diaryliodonium salts is somewhat difficult, they could be important alternatives to aryl halides and sulfonates as demonstrated by Chen's group [31]. The high efficiency of the Kharasch reaction involving this electrophiles is shown in the Table 1.

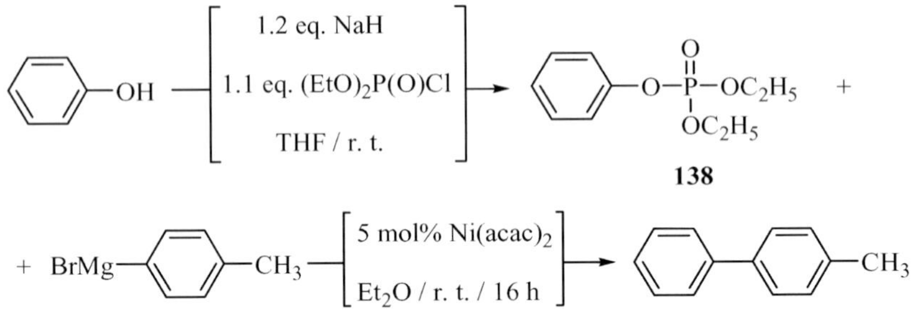

Scheme 9

Table 1. The yields in the cross-coupling reactions of diaryliodonium salts with aryl Grignard reagents [31]

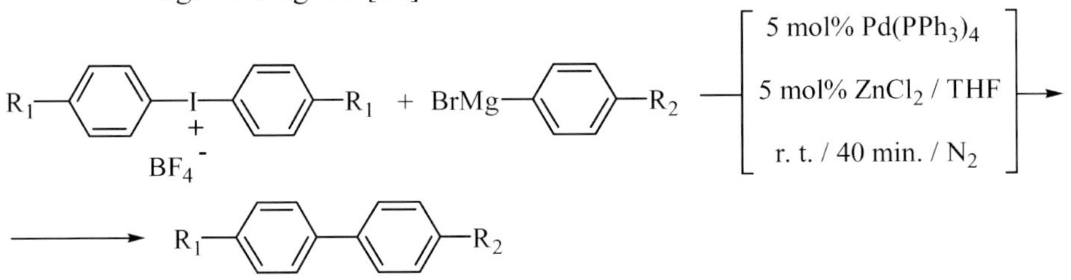

R_1	R_2	Yield (%)
H	H	96
4-CH$_3$O	H	86
3-O$_2$N	H	92
H	4-CH$_3$O	85
H	4-Cl	89

A recent extension of the Kharasch reaction to the arylnitriles is apparently an important contribution to the aryl-aryl bond forming methodology. Miller and coworkers [32] have found that arylmagnesium *t*-butoxides, phenoxides or thiophenoxides, *in situ* prepared from LiO*t*-Bu, LiOBHT, LiSPh, smoothly react with arylnitriles in the presence of Ni(PMe$_3$)$_2$Cl$_2$ as precatalyst to give the unsymmetrical biaryls in excellent yields. The results of this new reaction for synthesis of biaryls are presented in the Table 2. Although the use of suitable protecting groups, e.g. oxazolines as masked carboxylic acids [41], methyl-, *i*-propyl-, trityl-, benzyl-group(s)

[42] as well as other common ether-based protections for the phenols, alcohols or amines, etc. allows the cross-coupling reactions of relatively highly functionalyzed aryl halides with Grignard reagents, the Miller reaction conditions involving arylnitriles are compatible with ester function [32]. Herein, arylnitriles undergo the oxidative addition to the nickel(0) catalyst to generate the arylnickel(II) nitriles (**XVI**), which further react with arylmagnesium *t*-butoxide to give the diarylnickel(II) intermediate (**XVII**) and Mg(CN)O*t*-Bu. Compound **XVII** decomposes to the unsymmetrical biaryl and catalytically active parent nickel(0) complex, Scheme 10.

Table 2. Synthesis of biaryls *via* nickel-catalysed cross-coupling reaction of arylnitriles with aryl Grignard reagents [32]

$$Ar-MgZ \ + \ NC-Ar' \ \longrightarrow \left[\begin{array}{c} 5 \text{ mol\% Ni(PMe}_3)_2\text{Cl}_2 \\ \\ 60\ ^\circ C \ / \ THF \end{array} \right] \longrightarrow Ar-Ar'$$

Ar'-CN	Ar-MgZ	Time (h)	Yield (%)
4-MeOPh-	PhMgO*t*-Bu	2	91
4-MeOPh-	PhMgO*t*-Bu	2	88
4-MeOPh-	PhMgO*t*-Bu	2	97
4-MePh-	3-MeOPhMgO*t*-Bu	2	92
3-*t*-BuO$_2$CPh-	PhMgSPh	1	93
3-pyridyl-	PhMgOBHT [a]	6	80 [b]
2-furyl-	4-MePhMgO*t*-Bu	2	78

[a] BHT = 2,6-di-*t*-butyl-4-methylphenol
[b] The reaction was conducted at 25 $^\circ$C.

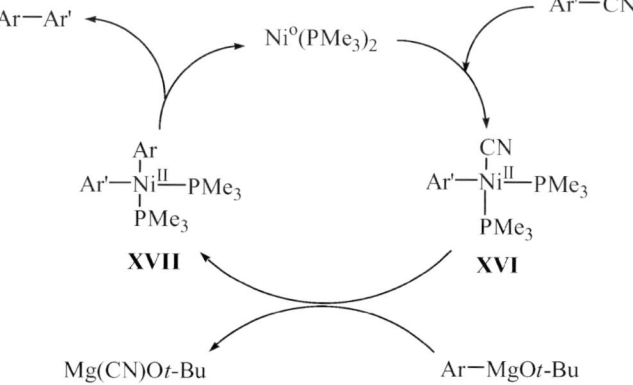

Scheme 10

An additional important application of the Kharasch reaction is in the synthesis of polyaryls. For example, 2,5-dibromopyridine (**140**) was transmetallated upon treatment with isopropylmagnesium bromide to give the corresponding Grignard reagent **141**, which subsequently underwent the Kharasch reaction affording the polypyridine (**PPy**) in 76% yield [43], respectively, Scheme 11.

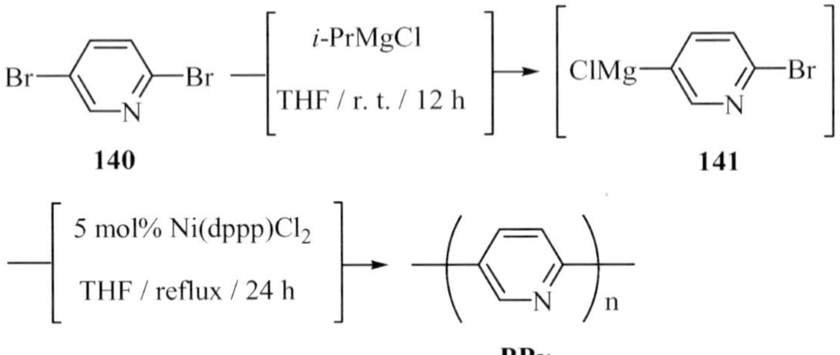

Scheme 11

Finally, the Kharasch reaction and its modern variants are valuable synthetic tools in the field of aryl-aryl bond formation.

4.4. The Negishi reaction

Arylzinc halides, as significantly less reactive organometallics than the Grignard reagents, are extremely convenient nucleophilic counterparts in the cross-coupling reactions with alkyl [44,45], alkynyl [46], benzyl [47], vinyl, and aryl halides [48], affording various valuable C-C bond formations. This reaction, discovered by Negishi's group, has became one of three the most popular approaches to biaryls. The Negishi synthesis of unsymmetrical biaryls includes the cross-coupling reaction of an arylzinc reagent with aryl iodides [48], bromides [48], chlorides [49,50], triflates [51], nonaflates [52], and tosylates [53] in the presence of nickel or palladium complexes (1-5 mol%) to give the biaryls in good to high yields. For instance, phenylzinc chloride was reacted with different monosubstituted iodo- and bromobenzenes to produce biaryls in high yields, Table 3.

Besides Ni(acac)$_2$ in the presence of triphenylphosphine, and Pd(PPh$_3$)$_2$Cl$_2$ as precatalysts [48], Pd(PPh$_3$)$_4$ [47,54,55], Pd(*t*-Bu)$_2$ [51], Pd(dba)$_2$ / dppf [52], Pd(dppf)Cl$_2$ [56], Pd(PPh$_3$)$_2$Cl$_2$ / CuI [53], nickel on charcoal / 4 PPh$_3$ [44], NiCl$_2$ / 2-4 PPh$_3$ [45], and Ni(PPh$_3$)$_2$Cl$_2$ [57], and Ni(acac)$_2$ / dppf [49], generally in amounts of 1-5 mol%, have been found to efficiently catalyse the cross-coupling reaction. Once again, the nickel(II) or palladium(II) precatalysts have to be reduced to the catalytically active zero-valent phosphine stabilized complexes, either by DIBAlH [48], *n*-BuLi [44,45], Zn-powder, etc. or by an excess of arylzinc reagent [54], prior to the Negishi reaction.

Table 3. The Negishi synthesis of unsymmetrical biphenyls [48]

R	X	Yield [b] (%)
MeO-	I	85
MeO-	I	87
NC-	Br	90
MeOOC-	Br	70
O₂N-	I	90

[a] A = Ni(PPh₃)₄ prepared *in situ* by the reaction of Ni(acac)₂, PPh₃, and (*i*-Bu)₂AlH (1 : 4 : 1); B = Pd(PPh₃)₂Cl₂ + (*i*-Bu)₂AlH (1 : 2); [b] GC yields

The arylzincs can be prepared by the following alternative reactions:
- metallation of aryl halides with the strong bases: *n*-, *s*-, *t*-BuLi [56-58], LDA [59,60], etc. followed by transmetallation with $ZnCl_2$ ($ZnBr_2$). A closely related method is based on the reaction of aryl Grignard reagents with $ZnCl_2$,
- metallation of relatively acidic arenes with the above mentioned strong bases followed by transmetallation with $ZnCl_2$,
- direct insertion of zinc dust or activated zinc to the aryl iodides and certain reactive aryl bromides.

While a background of the former two reactions is well established, the applications of the organozincs obtained directly from metallic zinc are quite recent. Although the alkylzincs are actually the oldest known organometallics, firstly prepared by Sir Frankland in 1840's, the related arylzincs appeared to be difficult to produce by simple Grignard-type reaction. However, Rieke's group have found that activated zinc, *in situ* prepared by reduction of zinc bromide with potassium in refluxing THF (4 h), reacts with bromobenzene (refluxing diglyme, 8 h) to give phenylzinc bromide in 83% yield [61]. In this manner, the formation of arylzincs also proceed with simple purchased zinc dust, although the polar solvents such as HMPA, 1,1,3,3-tetramethylurea (TMU), DMAc, or DMF are required [62]. The excellent results of Takagi's [62-64] and Knochel's [65,66] groups provided very convenient methods for the preparation of arylzincs from the ordinary zinc dust (3 eq.), activated with 1,2-dibromoethane (5-10 mol% to Zn) [65], trimethylsilyl chloride (3-12 mol% to Zn) [67] or with both reagents [66], in DMAc [65,66] or in common ethereal solvents: THF, diglyme or triglyme

[64]. Aryl iodides bearing an electron-donating substituents in *ortho-*, *meta-*, *para-* positions and electron-withdrawing substituents in the *ortho-*position to the iodine, furnished the arylzinc iodides in approx. 70-95% yields in a temperature rang between the room temperature and refluxing THF (65 °C). Yet, electron-withdrawing substituents in the *meta-* and *para-*position to the iodine do deactivate the aryl iodide. Higher reaction temperatures, approx. 100 °C, in diglyme, were essential to rich the high conversions [64]. Moreover, the 1,2-aryldizincs can be prepared by this methodology using TMU as the solvent at 90 °C from *o-*diiodo- or *o-*iodotriflates [63]. Apart from aryl iodides, a number of activated heteroaryl bromides [66,67] have reacted with activated zinc dust in THF, DMF or DMAc to yield the highly functionalyzed organozincs under very mild reaction conditions, r. t. to 70 °C, avoiding several difficulties connected with the usual work with hazard organolithiums, Grignard reagents, strong bases, etc. Recently, the convenient preparation of arylzincs by electroreduction of aryl iodides or bromides in an electrochemical cell fitted with a sacrificial zinc anode in DMF or acetonitrile as solvent in the presence of zinc bromide (30 mol%) and cobalt(II) bromide (10 mol%) has been developed [68]. The latter electrochemical method affords arylzincs in good yields, although with partial reductive dehalogenation, approx. 10-40%, as the only unwanted side event.

The Negishi reaction of thus prepared arylzincs with aryl halides or pseudohalides, e.g. triflates, in the presence of palladium- or nickel-catalyst led, to the 2-mono-, 2,2-di-, 2,2'-di-, tri- and tetra-*ortho-*substituted biarlys in good to high yields. However, the reaction of arylzincs with aryl halides bearing extremely bulky groups in the *ortho-* position(s) may fail even at reflux temperatures of 1,4-dioxane, 100 °C. Generally, the sterical encumbrances are more tolerated in an arylzinc counterpart than in an aryl halide. For extremely sterically hindered structures, the Suzuki-Miyaura reaction is a slightly more efficient alternative. The Negishi reaction of arylzincs with different aryl halides or pseudohalides has been accomplished by the following methods regarding the catalyst used, Table 4. As the reaction medium in all these methods serve all common dry ethers such as Et_2O, THF, DME, diglyme, triglyme, higher glymes, 1,4-dioxane, DIPE, MTBE as well as DMF, DMAc, HMPA, etc. In certain instances, even the mixtures of these solvents with hydrocarbons, e.g. toluene are used when the particular requirements concerning the solubility of reactants are needed. The reactions are generally conducted at a temperature ranging from room to about 120 °C depending on the reactivity of Negishi counterparts.

The *method A*, applying $Pd(PPh_3)_4$, and closely related *method B*, based on $Pd(PPh_3)_2$-catalysis, have been widely used since the original Negishi report, Table 4 [48]. $Pd(PPh_3)_4$ can be prepared separately, or *in situ* by reduction of $Pd(PPh_3)_2Cl_2$ with DIBAlH in the presence of triphenylphosphine (2 eq. to Pd) or by reaction of the latter (8 eq.) with $Pd_2(dba)_3$ [51]. The original Negishi protocol involving $Pd(PPh_3)_2$-

catalysis includes the previous reduction of $Pd(PPh_3)_2Cl_2$ with DIBAlH in the molar ratio 1 : 2. For example, the reaction of indolezinc chloride **142** with 2-bromo-4-methylpyridine (**143**) led to the formation of corresponding biaryl **144** in 73% yield [58], respectively, Scheme 12.

Table 4. Methodology of the Negishi reaction

Method	Actual catalyst (mol%)	Substrates Ar-X, X =	Lit. ref.
A	$Pd(PPh_3)_4$ (3 mol%)	I, Br, Cl, OTf, OTs	50, 52 53, 57
B	$Pd(PPh_3)_2$ (5 mol%)	I, Br, Cl	47, 54
C	$Pd(Pt\text{-}Bu_3)_2$ (2 mol%)	I, Br, Cl	49
D	Pd(dppf) (2 mol%)	I, Br, Cl	55
E	$Ni(PPh_3)_4$ (5 mol%)	I, Br, Cl	47
F	$Ni(PPh_3)_{2-3}$ (2 mol%)	I, Br, Cl	48, 56
G	$Ni(POi\text{-}Pr)_3$ (7.5 mol%)	Cl [a]	48
H	Ni(dppf) (2 mol%)	Cl [a]	48
D	Ni-C / 3-4 PPh$_3$ (5 mol% Ni)	Cl [a]	43

[a] Probably effective in the Negishi reactions of more reactive aryl iodides and bromides

Additionaly, less reactive aryl triflates react with arylzincs by *method A* furnishing the biaryls in excellent yields. Once again, the triflate involving reaction was accomplished in the presence of lithium chloride or bromide as mediators. These *in situ* generate appropriate more reactive aryl bromides or chlorides from the parent triflate, and thus accelerate the cross-coupling reaction with arylzinc. For instance, the reaction of 2-pyridylzinc chloride (**145**) with 3-methylpyridyl triflate (**146**) gave 3-methyl-2,2'-bipyridin (**147**) in 98% yield [51], respectively, Scheme 13.

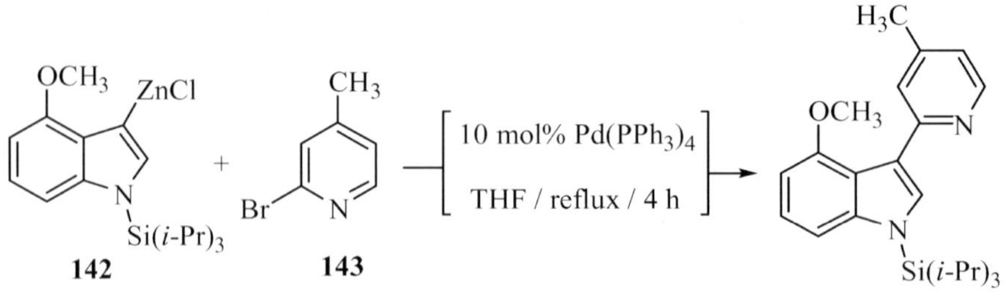

142 **143** **144**

Scheme 12

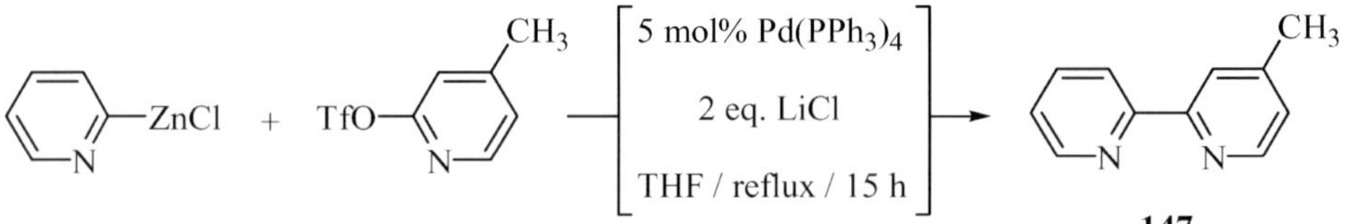

145 **146** **147**

Scheme 13

Method C, developed by Fu and Dai [46], offers the most efficient and general approach to a wide variety of substituted biaryls including the respectable synthetic targets, tetra-*ortho*-substituted biaryls. P*t*-Bu_3 as trialkylphosphine does form a more reactive palladium complex, capable of catalysing the Negishi reactions of arylzincs with aryl iodides or bromides even at room temperature. The reactions involving aryl chlorides proceed at 100 °C in the THF / NMP solvent mixture leading to the biaryls in excellent yields, often with very low palladium-catalyst loading. Thus an important pharmaceutical intermediate, 4'-methyl-2-cyanobiphenyl (**112**) was obtained from the 4-tolylzinc chloride (**148**) and 2-chlorobenzonitrile (**114**) in 97% isolated yield with only 0.03 mol% of Pd(P*t*-$Bu_3)_2$ as catalyst [50], Scheme 14.

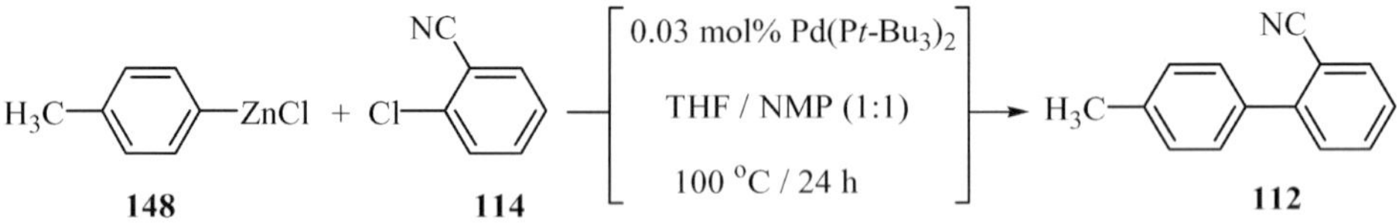

148 **114** **112**

Scheme 14

In *method D* an actual catalyst, Pd(dppf), was *in situ* prepared by reduction of a precatalyst, Pd(dppf)Cl_2, with an excess of an organozinc reagent followed by the Negishi reaction of aryl bromides and chlorides [56]. This method also works well in the reactions with higher order zincates, namely lithium triarylzincates, affording the corresponding biaryls at sub-stoichiometric zinc halide loading. Since all three zinc halides are quite hygroscopic solids, the removal of water present in commercial solutions (in THF) or bounded on the solid ZnX_2 is essential to avoid decreased yields. In addition, several difficulties may arise during the work-up of the reaction mixtures at stoichiometric ZnX_2 protocols. In this way, slow addition of PhLi (1 eq.) to the solution of $ZnCl_2$ (10 mol%), 2-chloropyridine (**149**, 1 eq.) and Pd(dppf)Cl$_2$ furnished the 2-phenylpyridine (**150**) with a 82% yield. To demonstrate the possibility of transfering all three aryl groups from higher order zincate to palladium, Gauthier and coworkers [56] have shown that the reaction of Ph$_3$ZnLi (0.35 eq.) and 2-chloro pyridine, under above mentioned conditions, led to the formation of 2-phenylpyridine in an essentially quantitative yield [56], Scheme 15.

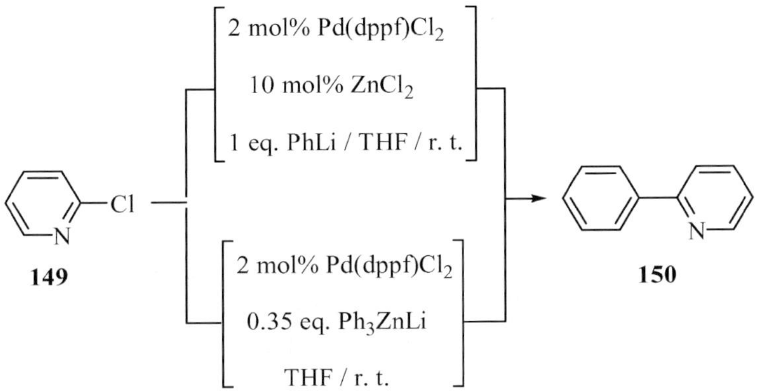

Scheme 15

Beside these higher order zincates, an ordinary diarylzincs are capable of undergoing the Negishi reaction with aryl halides, transfering both aryl groups.

Method E is the original Negishi nickel-catalysed protocol where the Ni(PPh$_3$)$_4$ is generated *in situ* by reduction of Ni(acac)$_2$ with an equimolar amount of DIBAlH in the presence of triphenylphosphine (4 eq. to Ni), Table 3.

An alternative nickel pre-catalyst is easily available Ni(PPh$_3$)$_2$Cl$_2$, and as the reductants, *n*-BuLi or zinc-dust have been employed. In the *method F* Ni(PPh$_3$)$_2$, obtained by *in situ* reduction of Ni(PPh$_3$)$_2$Cl$_2$ with an excess of an arylzinc reagent have been equally effective in the Negishi reactions of aryl bromides [57], as well as chlorides [49], although in particular instances at somewhat elevated temperatures, 50 °C to 65 °C (refluxing THF). However, phenylzinc chloride (**151**) was coupled with 2-chloro-3-amino pyridine (**152**) bearing a free-amino group, to give the 2-phenyl-3-aminopyridine (**153**) in 72% yield [45], respectively, Scheme 16.

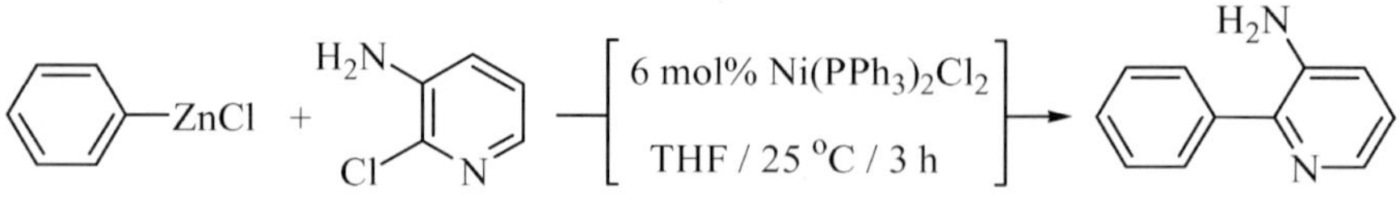

151
2.1 eq.

152

153

Scheme 16

Method G is close to the former two original Negishi methods, but uses the Ni[(PO*i*-Pr)$_3$]$_2$ as an actual catalyst *in situ* generated by the reaction of Ni(acac)$_2$ and P(O*i*-Pr)$_3$ (2 eq.). Thus catalysed Negishi reaction offers no significant advantages to the previously described nickel-based methods, except that the trialkylphosphites are somewhat less sensitive to the atmospheric oxygen.

Method H is based on the really efficient Ni(dppf)-catalysed high-yielding Negishi reaction system. For instance, biphenyl-2-carbonitrile (**154**) was prepared from phenylzinc chloride (**151**) and 2-chlorobenzonitrile (**114**) in 86% yield [49], Scheme 17.

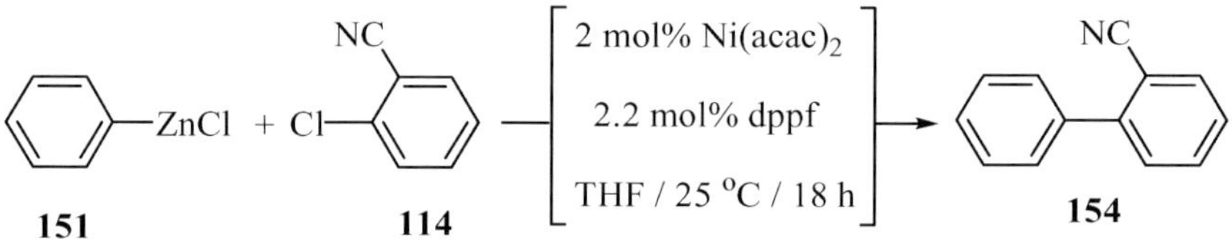

151

114

154

Scheme 17

Lipshutz has developed an original method using nickel on charcoal [44]. This valuable catalyst was prepared by *n*-BuLi-reduction of nickel(II) nitrate adsorbed on the activated charcoal with high surface area in the presence of triphenylphosphine. Thus the Negishi reaction of 3-tolylzinc chloride (**155**) with 2-chlorobenzonitrile (**114**) afforded 2-(3-tolyl)benzonitrile (**156**) in 92% yield [44], respectively, Scheme 18.

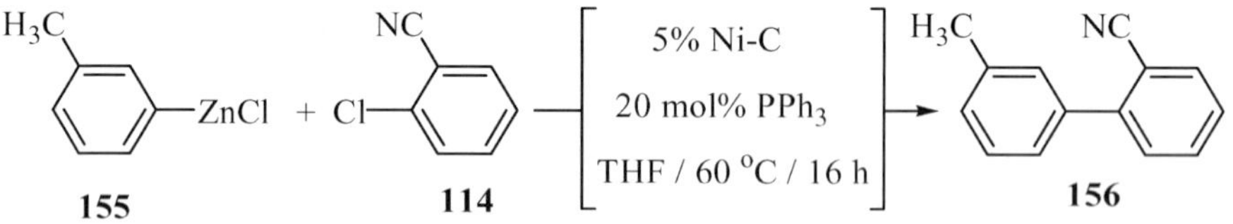

155

114

156

Scheme 18

This catalyst can be easily removed by simple filtration, in this way avoiding all common difficulties connected with expensive (Pd-complexes) or toxic (Ni-salts) unable to reuse homogeneous catalysts.

The most important side-reaction which occurs under the Negishi reaction conditions is the homo-coupling reaction of arylzinc reagent. Generally, the Negishi reaction is routinely conducted with arylzinc in slight excess, e.g. 10 mol%, since this unwanted side-event proceeds simultaneously with the major reaction. However, the extend of homo-coupling reaction is far less than in the Kharasch reaction. The homo-coupling side-reaction usually becomes important in the cases of sterically encumbered aryl halides where the Negishi reaction is significantly slower. Then a slightly higher excess of arylzinc, e.g. 1.2-1.5 eq., is necessary to rich the acceptable yield of desired product, although with considerable amounts of unwanted symmetrical biaryl. Its separation from the major product may cause difficulties and decreased yields of purified unsymmetrical biaryl.

The Negishi cross-coupling reactions have been employed wide-spread in the synthesis of different types of biaryls: wide variously substituted biphenyls [44,48,52,57], phenylpyridines [49,50,56,59,60,66,69], phenylthiophenes [49,50,56], phenylfurans [56], phenyl(N-alkyl)pyrroles [66], bipyridines [51,70], terpyridines [60], pyridylquinolines [70], bithiophenes [55], phenylthiazoles [52,67], naphthylthiazoles [67], thiophenylpyridines [55], pyridyl(N-alkyl)pyrroles [66], pyridylthiazoles [67], thiophenylthiazoles [67], pyridylindoles [54,58,71], pyrazinylindoles [71], thiophenylindoles [71], furylindoles [71], pyrimidylindoles [54], pyrazolylindoles [54], (N-t-BuMe$_2$Si, N'-Boc-protected)indoles [71], indol-6-yl-oxazoles [72], indol-5-yl-thiazoles [67], 4-pyrazolylthiazoles [67], and several other classes. An important characteristic of the Negishi reaction is the high compatibility with a number of common functional groups: nitrile, ester, arylarylketone, arylalkylketone, arylaldehyde group in the aryl halide counterpart [45], nitro group (in the palladium-catalysed reactions, see also Chapter 3 [48]), free arylamino group (with 2 eq. of arylzinc reagent) [49], N-triisopropyl- or t-butyldimethylsilylated amino groups, as well as indoles [69], ligand-capable heteroatoms in a wide variety of heteroaryl-moieties, etc. However, certain functional groups, e.g. O-tritylesters, do survive the Negishi reaction conditions, but are cleavaged during aqueous work-up by the influence of zinc halide (Lewis acids), liberated as a by-product. An electrochemical approach is a convenient way to prepare arylzincs bearing a free arylamino group [69]. Thus Gosmini and coworkers [68] obtained 4-H$_2$NC$_6$H$_4$ZnCl, which was readily coupled with iodobenzene by the original Negishi *method A* to the 4-aminobiphenyl in 85% yield, respectively. Among interesting applications, in the synthesis of natural occuring product steganone, 3,4-methylenedioxyphenylzinc chloride (**157**) and imine **158** were converted to the respective biaryl **159**, which, upon acid-catalysed cleavage of cyclohexylimino protecting group, gave the key-intermediate **160** with a 82% overall yield [73], respectively, Scheme 19.

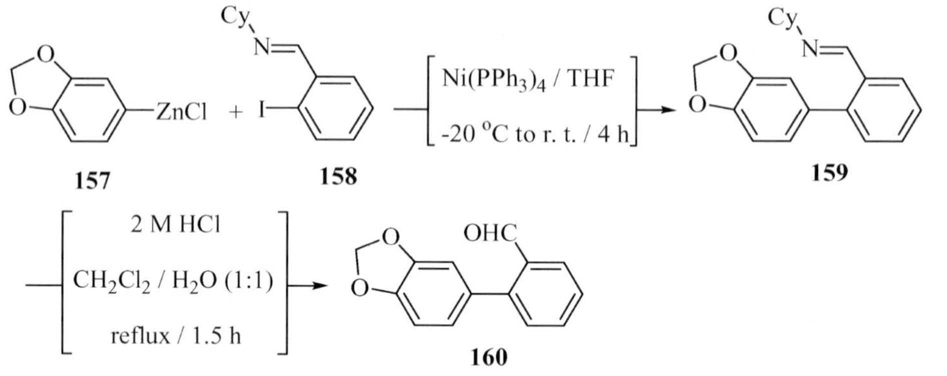

Scheme 19

In addition, among the synthesis of sterically hindered and bifunctionalized structures, 1,8-dichloroanthracene (**161**) was lithiated to **162**, which reacted with 3,4-dicyano-iodobenzene (**163**) to give the respective synthetic target, 1,8-diphenylanthracene (**164**) in 10% yield [74], Scheme 20.

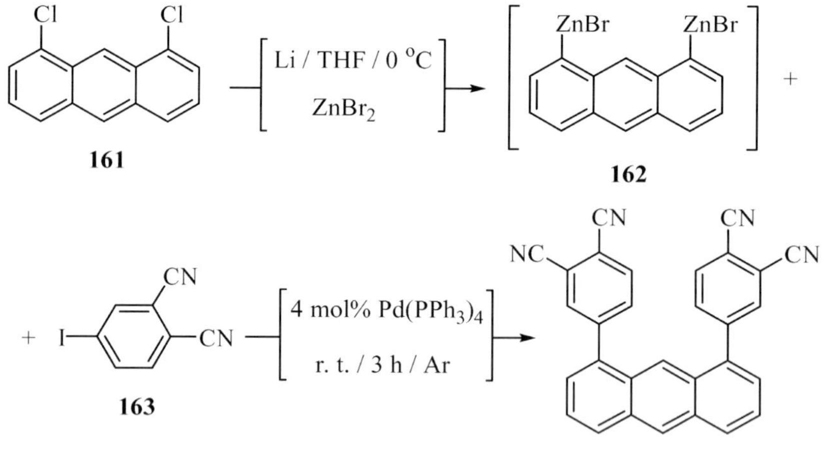

Scheme 20

In conclusion, the Negishi reaction is, beside the Stille and Suzuki-Miyaura approaches, apparently the most versatile method for aryl-aryl, C-C, bond formation.

4.5. The Stille reaction

Organostannanes are capable of undergoing the transmetallation process, transfering the hydrocarbon moieties to palladium (or copper). The reductive elimination of thus formed diarylpalladium(II) complex results with generation of the new C-C bond. The readiness of the organic groups to transfer from tin to palladium follows an order: alkyl < benzyl, allyl < aryl < alkenyl < alkynyl [6,10,75,76]. This is the key feature of an extremely versatile approach to the generation of almost all possible types of C-C bonds between above mentioned hydrocarbon groups, well known as the Stille reaction [6]. A nucleophilic part originates from the organostannanes, whereas as an electrophilic counterpart serves aryl halides or sulfonates. Herein the use of the Stille reaction in the synthesis of biaryls is focused. Thus arylstannanes type $ArSnR_3$, where R = lower alkyls, in the presence of palladium(0) catalysts, readily react with aryl halides, sulfonates or diazonium salts to give the corresponding biaryls in good to excellent yields, Scheme 21.

$$\text{Ar}-\text{SnR}_3 \ + \ \text{X}-\text{Ar}' \ \xrightarrow{\quad Pd^0 \ (or \ Cu^I, \ Mn^{II})\text{-catalysis} \quad} \ \text{Ar}-\text{Ar}'$$

R = lower alkyls, e.g. Me or n-Bu

X = $N_2^+ BF_4^-$, I, Br, Cl, OTf, OMs, OTs

Scheme 21

The aryltrialkylstannanes consisting of a wide variety of substituted phenyls, O, S, N-containing heteroaryls have been affecting the Stille cross-coupling reactions with aryl iodides [77], bromides [77], chlorides [78], triflates [75,76,79-82], mesylates [83], and tosylates [83]. Whereas the major starting materials in the synthesis of the parent aryltrialkylstannanes are readily available trimethyltin chloride, hexamethyldistannane, and tri-n-butyltin chloride, as the alkyl-moieties serve methyl- or n-butyl groups. The Stille reaction is catalysed by different palladium complexes including: $Pd(OAc)_2$ [1], $Pd(dba)_2$ [75], $BnPd(PPh_3)_2Cl$ [6], $Pd(PPh_3)_4$ [75,76], $Pd(PPh_3)_2Cl_2$ [75-77], $Pd(MeCN)_2Cl_2$ [75], $[(\eta^3\text{-}C_4H_7)PdCl]_2$ [75], $Pd(dppf)Cl_2$ [76], $Pd_2(dba)_3$ with 4 eq. of Pt-Bu_3 [78], tri(2-furyl)phosphine [79], tri(2-tolyl)phosphine [79], triphenylarsine [79], tri(2,4,6-trimethoxypheny)phosphine [79], dppe [82], $Pd(PPh_3)_2Cl_2$ / dppf [80], and certain supported palladium-catalysts. Generally, among the palladium-phosphine complexes, the monodentate ligands, e.g. PPh_3, have worked significantly better than the bidentate ones, e.g. dppf, dppp, dppe [80]. Apart from the palladium-phosphine complexes, certain N-heterocyclic carbenes have been employed as a versatile ligands that stabilize the palladium(0) species effectively [84]. The catalysts are usually employed in amounts between 1-10 mol%, and most commonly around 5 mol%. Once again, an active palladium(0) catalyst is added in required zero-valent form or the reaction is conducted with a different palladium(II) precatalysts which are reduced *in situ* to the former by an excess of organostannane. In addition, it has been found that copper(I) salts such as CuCl, CuBr or CuI, added in 2-4 eq. relative to Pd, act as co-catalysts accelerating the sluggish Stille cross-coupling reactions [81,85].

Moreover, copper(I) halides or manganese(II) bromide efficiently catalysed the Stille reaction of arylstannanes with aryl halides without any palladium catalyst added [87,88]. In contrast to $MnBr_2$, manganese(II) chloride was proved as a less effective catalyst, whereas manganese(II) iodide-mediated reactions failed to give any cross-coupled products [88]. The latter achievements dramatically extend the economy of the Stille reaction since the use of, generally very expensive, palladium chemicals is completely avoided.

The palladium-catalysed Stille reactions involving aryl iodides and bromides have been carried out mainly in ethereal solvents [77,78]: THF, 1,4-dioxane, or DME, whereas those with aryl sulfonates, except ethers, in DMF [75,76,80], NMP [79], acetone [82], HMPA [82], and DMSO [82]. However, less polar solvents such as toluene or xylenes have also been successfully used as reaction medium [83]. The reactions involving aryl sulfonates require the addition of lithium salts, e.g. LiCl (2-4 eq.) providing a soluble chloride source to mediate the Stille-catalytic process [75,76,88-91]. So, the prefered reaction medium for the couplings with the aryl sulfonates are polar, good coordinating solvents, which solubilize LiCl well. On the contrary, polar solvents such as acetonitrile where LiCl is not soluble enough or which do not act as good ligands, e.g. sulfolane, lead to catalyst decomposition prior to the completion of the reaction. The copper(I)- and manganese(II)-catalysed Stille reactions were accomplished in DMF or the slightly more effective NMP [85-88].

Alternatively, the Stille reactions have been accelerated by the microwave heating [92], as well as with tetra-*n*-butylammonium fluoride [93]. The reaction temperature ranged from room temperature to about 150 °C (refluxing DMF). Generally, the reactions have been performed in an inert atmosphere because the presence of oxygen diminishes the yields of desired cross-coupling products. Since the aryltrialkyl stannanes are quite insensitive to the presence of water, the Stille reactions can be conducted in solvents containing usual amount of water (e.g. 0.1%). Moreover, the reaction has been successfully carried out in acetonitrile / water mixture at 80 °C, nicely illustrating the stability of Stille-stannanes to hot water.

The aryltrialkylstannanes as the key-Stille intermediates can be obtained by the following alternative reactions:

-   metallation of aryl halides or relatively acidic arenes with the strong bases, e.g. *n*-BuLi or LDA followed by reaction with trimethyltin- [94-96] or tri-*n*-butyltin chloride [96,97],

- palladium-catalysed reaction of aryl iodides, bromides or triflates with hexymethyldistannane (Me_6Sn_2) [98,99] or hexa-*n*-butyldistannane (*n*-Bu_6Sn_2) [99],
- palladium-catalysed reaction of tributyltin hydride (*n*-Bu_3SnH) with aryl iodides [101], or
- photostimulated nucleophilic substitution of aryl halides with sodium trimethylstannate ($NaSnMe_3$) [102,103].

Whereas the quenching of lithiated arene with Me_3SnCl or *n*-Bu_3SnCl is widely used and a well established access to aryltrialkylstannanes [94-97], the other three methods are more recent and relatively less known. The palladium-catalysed reactions of aryl iodides [98], bromides [99], or triflates [100] with $Me_3SnSnMe_3$ [98-100] or $Bu_3SnSnBu_3$ [100] (1.2-1.5 eq. to ArX) proceed smoothly at elevated temperatures in THF [99] or toluene [98] in the presence of $Pd(PPh_3)_2(OAc)_2$ [98], $Pd(PPh_3)_4$ [99], or $Pd(PPh_3)_2Cl_2$ [100] (5 mol%) leading to the corresponding aryltrialkylstannane in good to excellent yields. For instance, the $Pd(PPh_3)_4$-catalysed reaction from 2-nitro-4,5-disubstituted bromobenzenes afforded the trimethylstannanes in good yields [99], Table 5.

Table 5. The yields of arylstannanes in the palladium-catalysed coupling of aryl bromides with hexamethyldistannane [99]

R_1	R_2	Yield (%)
CH_3O	CH_3O	70
H	H	75
H	CH_3O	55
H	CF_3	60

In similar manner, hexaalkyldistannanes allow the homo-coupling reactions of aryl iodides, bromides [104] and triflates [105] to the symmetrical biaryls. First, the aryltrialkylstannane is generated by reaction of the aryl halides with Me_6Sn_2 or *n*-Bu_6Sn_2, followed by the *in situ* Stille reaction with an excess of electrophilic reactant, since both reactions are catalysted by the same complexes and proceed at comparable reaction rates. For instance, 2-naphthyl triflate (**165**) was coupled to 2,2'-binaphthalene (**166**) in 69% yield [105], respectively, Scheme 22.

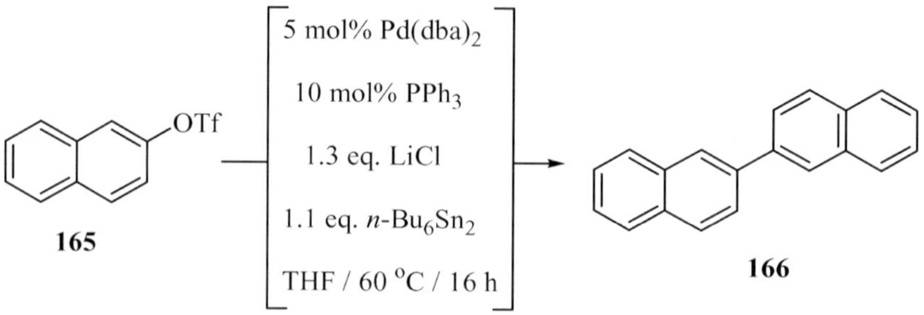

Scheme 22

Moreover, since the cross-coupling reaction of hexaalkyldistannane with aryl halides strongly depends on the reactivity of the latter, a chemoselective cross-coupling of two different aryl halides has become of practical value. Thus Zhang described a simple and versatile method for synthesis of unsymmetrical biaryls from 2-bromopyridines (more reactive) and various phenyl- and 3-pyridyl bromides (less reactive) to afford the appropriate biaryls in moderate to high yields [106]. For example, 5-cyano-2-bromopyridine (**167**) was cross-coupled with 4-nitro-bromobenzene (**168**) to give the biaryl **169** in 67% yield [106], respectively, Scheme 23.

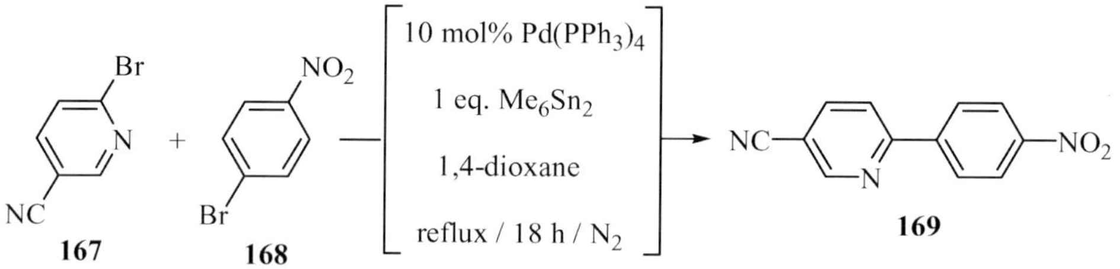

Scheme 23

Analogously to pinacolboranes, Masuda's group has developed an important alternative route to arylstannanes by palladium-catalysed reaction of aryl iodides with tri-*n*-butyltin hydride [101]. In this manner, Pd(PMePh$_2$)$_2$Cl$_2$-catalysed reactions of several substituted iodobenzenes with *n*-Bu$_3$SnH in the presence of potassium acetate as the base furnish the aryltri-*n*-butylstannanes in good to excellent yields. Interestingly, the presence of the free-amino group is well tolerated, as can be seen from the reaction of 4-iodoaniline (**170**) which gave 4-aminophenyltri-*n*-butyl stannane (**171**) in 73% isolated yield [101], Scheme 24.

The latest approach to arylstannanes involves the reaction of aryl chlorides with sodium trimethylstannate under irradiation. Rossi's group has found that by simple irradiation (350 nm; 250 W, 1-2 h) a mixture of Me_3SnCl (1.1 eq.), metallic sodium (2.2 eq.) and desired aryl chloride (1 eq.) in liquid ammonia led to the aryltrimethyl stannane in high to quantitative yields [102]. Alternatively, $NaSnMe_3$ obtained in liquid ammonia, after evaporation of the latter, is reacted with aryl chloride in diglyme to give aryltrimethylstannane with similar success [103].

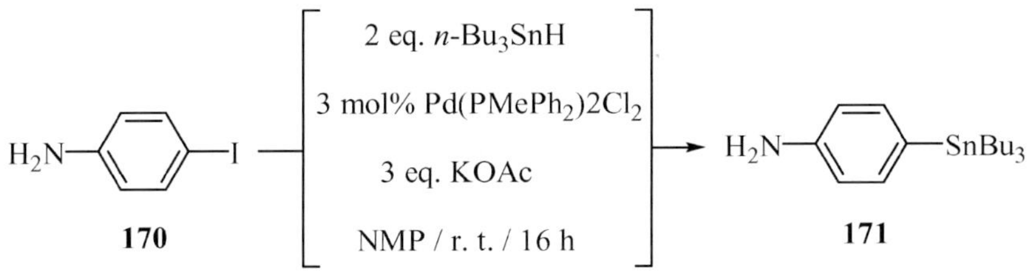

170   2 eq. *n*-Bu_3SnH
3 mol% Pd($PMePh_2$)$2Cl_2$
3 eq. KOAc
NMP / r. t. / 16 h   **171**

Scheme 24

The Pd(PPh_3)$_2Cl_2$ and Pd(PPh_3)$_4$ catalysts in THF or dioxane as a reaction medium are the most common Stille reaction systems. Alternative methods for performing the reaction of aryltrialkylstannanes with different electrophilic substrates are given in the Table 6.

Table 6. Methodology of the Stille reaction [1,9,10]

Ar-X, X =	Catalyst [a]	Solvent [b]	Additives	Lit. ref.
$N_2^+ BF_4^-$	A, B	1	-	107
I, Br	C-H, M,N	2, 3, 4, 6	NaCl (1 eq.) [c]	1, 6, 77
Cl	C, D, I	2, 3	CsF (2.2 eq)	78
OTf, OTs, OMs	C, D, J (K, L) [d]	6, 5, (7-10)	LiCl (3 eq.)	75, 76, 79-83

[a] Catalysts (3-5 mol%): A = Pd(OAc)$_2$; B = Pd(dba)$_2$; C = Pd(PPh_3)$_4$; D = Pd(PPh_3)$_2Cl_2$; E = BnPd(PPh_3)$_2$Cl; F = Pd(MeCN)$_2Cl_2$; G = [(η^3-C_4H_7)PdCl]$_2$; H = Pd(dppf)Cl_2; I = Pd$_2$(dba)$_3$ / P*t*-Bu$_3$ (Pd : P = 1 : 2); J = Pd$_2$(dba)$_3$ / AsPh$_3$ (Pd : As = 1 : 4); K = CuCl; L = CuBr; M = CuI; N = MnBr$_2$.
[b] Solvents: 1 = MeCN, 2 = THF; 3 = 1,4-dioxane; 4 = DME; 5 = DMF; 6 = NMP; 7 = acetone; 8 = HMPA; 9 = DMSO; 10 = xylene.
[c] Sodium chloride has been used as an additive in the copper- and manganese-catalysed Stille reactions with aryl iodides in NMP [88].
[d] CuCl and CuBr accelerate the Pd-catalysed Stille reactions in DMF [85,86].

The aryltrialkylstannanes obtained by above described methods affected the Stille cross-coupling reactions whose efficacy is presented through the following examples where arylstannanes **172-176** were reacted with aryl halides **25, 177, 178** and iodobenzene giving biaryls **179-183** in respective yields [77], Scheme 25.

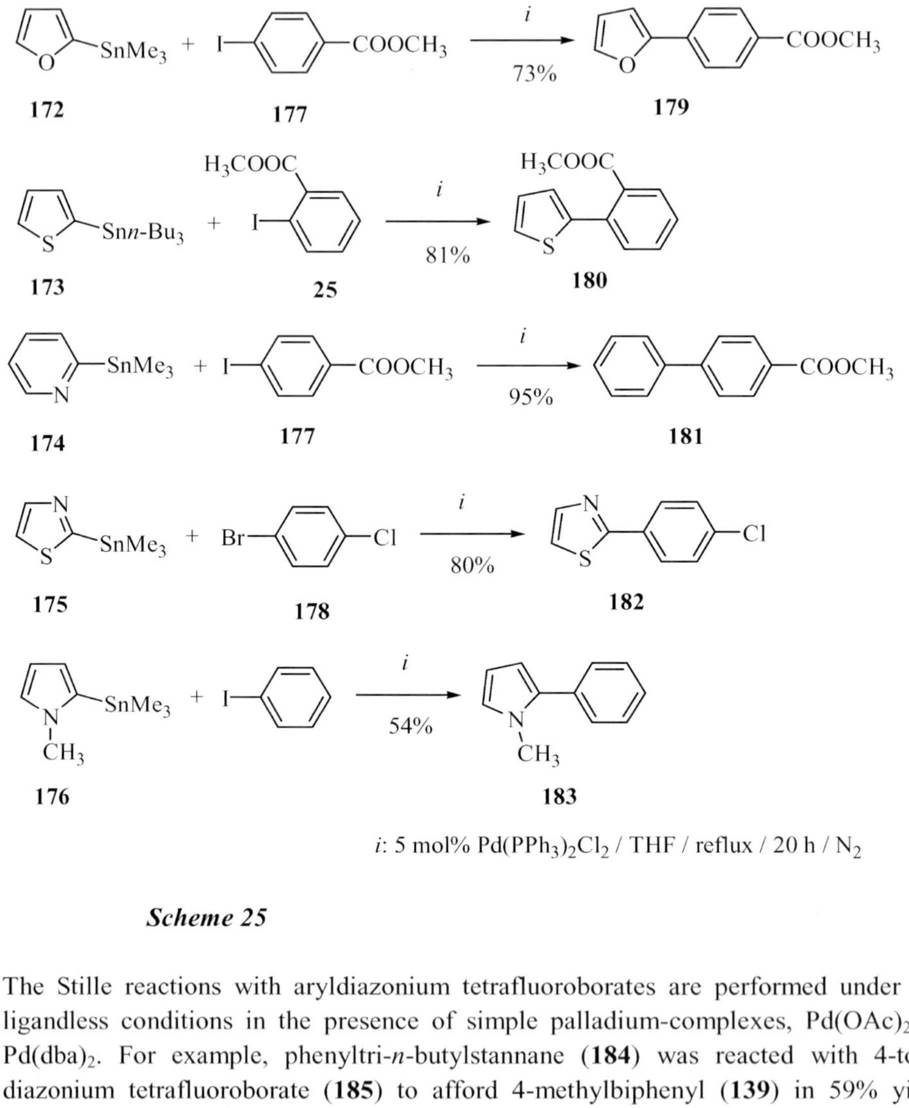

i: 5 mol% Pd(PPh₃)₂Cl₂ / THF / reflux / 20 h / N₂

i: 5 mol% Pd(PPh$_3$)$_2$Cl$_2$ / THF / reflux / 20 h / N$_2$

Scheme 25

The Stille reactions with aryldiazonium tetrafluoroborates are performed under the ligandless conditions in the presence of simple palladium-complexes, Pd(OAc)$_2$ or Pd(dba)$_2$. For example, phenyltri-*n*-butylstannane (**184**) was reacted with 4-tolyl diazonium tetrafluoroborate (**185**) to afford 4-methylbiphenyl (**139**) in 59% yield [107], Scheme 26.

Among different palladium-phosphine complexes, simple triphenylphosphine-based complexes were the most effective in the reactions with aryl iodides and bromides.

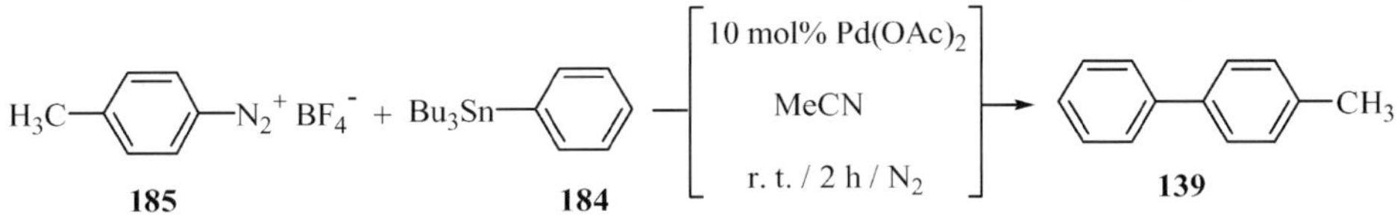

Scheme 26

However, the copper- and manganese-catalysed Stille reactions with aryl iodides have become the most important approaches to the synthesis of biaryls involving organostannanes. The efficacy of Cu- and Mn-catalysed Stille reactions can be seen from the examples in the Table 7.

Table 7. The reaction yields in copper- and manganese-catalysed Stille reactions with aryl iodides [88]

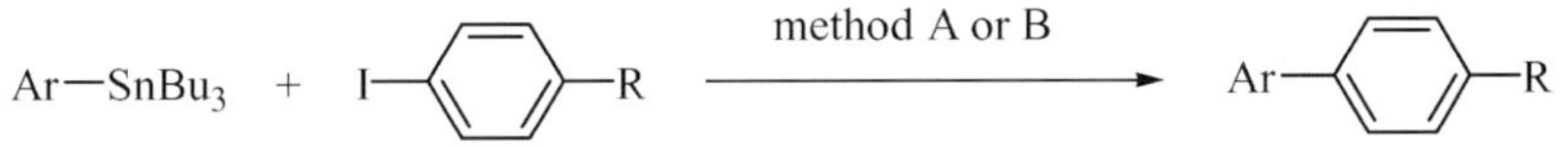

Ar	R	Method [a]	Temp. (°C)	Time (h)	Yield (%)
Phenyl	CH_3	A	110	10	81
		B	120	10	76
2-Thiophenyl	H	A	100	12	92
		B	100	12	92
2-Thiophenyl	CH_3O	A	90	8	85
		B	100	8	86
2-Thiophenyl	I	A	100	6	86 [b]
		B	110	10	80 [b]
2-Furyl	CH_3O	A	90	8	90
		B	90	8	90

[a] Method: A = CuI (10 mol%), B = $MnBr_2$ (10 mol%); NMP, NaCl (1 eq.)
[b] 1,4-Bis(2-thiophenyl)benzene was obtained.

The Stille reactions of aryl chlorides, as the least reactive electrophilic substrates can be successfully performed by applying a palladium-complex of electron-rich and sterically encumbered Pt-Bu$_3$ as Pd(0)-stabilizing ligand [78]. The latter increases the electron-density at palladium metallic centre and thus facilitates the oxidative addition step of unreactive electron-rich aryl chlorides to the Pd(Pt-Bu$_3$)$_2$. For instance, even 4-chloroanisole (**76**), among the least reactive chlorides, was reacted with phenyltri-*n*-butylstannane (**184**) to give 4-methoxybiphenyl (**78**) in 94% yield [78], respectively,

Scheme 27. Addition of cesium fluoride, as the most suitable fluoride source, was necessary to rich the high yields. The fluoride ion reacts with aryltrialkylstannane to form a hypervalent, pentacoordinative, organotin species typically more nucleophilic than its tetravalent precursor. In this manner, the fluoride ion does co-catalyse the Stille reaction.

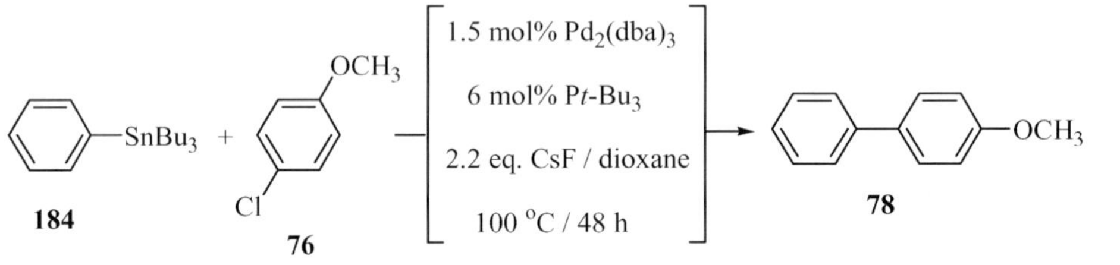

Scheme 27

Moreover, the accelerating influence of the fluorides, e.g. CsF, n-Bu$_4$NF, KF, etc., was found to be a general phenomena connected with the all processes involving tin to palladium transmetallation step [78,93]. The Stille reactions of aryl iodides and bromides with diaryldialkylstannanes and tetraarylstannanes have been also facilitated by n-Bu$_4$NF (2-6 eq.) transfering all aryl groups from tin to palladium [93]. Apart from phosphine-based palladium complexes, variuos N-heterocyclic carbenes have been proved as equally effective palladium(0)-stabilizers. Pd(OAc)$_2$ / imidazolium chloride **186** in the presence of n-Bu$_4$NF was found to be very active catalytic system for the Stille reactions of aryl bromides and chlorides. For example, the reaction of phenyltrimethylstannane (**187**) with mesityl bromide (**188**) led to 2,4,6-trimethyl biphenyl (**189**) in 86% yield [80], respectively, Scheme 28.

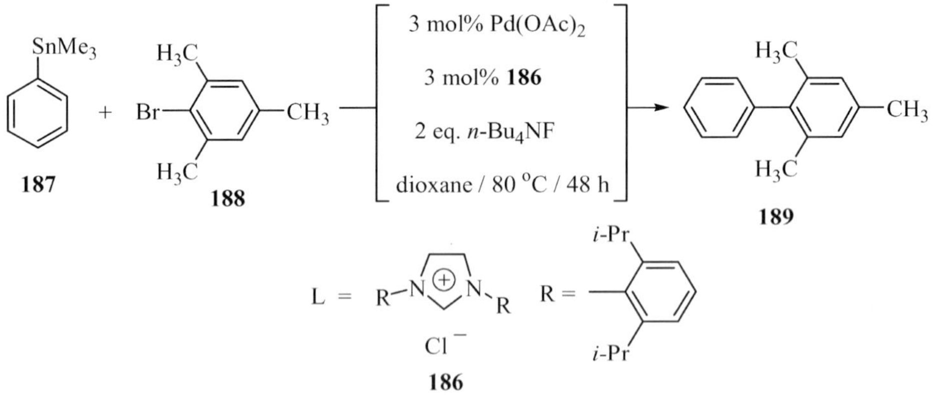

Scheme 28

The Stille reactions of aryl sulfonates are an extremely valuable alternative approach to generate the aryl-aryl bond, since high yields can be obtained even with quite unreactive polymethoxylated-phenyl triflates [80,81]. Once again, the cross-coupling reactions with aryl sulfonates require the presence of lithium chloride (2-3 eq.), otherwise the decomposition of the catalyst occurs and reactions often failed or lower conversions were achieved. Other lithium halides, e.g. LiBr, LiI, accelerate the Stille reaction more strongly than LiCl in the initial stage, however, as the reaction proceeds, the catalyst decomposition was much faster leading to longer reaction times and uncomplete conversions [82]. The reactions with aryl triflates have been generally conducted in polar, coordinating solvents, capable of solubilizing lithium chloride enough to provide a required concentration of the chloride anion, e.g. NMP or DMF. For instance, sterically hindered 2,6-dimethoxyphenyl triflate (190) was reacted with phenyltri-*n*-butylstannane (184) to give the 2,6-dimethoxybiphenyl (191) in 74% yield [81], respectively, Scheme 29.

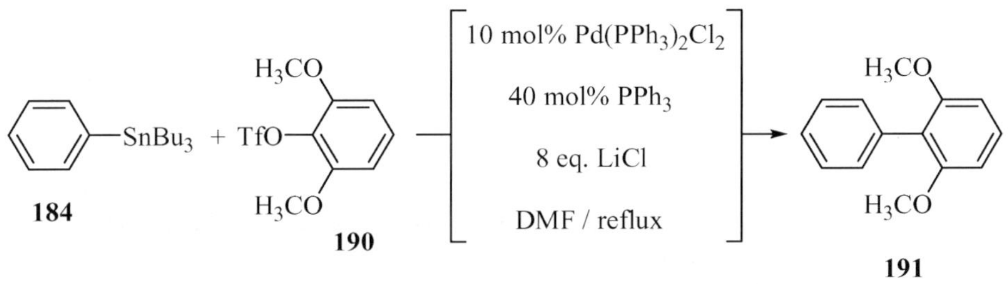

Scheme 29

However, aryl triflates have been coupled with arylstannanes in even less polar solvents in the absence of LiCl. The reactions of arylstannane 192 with pyridyl-2-sulfonates 193a-c furnish the corresponding 2,2'-bipyridines 194 and 195 in moderate to good yields, whereas the amount of the homo-coupling products derived from the parent arylstannane decreases in order: Ar-OMs > Ar-OTs > ArOTf [83], Table 8.

In addition, the Stille reactions of aryl triflates can be co-catalysed with copper(I) bromide or iodide (2-4 eq. to Pd) in refluxing DMF under standard conditions (Pd(PPh$_3$)$_4$ or Pd(PPh$_3$)$_2$Cl$_2$ / 2PPh$_3$ catalysis + LiCl) [86].

The major side-reactions which occur under the Stille reaction conditions are:

- homo-coupling reaction of aryltrialkylstannane and
- alkyl-transfer reaction from tin to palladium.

The former reaction proceeds by the oxidative addition of the aryl-tin bond to the palladium(0) catalyst to form a putative Ar-Pd-SnR$_3$ intermediate. This undergoes the free-radical oxidation at tin with traces of oxygen to produce R$_3$SnOSnR$_3$ as well as homo-coupling product derived from the parent aryltrialkylstannane. Although this mechanism is far from being clear, the oxygen apparently affects the homo-coupling reaction [79]. Moreover, the well known radical scavenger 2,6-di-*tert*-butyl-4-methyl phenol (BHT) is inhibiting this side-reaction significantly. Thus when the Stille reaction mixtures have been exhaustively degassed (by at least six repeated freeze-thaw cycles), this unwanted side-event was completely shut off. Since the ordinarily "degassed" reaction mixtures usually still contain a small amount of dissolved oxygen, especially in more polar solvents such as DMF or NMP, the homo-coupling reaction always proceeds in a range between 5-15%. The copper(I) iodide and manganese(II) bromide-catalysed Stille reactions of aryl iodides led to significant amount of arylstannane homo-coupling side-products only in the absence of sodium chloride as an additive.

Table 8. The selectivity of the Stille reactions of pyridyl-2-sulfonates (**193a-c**) with 6-methylpyridin-2-yltri-*n*-butylstannane (**192**): cross- vs. homo-coupling [83]

193, X =	Yield 194 (%)	Yield 195 (%)
OMs (a)	31	36
OTs (b)	53	40
OTf (c)	78	0

The alkyl-transfer process from tin to palladium causes the formation of methylarenes when ArSnMe$_3$ was used as the parent stannane, or *n*-butylarenes in the case of ArSnBu$_3$, usually in amount up to 30% [79]. Ordinarily preferential transfer of an aryl-over alkyl-group from tin to palladium became less favourable in the case of arylstannanes bearing coordinating substituents at the *ortho*-benzylic positions. For example, the reaction of stannane **196** with 4-acetylphenyl triflate (**197**) gave the expected biaryl **198** in 50% yield, but also 4-*n*-butylacetophenone (**199**) with a 24% yield [79], Scheme 30.

Although the Stille reactions with aryltrimethylstannanes usually proceed faster, the reactions involving aryltri-*n*-butylstannanes have given a lower amount of side-products from alkyl-transfer.

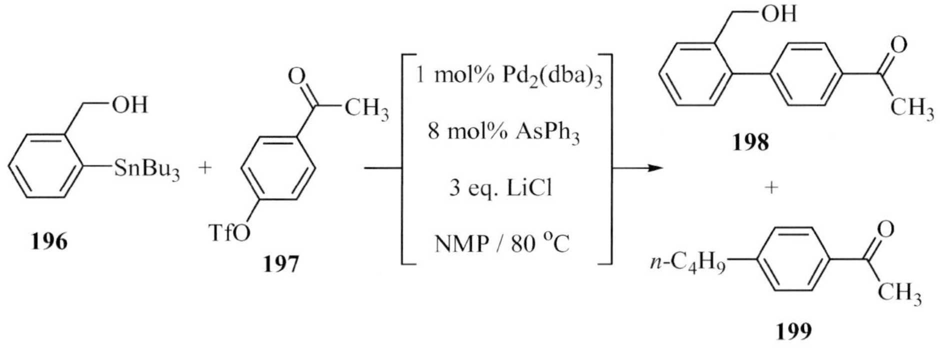

Scheme 30

Among the recent achievements in the Stille reaction chemistry, Kang's group has introduced the first polymer supported palladium-catalyst which has been proved as efficient and reusable [108]. A polymeric palladium-complex of modified silica bearing sulfide and cyano groups catalysed the reaction of 2-thiophenyltri-*n*-butyl stannane (**173**) with various iodobenzenes in aqueous media to give biaryls in high yields. Thus the reaction with iodobenzene itself gave 2-phenylthiophene (**71**) in 80% yield [108], respectively, Scheme 31.

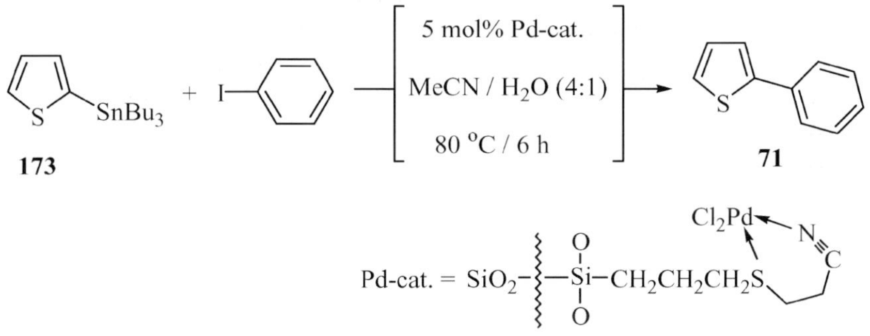

Scheme 31

The Stille reaction is somewhat more sensitive to the steric hindrances than the Negishi reaction involving organozincs. According to this, 2,2-, 2,2'-di-, and 2,6,2'-trisubstituted biaryls have been obtained in good yields [80], however the fully *ortho*-substituted biaryls, if any, were prepared in poor yields. The yields of extremely encumbered tetra-*ortho*-substituted biaryls can be somewhat improved by addition of copper(I) bromide or copper(I) iodide as cocatalysts (2-4 eq. to Pd) [85,86].

The Stille reaction have been successfully employed in the synthesis of a great number of variousy substituted biphenyls [75,78,80,81,84-86,92,93,96,99], phenylthiophenes

[77,87], phenylfurans [77,87,97], phenylpyridines [77], phenylthiazoles [77], phenyl-
N-alkylpyrroles [77], phenylbenzo[*b*]furans [98,109], phenylquinolines [75],
bipyridines [83,104], bithiazoles [94], phenylnaphtalenes [110], dihydrophenanthrenes
[111], phenanthro[9,10-*d*]heterocycles [112], terphenyls [113], terpyridines [114-116],
oligopyridines [117,118], cyclic-sexipyridines [119], heteroarylpyridinium salts [120],
phenyl-, thiophenyl-, and pyridyl-pyridazines [121], bithiophenes [122], various
polythiophenes [95,123], polyfurylthiophenes [123], phenyl- and thiophenyl-oxazoles
[124], thiophenylferrocenes [125,126], pyridyl- and quinolinyl-ferrocenes [127], as
well as several other biaryl structures. The Stille reaction conditions are well tolerating
several common functional groups: ester, nitrile, aldehyde, ketone, free alkyl-hydroxy,
primary amino group, etc. The versatility of the Stille reaction can be judged from
several successful cross-coupling reactions of cephalosporin derived alkenyl triflates
with several aryl and heteroarylstannanes where the appropriate products were
obtained with no disturbance of otherwise very sensitive cephem-structures [128,129].
This fact is very important advantage of this reaction over the Kharasch and Negishi
reaction. However, the requirement of extremely toxic organostannanes in
stoichiometric amount is the most serious limit for wide-spread use of the Stille
reaction. Otherwise this is a very powerful tool in the synthetic organic chemistry.

4.6. Cross-coupling reactions of arylsilanes with aryl halides

The cross-coupling reaction of aryl halides with arylsilane reagents is an
important alternative aryl-aryl bond forming strategy. Thus the reaction of various
aryltrialk(ox)ylsilanes readily proceeds with aryl iodides [130] and bromides [131] in
the presence of palladium(0)-catalyst to give the biaryls in good to excellent yields,
Scheme 32.

$$\text{Ar—SiR} \quad + \quad \text{X—Ar'} \xrightarrow{\text{Pd-catalysis}} \text{Ar—Ar'}$$

X = I, Br

R = $(OMe)_3$, $(OEt)_3$, $(OCH_2CF_3)_3$, $(Me)_2OH$, $Me(OH)_2$, $Et(OH)_2$

Scheme 32

Among the arylsilane reagents, $ArSi(OR)_3$, where R = Me [130], Et [131], CH_2CF_3
[130], are the most convenient and easily available. The aryldialkylhydroxy- and
arylalkyldihydroxy-silanes, $ArSiR_2OH$ and $ArSiR(OH)_2$, where R = Me, Et, have also
been used in the synthesis of biaryls [132]. Moreover, polysiloxanes bearing a phenyl

group, e.g. $(Si(Ph)(Me)O)_n$, have been successful aryl-donors in the palladium-catalysed reactions with aryl iodides [133]. Once again, the oxidative addition of aryl halide (**I**) to the catalytically active palladium(0) complex occurs to form arylpalladium(II) intermediate (**XIII**). The crucial step is the transmetallation process of an aryl group from arylsilane to the arylpalladium(II) intermediate to produce a diarylpalladium(II) complex (**XIV**) which rapidly eliminates the biaryl (**II**) (see Chapter 3), Scheme 33. Analogously to the Stille and Suzuki-Miyaura reaction (see Chapter 5), the transmetallation from silicon to palladium can be strongly accelerated with fluoride, as well as oxygen anions. The latter react with tetracoordinative silanes **XVIII** to form a pentacoordinative silanolates **XIX**, which smoothly undergo aryl-transfer process to palladium.

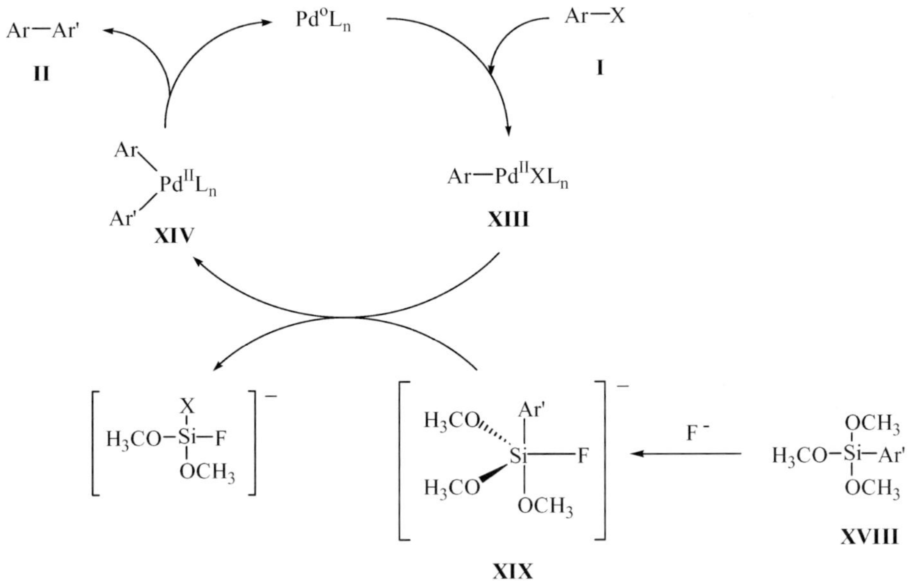

Scheme 33

As the reaction (pre)catalysts (5-10 mol%): $Pd(dba)_2$ [130], $Pd(OAc)_2$ [131], $Pd(MeCN)_2Cl_2$ [131], $Pd(PPh_3)_2Cl_2$ [131], $Pd(PPh_3)_4$ [132], $Pd_2(dba)_3$ [132], $(\eta^3\text{-}C_3H_5PdCl)_2$ [134], $Pd(OAc)_2$ / $P(2\text{-tolyl})_3$ [134] have been effective. As cocatalysts, tetra-*n*-butylammonium fluoride (TBAF) [130], potassium fluoride [134], aqueous sodium or potassium hydroxide [131], and silver(I) oxide [132] have been used. The reactions are typically conducted in DMF [130], but THF [130,132,133] and 1,4-dioxane [131] have also been applied as an alternative reaction medium at 65 °C (refluxing THF) to about 100 °C (refluxing dioxane) under an inert atmosphere.

The arylsilanes can be obtained by:
- reaction of aryllithiums or aryl Grignard reagents with $Si(OR)_4$, $ClSi(OR)_3$, or $SiCl_4$ followed by the treatment with ROH; R is usually Me or Et [135],
- palladium(0)-catalysed silylation of aryl iodides and bromides with triethoxysilane, $((EtO)_3SiH)$ [135], or by
- palladium(0)-catalysed silylation of aryl iodides, bromides and certain electron-deficient aryl chlorides with hexamethyldisilane (Me_6Si_2) [136].

The first approach is well established and widely used, even on an industrial scale. Herein, rather practical access to aryltriethoxysilanes *via* Barbier-type reaction of aryl halides, iodides or bromides, and tetraethyl orthosilicate is discussed. So, when the latter (2 eq.) and 4-bromotoluene (**113**, 1 eq.) were added successively to magnesium turnings in diethyl ether, followed by 6 h-refluxing, filtration and distillation, 4-tolyl triethoxysilane (**200**) was produced in 72% yield [131], see section 4.10.8. However, this method is restricted to the substrates lacking sensitive functional groups. Analogously to the borylation with pinacolborane (see Chapter 5), triethoxysilane reacts with aryl iodides and bromides to afford the aryltriethoxysilanes. Thus a palladium(0) complex, *in situ* generated from $Pd(dba)_2$ and sterically encumbered and electron-rich aryldialkylphosphine **201**, efficiently catalyses the reaction of triethoxysilane with various aryl iodides and bromides to give the aryltriethoxysilanes in moderate to good yields. For instance, 4-iodoaniline (**170**) was converted to 4-triethoxysilylaniline (**202**) in 77% yield [135], respectively, Scheme 34.

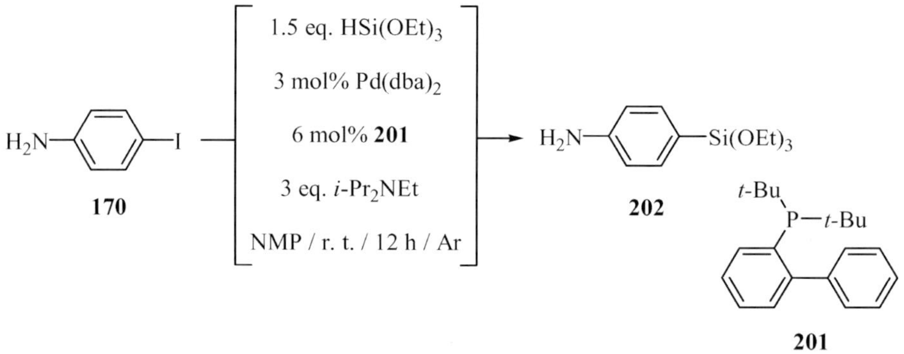

Scheme 34

Although the method is well tolerating to the presence of free-amino, hydroxy, acetamido, ketone, and acetate ester groups, in several cases yields were rather low. Moreover, the strong deleterious effect of the *ortho*-substituent to the halide was found, e.g. 2-iodotoluene and 2-iodopyridine were recovered quantitatively. Masuda's group has developed a similar method based on $Pd(dba)_2$ / $P(2-tolyl)_3$-catalysts suitable for electron-rich *para*-substituted aryl iodides. The most serious side-reaction is the reductive dehalogenation which, under certain reaction conditions, may become a major process.

The last alternative access to arylsilanes is based on palladium(0)-catalysed silylation reaction of aryl iodides and bromides (and certain chlorides) with hexamethyldisilane in HMPA, DMF, NMP, or the most effective DMPU. In this manner, 4-trimethylsilyl acetophenone (**203**) was prepared from 4-bromoacetophenone (**73**) in 83% isolated yield [136], Scheme 35.

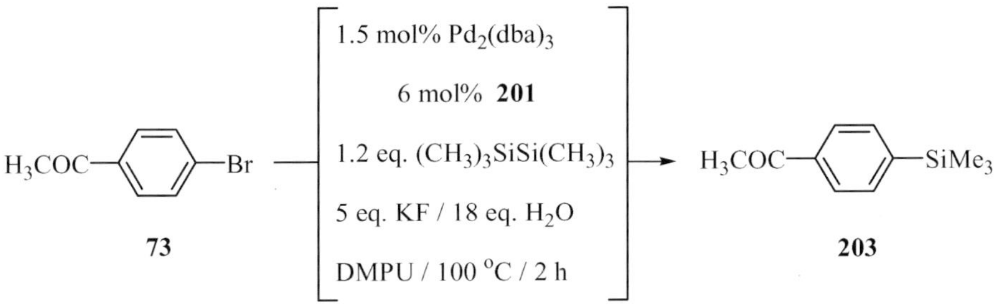

$$
H_3COC{-}\!\!\left\langle\ \right\rangle\!\!{-}Br \quad
\begin{bmatrix}
1.5\ \text{mol\% Pd}_2(\text{dba})_3 \\
6\ \text{mol\% } \mathbf{201} \\
1.2\ \text{eq. } (CH_3)_3SiSi(CH_3)_3 \\
5\ \text{eq. KF / 18 eq. } H_2O \\
DMPU\ /\ 100\ ^{\circ}C\ /\ 2\ h
\end{bmatrix}
\longrightarrow \quad
H_3COC{-}\!\!\left\langle\ \right\rangle\!\!{-}SiMe_3
$$

73   **203**

Scheme 35

During the silylation with hexamethyldisilane, the reductive dehalogenation and the formation of a biaryl are common side-reactions. However, these can be minimized by using potassium carbonate as the base in the case of electron-rich substrates, or with potassium fluoride in the reactions with electron-poor substrates. The arylsilanes produced by any of above mentioned reactions can be cross-coupled with aryl halides by employing the following methods, Table 9.

The most convenient is the second, Masuda's method involving the easily available aryltriethoxysilanes which have been cross-coupled with a wide variety of aryl bromides to afford the biaryls in good to excellent yields. The efficacy of Masuda's method is presented in the Table 10.

The palladium complex $Pd(MeCN)_2Cl_2$ is actually decomposed under the reaction conditions to form catalytically active nano-sized palladium clusters. However, in the case of less reactive aryl halides bearing electron-donating groups, e.g. 4-bromo toluene, addition of 10 mol% of $P(Oi\text{-}Pr)_3$ as stabilizing ligand for zero-valent palladium was necessary to rich high conversions.

The only side-reaction of considerable interest is the homo-coupling reaction of the parent arylsilane. For that reason, Masuda's method is usually conducted with a slight excess, e.g. 20 mol%, of arylsilane. Once again, methods involving significantly higher excesses of arylsilanes cause several difficulties during the purification procedure of the final biaryl product.

Table 9. Methodology of the arylsilane involving synthesis of biaryls

ArSiR, R =	Catalyst (mol%)	Base (eq.)	Reaction conditions	Lit. ref.
Me_3, OMe_3, $(OCH_2CF)_3$	$Pd(dba)_2$ (10)	n-Bu_4NF (2-3)	THF, DMF 60-100 °C	130
$(OEt)_3$	$Pd(MeCN)Cl_2$ (2)	aq. NaOH (3)	dioxane or DME - H_2O / 80 °C	131
$(Me)_2OH$	$Pd(PPh_3)_4$ (5)	Ag_2O (1)	THF / 60 °C	132
$(X)_mA_{3-m}$ X = Cl, F, m = 1, 2 A = Me, Et, i-Pr	$(\eta^3$-$C_3H_5PdCl)_2$ (2.5)	n-Bu_4NF or KF (2)	DMF / 60 °C	134

Table 10. The yields of biaryls obtained by Masuda's cross-coupling reaction of aryl triethoxysilanes with aryl bromides [131]

R_1	R_2	Time (h)	Yield (%)
H	CH_3	12	89
CF_3	CH_3CO	4	81
CF_3	CH_3O	8	84
Cl	OH	4	80
CH_3O	CH_3CO	4	94

Synthesis of biaryls using arylsilanes is apparently the focus of current research, and is going to be an important, environmental friendly, and valuable alternative to the Suzuki-Miyaura and Negishi reactions.

4.7. Cross-coupling reaction of diarylmercurials with aryl halides

Diarylmercurials, Ar_2Hg, undergo the palladium-catalysed cross-coupling reactions with aryl halides, preferably iodides, to give biaryls in high yields, Scheme 36.

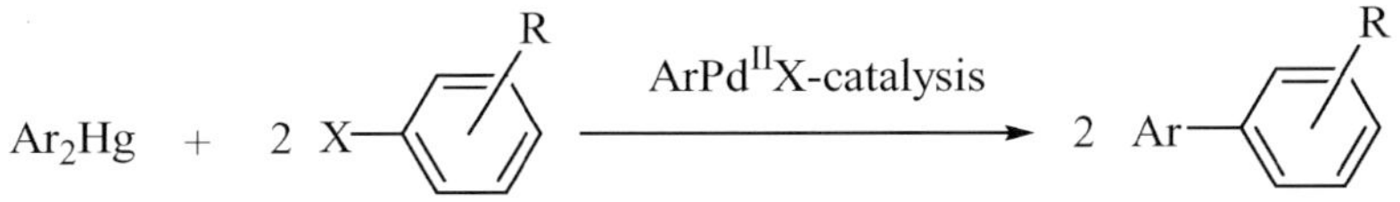

$$Ar_2Hg \quad + \quad 2 \; X \text{—} \underset{}{\overset{R}{\bigcirc}} \quad \xrightarrow{\text{ArPd}^{II}\text{X-catalysis}} \quad 2 \; Ar \text{—} \underset{}{\overset{R}{\bigcirc}}$$

$$X = I, (Br, Cl)$$

Scheme 36

The reaction proceeds by the mechanism closely related to other organometallics (see above), and both aryl groups are incorporated into the biaryl product indicating that the transmetallation actually involves two steps: the reaction of Ar'PdX with Ar_2Hg to give ArAr'Pd plus ArHgX, and the reaction of the latter and the second molecule of Ar'PdX to produce an additional molecule of ArAr'Pd and HgX_2. The unsymmetrical *cis*-diarylpalladium intermediates decompose to unsymmerical biaryls with regeneration of catalytically active palladium(0)-complex. Despite the high toxicity and ecological problems, diarylmercurials exhibit good stability towards water, air and alcohols. The diarylmercurials can be obtained by the following convenient alternative reactions:

- transmetallation of aryllithiums or aryl Grignard reagents with $HgCl_2$,
- direct mercuration of activated arenes with HgX_2, X = Cl, OAc, OOCCF₃, etc., and
- mercuration of aryldiazonium salts with $HgCl_2$ in the presence of copper metal [137].

Although the first access is general, an organic group must lack sensitive functional group(s). Mercuration with mercury(II) halides or acetate is a well known electrophilic aromatic substitution reaction which proceeds smoothly with arenes that are reactive at least as benzene itself. For example, the latter reacts with mercury(II) acetate at reflux temperature, about 80 °C, to give phenylmercuric acetate in high yield. However, more reactive arenes such as phenols, *N,N*-dialkylanilines, or reactive electron-rich heterocycles, e.g. thiophene or furan, react with mercuric salts at very mild conditions, e.g. room temperature [138]. As distinguished from arylzincs, arylboronic acids, and several other organometallics, organomercurials can be readily prepared from anilines *via* aryldiazonium salts. The latter approach, well known as the Nesmeyanov reaction, significantly extends the scope of cross-coupling reactions involving diarylmercurials, since anilines are more widely available than aryl halides. Moreover, the Nesmeyanov reaction can be applied in the synthesis of diarylmercurials derived from electron-poor arenes bearing a wide variety of sensitive and protic functional groups. Thus double salts of aryldiazonium chlorides and mercury(II) chloride are decomposed by copper metal in acetone at -5 to -15 °C followed by reaction with aqueous ammonia furnishing the symmetrical diarylmercurials in moderate to good yields. The efficiency of the Nesmeyanov reaction is shown in the Table 11.

The diarylmercurials prepared by any of these reactions are the most efficiently coupled with aryl iodides in the presence of a catalytic amount (1 mol%) of aryl(diphenylphosphino)palladium(II) iodides to give the corresponding biaryls in good to excellent yields under very mild reaction conditions. This method was developed by Beletskaya's group [138], Table 12.

Table 11. The yields of symmetrical diarylmercurials obtained by Nesmeyanov reaction [137]

R	Yield (%) [a]
H	65
4-I	70
4-O$_2$N	10
2,5-Cl$_2$	20
2-CH$_3$O	60
2-CH$_3$OOC	23
4-CH$_3$OOC	68

[a] The copper for performing the Nesmeyanov reaction is prepared by reaction of aqueous CuSO$_4$ and zinc dust.

The only side-reaction is the homo-coupling reaction of the parent diarylmercury which occurs as a result of oxidative addition of diarylmercury to palladium(0) followed by oxidative demercuration reaction. However, the selectivity can be enhanced by adding a nucleophilic catalyst, e.g. iodide ion. The role of iodide ion is apparently crucial as it significantly reduces the amount of symmetrical biaryl from the oxidative demercuration process. Since diarylmercurials act as two aryl-groups donors, only a half equivalent of aryl iodide is required for the cross-coupling reaction.

Table 12. The selectivity of the Bumagin's method for cross-coupling of diarylmercurials with aryl iodides [138]

$$\left(R\underset{Z}{\diagdown}\right)_2 Hg \quad + \quad I-Ar \quad \longrightarrow \quad \left[\begin{array}{c} 1 \text{ mol\% ArPdI} \\ 2 \text{ eq. NaI} \\ DMF / Ar \end{array}\right] \quad \longrightarrow \quad R\underset{Z}{\diagdown}Ar$$

Z, R	Ar	Time (min)	Temp. (°C)	Yield (%) [a]
S, H	4-O$_2$NPh-	10	20	80
S, H	2-pyridyl-	10	20	91 (8)
O, CH$_3$	2-pyridyl-	10	20	75 (10)
O, CH$_3$	2-thienyl-	120	80-90	60 (20)
O, H	2-nitropyridin-5-yl-	120	80-90	55 (15)

[a] The yields of the homo-coupling by-products are given in parenthesis.

The Bumagin's method has been successfully used in the synthesis of arylferrocenes [139].

4.8. Cross-coupling reaction of arylcopper reagents with aryl halides

The Ullmann-type reaction of well defined arylcopper(I) reagents, ArCu, with aryl iodides to give biaryls was introduced by Nilsson's group [140]. The respective organometallics are prepared by transmetallation of aryllithiums and copper(I) iodide, solubilized with dimethyl sulfide. Arylcopper(I) reagents are relatively stable compounds which can be stored under nitrogen or wetted with dry ethereal solvents at -18 °C, or even at room temperature for a few days without significant decomposition. Relatively low solubility of these organometallics in toluene, which is common reaction solvent, can be improved by adding a slight excess of an equimolar amount of triphenylphosphine (PPh$_3$). The latter forms a complex such as ArCu(PPh$_3$), which is far more soluble in nonpolar solvents, and thus provides clean cross-couplings, as reactions proceed in a homogeneous medium. Nilsson's group has shown that unsymmetrical biaryls can be obtained in good yields, even with respect to both steric encumbrances and a variety of substituents. As an illustrative example, 2-pyridyl copper (**203**), *in situ* produced from 2-bromopyridine (**204**) *via* 2-pyridyllithium (**205**), in the presence of triphenylphosphine was reacted with mesityl iodide (**206**) to produce the respective biaryl **207** in 57% yield [140], Scheme 37.

In contrast to Ziegler's method (see Chapter 2), this reaction can be affected with aryl halides bearing no *ortho*-coordinating substituents.

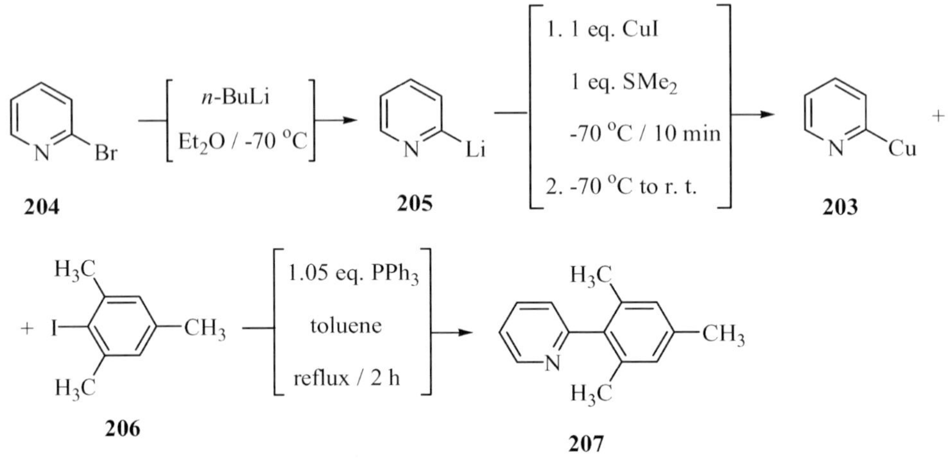

Scheme 37

Although this is an interesting and obviously perspective reaction, it has not been studied with other classes of electrophilic counterparts.

4.9. Cross-coupling reactions of miscellaneous arylmetallic: Mn, Ti, In, and Ge reagents with aryl halides and sulfonates

Arylmanganese reagents, ArMnX, are slightly less reactive than the corresponding arylzincs, and thus exhibit a higher selectivity in the cross-coupling reactions with aryl halides and triflates affording the biaryls in good to excellent yields [141], Scheme 38.

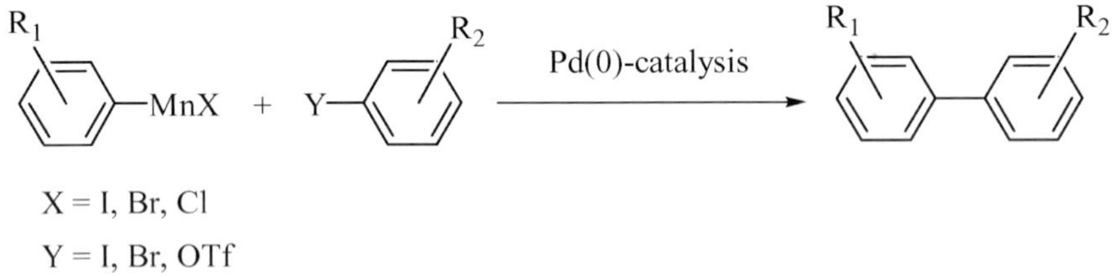

X = I, Br, Cl
Y = I, Br, OTf

Scheme 38

The arylmanganese halides can be prepared by:
- reaction of aryllithiums or aryl Grignard reagents with MnX_2 (X = I, Br, Cl),
- reaction of organomercurials with manganese, and
- reaction of Rieke manganese with aryl iodides or bromides [142].

Arylmanganese halides obtained this way are smoothly cross-coupled with aryl iodides, bromides and triflates in the presence of palladium catalysts (1-10 mol%): $Pd(PPh_3)_2Cl_2$ or $Pd(PPh_3)_4$ under very mild reaction conditions to give the biaryls in the range of 70% to almost quantitative yields, tolerating several sensitive functional groups such as ester, nitrile, ketone, free-amino group, etc. For example, the reaction of Rieke manganese with 3,4-dibromothiophene (**208**) was affected in THF at room temperature furnishing the 4-bromothiophene-3-ylmanganese bromide (**209**). The latter was subsequently coupled with ethyl 4-iodobenzoate (**210**) to produce the biaryl **211** with a 86% yield [141], respectively, Scheme 39.

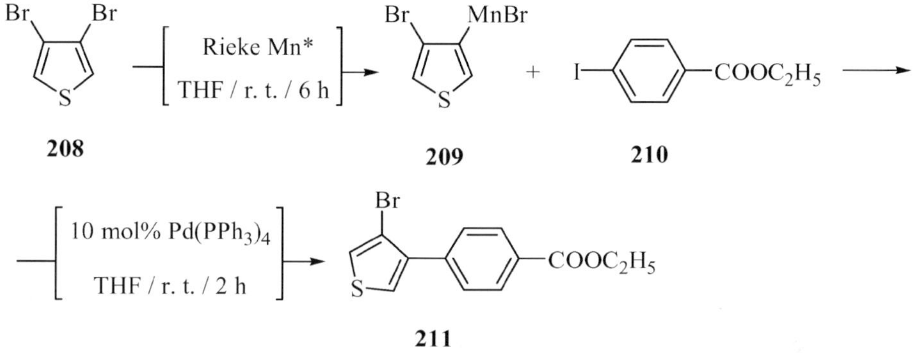

Scheme 39

The synthesis of biaryls involving the arylmanganese halides seems to be a very perspective approach and will become an extremely valuable aryl-aryl bond forming reaction.

In addition, useful organometallics that have been successfully employed in the aryl-aryl bond forming reactions are aryltrialkoxytitanium $(ArTi(OR)_3)$, diaryldialkoxy titanium reagents $(Ar_2Ti(OR)_2)$ where R is usually *i*-Pr [143,144], and triarylindium reagents (Ar_3In) [145-147]. The former are conveniently obtained by reaction of aryllithiums or aryl Grignard reagents with $Ti(O\text{-}i\text{-Pr})_4$ or $ClTi(O\text{-}i\text{-Pr})_3$ at -100 °C to 0 °C [143]. The triarylindiums have been prepared by reaction of aryllithium or aryl Grignard reagents with indium(III) chloride, transmetallation between organomercurials and indium metal, and direct oxidative addition of reactive aryl halides to Rieke indium [145].

Aryltrialkoxy- and diaryldialkoxytitaniums are relatively stable (some of them are even distillable) compounds which readily undergo palladium-catalysed reaction with aryl iodides, bromides and chlorides, as well as triflates to furnish the biaryls in very high yields. For instance, the reaction of 2-naphthyl triflate (**165**) with phenyltitanium triisopropoxide (**212**, in 20 mol% excess) catalysed by palladium-complex of *N,P*-bidentate ligand **213** gave 2-phenylnaphthalene (**214**) in 98% yield [143], respectively, Scheme 40.

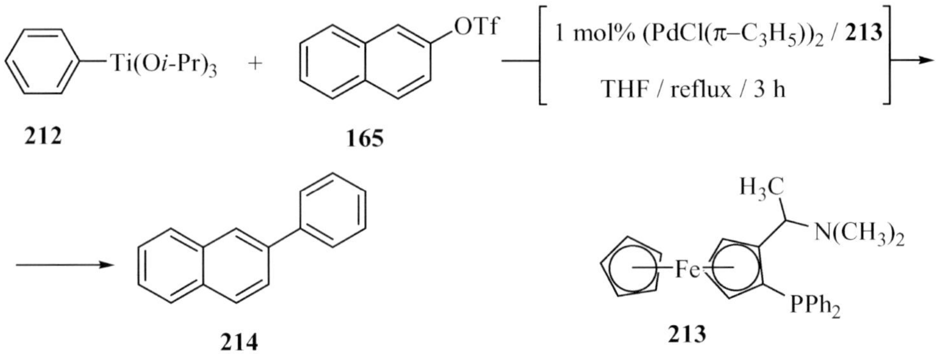

Scheme 40

The diaryltitanium diisopropoxides behave analogously acting as donors of two aryl-groups. Recently, the triarylindiums have been introduced as extremely efficient organometallics capable of transfering all three aryl groups from indium to palladium or nickel. In this manner, the reaction of aryl iodides, bromides and triflates (1 eq.) with triarylindiums (0.34 eq.) was readily effected in the presence of a catalytic amount of either palladium, e.g. $Pd(PPh_3)_2Cl_2$ [145,147], $Pd(MeCN)_2Cl_2$ [145], $Pd_2(dba)_3$ [145], $Pd(dppp)Cl_2$ [145], $Pd(PPh_3)_4$ [145], $Pd_2(dba)_3$ / $P(2\text{-furyl})_3$ [146], $Pd(dppf)Cl_2$ [147], or nickel, e.g. $Ni(PPh_3)_4$ [145], catalysts to produce the biaryls in excellent yields. Among mentioned catalysts, easily available $Pd(PPh_3)_2Cl_2$ and $Ni(PPh_3)_4$ have been found as the most versatile. The reactions of aryl iodides, bromides and triflates were conducted with the palladium-catalysts, whereas the reactions of less reactive aryl chlorides have been performed with $Ni(PPh_3)_4$ [145]. In this manner, triphenylindium (**215**, 0.34 eq.) reacts with 4-acetylphenyl triflate (**197**, 1 eq.) to afford 4-acetylbiphenyl (**68**) in 95% yield [145], Scheme 41.

Apart from triarylindiums, diarylindium halides and arylindium dihalides, stable in an aqueous media, smoothly react with aryl iodides and bromides at a very low palladium-catalyst loading in a THF / water mixture (8:1 to 6:1, v/v) to give the biaryls in very high yields tolerating free-amino, hydroxy, carboxy groups, as well as aldehyde, nitro group, etc. The presence of a polar cosolvent (water) has been essential to promote the cross-coupling reactions [145]. The homo-coupling of aryl halides with indium metal is described in the Chapter 3.

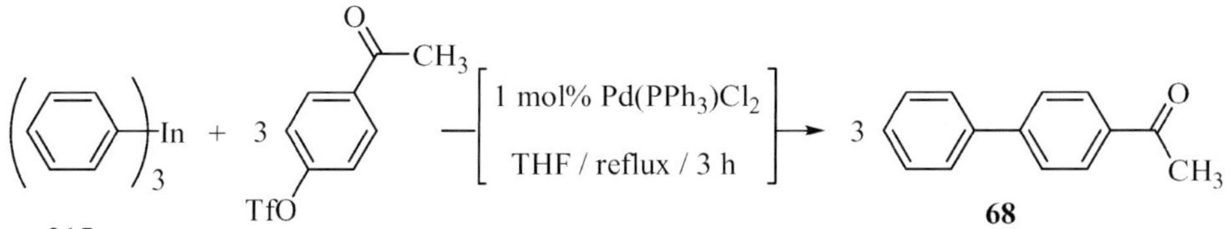

215 **197** **68**

Scheme 41

The synthesis of biaryls using aryltitanium and arylindium reagents are under extensive investigations which will obviously be a significant contribution to the aryl-aryl bond forming methodology.

Analogously to aryltri-*n*-butylstannanes, corresponding germanes can affect the transmetallation to palladium to participate in the cross-coupling reactions with aryl halides. However, the synthetically useful yields were obtained by using 1-aza-5-germa-5-aryl-bicyclo[3.3.3]undecanes (**XX**). The latter are prepared by reaction of 1-aza-5-germa-5-chloro-bicyclo[3.3.3]undecane (**216**) with aryl Grignard reagents. For example, compound **217** was converted to 4-methylbiphenyl (**139**) in the reaction with 4-bromotoluene (**113**) in 85% yield [148], Scheme 42.

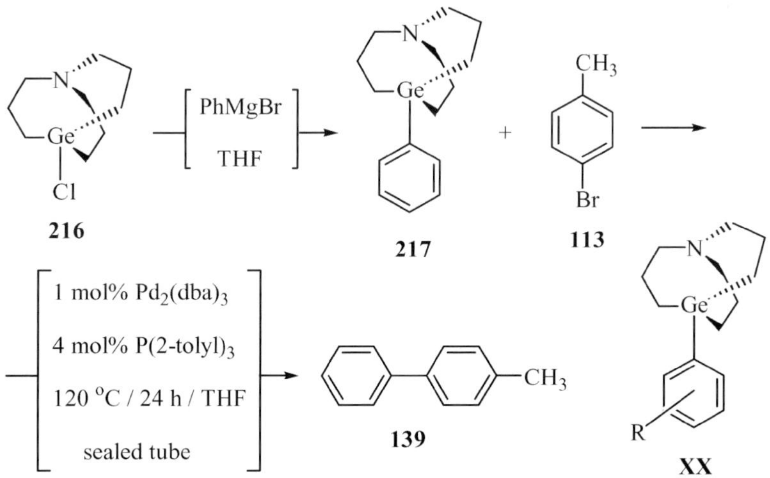

216 **217** **113** **139** **XX**

Scheme 42

Reagent **216** is easily accessible by successive treatment of triallylamine with zirconocene hydride (*in situ* generated from Cp_2ZrCl_2 and $LiAlH_4$ / THF / 0 °C) and germanium(IV) chloride.

4.10. Selected synthetic procedures

4.10.1. The Kharasch reaction: Preparation of 2-methoxybiphenyl (91) [12]

To a solution of bromobenzene (0.78 g, 5 mmol) in THF (5 ml) palladium catalyst $Pd(PPh_3)_4$, (0.115 g, 0.1 mmol) was added under nitrogen. A solution of the Grignard reagent **124**, prepared by the usual method, from 2-bromoanisole (**218**, 1.40 g, 7.5 mmol, 1.5 eq.) and magnesium turnings (0.20 g, 8 mmol, 1.07 eq.) in THF (5 ml), was added dropwise to the mixture under reflux. The reaction mixture was heated under reflux for 20 h. The resulting slurry was evaporated under reduced pressure to remove the THF. Diethyl ether (20 ml) and 1 M hydrochloric acid (10 ml) were added until the solids just dissolved. The aqueous layer was separated and extracted with diethyl ether (2x10 ml). The combined ether layers were washed with water until the washings were neutral, dried ($MgSO_4$), and the ether was removed under reduced pressure. From the crude product, 847 mg (92%) of pure 2-methoxybiphenyl (**91**) was obtained by preparative chromatography over silica gel as a colourless oil.

4.10.2. The Kharasch reaction of arylnitriles: Preparation of tert-butyl biphenyl-3-carboxylate (219) [32]

A solution of lithium thiophenoxide (**220**) in THF (4.0 ml of 1 M solution, 4.0 mmol) was treated at room temperature with phenylmagnesium chloride (**221**, 2.8 ml, 1.4 M in THF, 4.0 mmol) and the resulting solution heated at 60 °C for 1 h. After cooling to room temperature, the reaction solution containing **222** was treated with a solution of 3-*t*-butylcarbonylbenzonitrile (**223**, 410 mg, 2.0 mmol), and bis(trimethylphosphine) nickel(II) chloride (28 mg, 5 mol%). The reaction mixture was stirred at room temperature for 1 h. The THF was removed under reduced pressure, and 5% hydrochloric acid (10 ml) and dichloromethane (20 ml) were added to the residue. The organic layer was separated, whereas the water layer was extracted with additional dichloromethane (2x20 ml). The combined extracts were dried (Na$_2$SO$_4$), filtered and evaporated. The 470 mg (93%) of pure *t*-butyl biphenyl-3-carboxylate (**219**) was isolated by preparative chromatography.

4.10.3. The Negishi reaction catalysed by palladium: Preparation of 4-nitro biphenyl (224) [48]

To a solution of ZnCl$_2$ (2.04 g, 15 mmol) in THF (25 ml), a solution of phenyllithium prepared by reaction of bromobenzene (2.35 g, 15 mmol) and an excess of lithium (0.28 g, 40 mmol) in diethyl ether (15 ml) was added to generate phenylzinc chloride (**151**). The Pd-catalyst was prepared in a separate flask by treating of Pd(PPh$_3$)$_2$Cl$_2$ (350 mg, 0.5 mmol, 5 mol%) in THF (15 ml) with diisobutylaluminum hydride (1 ml, 1 mmol, 1 M in *n*-hexane). To this catalyst, 4-nitro-iodobenzene (**225**, 2.49 g, 10 mmol) and the supernatant solution of phenylzinc chloride prepared above were added. The reaction mixture was stirred for 1 h at room temperature. The reaction mixture was quenched with 10% aqueous hydrochloric acid (100 ml) and extracted with diethyl ether (3x50 ml). The organic layers were collected, dried (Na$_2$SO$_4$), filtered, and evaporated to dryness. The crude product was recrystallized from methanol to give 1.47 g (74%) of pure 4-nitrobiphenyl (**224**), m.p. 112-114 °C.

4.10.4. The Negishi reaction catalysed by nickel: Preparation of 2-cyano phenyzinc bromide (226) and its application in the synthesis of 4'-methylbiphenyl-2-carbonitrile (112) [49]

To a suspension of zinc dust (2.45 g, 37.5 mmol) in dry THF (5 ml), 1,2-dibromo ethane (0.32 ml, 0.70 g, 3.7 mmol, 10 mol% to Zn) was added and the reaction mixture was vigorously stirred at reflux temperature for 5 min. The reaction mixture was allowed to cool to room temperature. This was repeated for three times. Then, trimethylsilyl chloride (0.57 ml, 0.49 g, 4.5 mmol, 12 mol% to Zn) was added and the mixture was refluxed for 5 min. The reaction mixture was allowed to cool to room temperature. The latter was repeated for three times. To the activated zinc suspension 2-bromobenzonitrile (**116**, 2.27 g, 12.5 mmol, 1.25 eq.) was added. The reaction mixture was refluxed for 5 h. After beeing cooled, the clear supernatant containing 2-cyanophenylzinc bromide (**226**) was taken by syringe and added to the solution of 4-chlorotoluene (**227**, 1.27 g, 10 mmol), Ni(PPh$_3$)$_2$Cl$_2$ (0.49 g, 0.75 mmol, 7.5 mol%) and triphenylphosphine (0.39 g, 1.5 mmol, 15 mol%) in THF (20 ml). The reaction mixture was stirred at 55 °C for 5 h, quenched with 1 M aqueous hydrochloric acid (50 ml) and extracted with diethyl ether (3x30 ml). The combined extracts were dried (Na$_2$SO$_4$), filtered, and evaporated to remove ether. From the crude product, 1.45 g (75%) of 4'-methyl-2-cyanobiphenyl (**112**) was isolated by means of preparative chromatography over silica gel (*n*-hexane / dichloromethane).

4.10.5. The Stille reaction catalysed by palladium: Preparation of methyl 2-(2-thienyl)benzoate (180) [77]

A suspension of methyl 2-iodobenzoate (**25**, 2.62 g, 10 mmol), 2-thiophenyltri-*n*-butyl stannane (**173**, 4.21 g, 11.3 mmol), and Pd(PPh$_3$)$_2$Cl$_2$ (0.35 g, 5 mol%) in dry THF (30 ml) was refluxed under a nitrogen atmosphere for 20 h. During the course of the reaction, the colour changed from yellow to black as Pd^0 was formed. The reaction mixture was cooled, diluted with diethyl ether, and filtered through a pad of neutral alumina. The filtrate was washed several times with water, dried (anhydrous K_2CO_3), concentrated, and chromatographed using *n*-hexane / ethyl acetate (9:1) as an eluent to afford 1.77 g (81%) of pure methyl 2-(2-thienyl)benzoate (**180**).

4.10.6. The Stille reaction catalysed by copper(I) iodide: Preparation of 2-phenylthiophene (71) [88]

173   **71**

To a stirred solution of iodobenzene (175 mg, 0.85 mmol) and sodium chloride (50 mg, 0.85 mmol) in NMP (2 ml), copper(I) iodide (16.2 mg, 10 mol%) was added, the mixture was heated to 90 °C, and then 2-thiophenyltri-*n*-butylstannane (**173**, 320 mg, 0.85 mmol) in NMP (2 ml) was added slowly for 1 h *via* a syringe pump. The reaction mixture was stirred at 90 °C for 8 h, and cooled to room temperature, and saturated potassium fluoride solution was added. The mixture was extracted with diethyl ether (3x20 ml), and the combined organic layers were dried (MgSO₄), filtered, and evaporated *in vacuo*. The crude product was separated by preparative chromatography to afford 125 mg (92%) of pure 2-phenylthiophene (**71**).

4.9.7. Synthesis of aryltrialkylstannanes by reaction of aryl bromides with hexaalkyldistannanes: Preparation of 2-nitrophenyltrimethylstannane (228) [99]

94   **228**

A mixture of hexamethyldistannane (3.00 g, 9.2 mmol, 1.5 eq.), 2-nitro-bromobenzene (**94**, 1.23 g, 6.1 mmol), and Pd(PPh₃)₄ (0.20 g, 0.17 mmol, 2.8 mol%) in anhydrous THF (30 ml) was heated to reflux under nitrogen for 10 h. After the mixture was cooled to room temperature, THF was evaporated and dichloromethane (30 ml) was added to the residue. To this mixture, aqueous potassium fluoride (7.0 M, 2 ml) with vigorous stirring was added dropwise. The organic layer was separated, dried and evaporated to dryness The residue was chromatographed over silica gel (100 g) using ethylacetate / *n*-hexane (1:6) as an eluent to give 1.30 g (75%) of pure 2-nitrophenyl trimethylstannane (**228**).

4.10.8. Synthesis of biaryls involving aryltriethoxysilanes: Preparation of 4-tolyl triethoxysilane (200) and its application in the synthesis of 3-(4-tolyl) pyridine (229) [131]

113   **200**   **230**

229

Tetraethyl orthosilicate (1.3 ml, 5.8 mmol) and 4-bromotoluene (**113**, 447 mg, 2.6 mmol) were added successively to magnesium turnings (80 mg, 3.3 mmol) in dry diethyl ether (5 ml). The reaction mixture was refluxed for 6 h and then filtered to remove the salt. The filtrate was purified by Kugelrohr distillation to give 398 mg (72%) of 4-tolyl triethoxysilane (**200**).

To a solution of Pd(MeCN)$_2$Cl$_2$ (5 mg, 0.02 mmol) and 3-bromopyridine (**230**, 158 mg, 1.0 mmol) in dioxane (4 ml), aqueous sodium hydroxide solution (2 M, 1.5 ml, 120 mg, 3 mmol, 3 eq.) and 4-tolyltriethoxysilane (**200**, 305 mg, 1.2 mmol, 1.2 eq.) were added. After being stirred for 4 h at 80 °C, the mixture was extracted with diethyl ether (3x20 ml). The extract was washed with water, dried (MgSO$_4$), and concentrated. The residue was purified by column chromatography over silica gel by using *n*-hexane / ethylacetate (4:1) to give 175 mg (80%) of 3-(4-tolyl)pyridine (**229**).

4.10.9. Synthesis of biaryls using diarylmercurials: Preparation of 2-(2-thienyl) pyridine (231) [138]

233   **232**   **231**

To a mixture of sodium iodide (0.30 g, 2 mmol, 2 eq.), 2-iodopyridine (**232**, 205 mg, 1 mmol), di(2-thienyl)mercury (**233**, 202 mg, 0.55 mmol) in DMF (2.5 ml), was added 2-pyridylpalladium(II) iodide (3.1 mg, 1 mol%) and the reaction mixture was stirred at room temperature for 10 min under argon. The reaction mixture was diluted with water (10 ml) and extracted with diethyl ether (3x5 ml). The extracts were combined, and washed with an aqueous solution of sodium iodide. The ether layer was dried (MgSO$_4$), and then evaporated *in vacuo* to give the crude product. From the latter, 147 mg (91%) of pure 2-(2-thienyl)pyridine (**231**) was obtained by preparative chromatography, as well as 13 mg (8%) of 2,2'-bithienyl.

4.10.10. Synthesis of biaryls from triarylindiums: Preparation of 4-acetyl biphenyl (68) [145]

221a   **215**   **197**

68

To a solution of indium(III) chloride (82 mg, 0.37 mmol) in THF (4 ml), a solution of phenylmagnesium bromide (**221a**, 1.1 mmol in THF or Et$_2$O) was slowly added (15-30 min). The mixture was stirred for 30 min, the cooling bath was removed, and the reaction mixture was warmed to room temperature. A solution of triphenylindium (**215**, 0.34 mmol) in THF was added to a refluxing mixture of 4-acetylphenyl triflate (**197**, 268 mg, 1 mmol) and Pd(PPh$_3$)$_2$Cl$_2$ (7 mg, 1 mol%) in THF (4 ml). The resulting mixture was refluxed for 3 h. The reaction mixture was quenched by the addition of a few drops of methanol. The mixture was concentrated *in vacuo* and diethyl ether (30 ml) was added. The organic phase was washed with aqueous HCl (5%, 10 ml), saturated aqueous NaHCO$_3$ (15 ml), and brine (15 ml), dried (Na$_2$SO$_4$), filtered and evaporated to dryness. The residue was purified by flash chromatography to afford, after concentration and high-vacuum-drying, 186 mg (95%) of pure 4-acetylbiphenyl (**68**).

4.10.11. Synthesis of biaryls based on aryltitanium triisopropoxides: Preparation of 2-phenylnaphthalene (214) [143]

To a solution of phenyllithium (5.56 ml, 1.8 M in cyclohexane-ether, 12 mmol, 1.2 eq.), chlorotitanium triisopropoxide (10 ml, 1.0 M solution in *n*-hexane, 12 mmol, 1.2 eq.) was added dropwise at -78 °C under nitrogen to obtain **212**. The reaction mixture was warmed to 0 °C over 2 h. Then, the solution of [PdCl(π-C$_3$H$_5$)]$_2$ (18 mg, 1 mol%), *N,N*-dimethyl-1-[2-(diphenylphosphino)ferrocenyl]ethylamine (**213**, 44 mg, 1 mol%), and 2-naphthyl triflate (**165**, 2.76 g, 10 mmol) in THF (30 ml) was added dropwise and the resulting mixture was refluxed under nitrogen for 3 h. The reaction mixture was evaporated to dryness. To the residue, water (50 ml) was added and extracted with diethyl ether (3x50 ml). The combined organic layers were dried (Na$_2$SO$_4$), filtered, and evaporated. From the crude product, 2.00 g (98%) of pure 2-phenylnaphthalene (**214**) was isolated by preparative chromatography over silica gel (*n*-hexane / ethyl acetate).

4.10. Conclusion

This Chapter focuses on synthetically useful cross-coupling reactions of aryl halides or sulfonates with various arylmetallic reagents (ArM). Among the latter, aryl Grignard reagents (M = MgX) in the Kharasch, arylzincs (M = ZnX) in the Negishi, and aryltrialkylstannanes (M = SnR$_3$) in the Stille reaction are the most important organometallics for the synthesis of biaryls. These reactions are catalysed by zero-valent palladium or nickel complexes, whilst only in the case of the Stille reaction, the copper(I) salts exhibit a profound catalytic activity. Recently, impressive improvements in the chemical methodology have been introduced by the new cross-coupling reactions involving arylsilane, arylmanganese, aryltitanium, and arylindium reagents. Of course, the complementare methods using diarylmercurials and arylcopper reagents are also valuable methods for small-scale synthesis of unsymmetrical biaryls. Diarylmercurials, despite their toxicity, are respective intermediates since they can be easily prepared from widely available anilines, *via* the Nesmeyanov reaction, tolerating several sensitive functional groups.

4.11. References

1. S. P. Stanford, Tetrahedron 54 (1998) 263.
2. K. Tamao, K. Sumitani and M. Kumada, J. Am. Chem. Soc. 94 (1972) 4374.
3. K. Tamao, K. Sumitani, Y. Kiso, M. Zemabayashi, A. Fujioka, S. Kodama, I. Nakajima, A. Minato and M. Kumada, Bull. Chem. Soc. Jpn. 49 (1976) 1958.
4. E. Negishi, Pure & Appl. Chem. 53 (1981) 2333.
5. T. Hayashi, M. Konishi, Y. Kobori, M. Kumada, T. Higuchi and K. Hirotsu, J. Am. Chem. Soc. 106 (1984) 158.
6. D. Milstein and J. K. Stille, J. Am. Chem. Soc. 101 (1979) 4992.
7. C. H. Cho, H. S. Yun and K. Park, J. Org. Chem. 68 (2003) 3017.
8. A. Minato and K. Suzuki, Tetrahedron Lett. 25 (1984) 83.
9. J. K. Stille, Angew. Chem., Int. Ed. Engl. 25 (1986) 508.
10. J. K. Stille, Pure & Appl. Chem. 57 (1985) 1771.
11. A. Gillie and J. K. Stille, J. Am. Chem. Soc. 102 (1980) 4933.
12. D. A. Widdowson and Y. Z. Zhang, Tetrahedron 42 (1986) 2111.
13. M. Schlosser (Editor), L. Hegedus, B. Lipshutz, H. Nozaki, M. Reetz, P. Rittmeyer, K. Smith, F. Totter and H. Yamamoto, Organometallics in Synthesis, John Wiley & Sons, New York, 1994.
14. S. Gronowitz, B. Cederlund and A.-B. Hörnfeldt, Chem. Scripta 5 (1974) 217.
15. M. Schlosser, J. Organomet. Chem. 8 (1967) 9.
16. A. V. Rao, K. L. Reddy and M. M. Reddy, Tetrahedron Lett. 35 (1994) 5039.
17. B. J. Brisdon and P. Nair, Tetrahedron 37 (1981) 173.
18. S.-I. Murahashi, Y. Tamba, M. Yamamura and N. Yoshimura, J. Org. Chem. 43 (1978) 4099.
19. M. S. Kharasch and E. K. Fields, J. Am. Chem. Soc. 63 (1941) 2316.
20. E. Wenkert, E. L. Michelotti and C. S. Swindell, J. Am. Chem. Soc. 101 (1979) 2246.
21. E. Wenkert, E. L. Michelotti, C. S. Swindell and M. Tingoli, J. Org. Chem. 49 (1984) 4894.
22. A. Minato, K. Tamao, T. Hayashi, K. Suzuki and M. Kumada, Tetrahedron Lett. 21 (1980) 845.
23. T. Katayama and M. Umeno, Chem. Lett. (1991) 2073.
24. Y. Ikoma, F. Taya, E. Ozaki, S. Higuchi, Y. Naoi and K. Fuji-i, Synthesis (1990) 147.
25. R. J. P. Corriu and J. P. Masse, J. Chem. Soc., Chem. Commun. (1972) 144.
26. R. L. Clough, P. Mison and J. D. Roberts, J. Org. Chem. 41 (1976) 2252.
27. M. Kumada, Pure & Appl. Chem. 52 (1980) 669.
28. S. Miyano, S. Okada, T. Suzuki, S. Handa and H. Hashimoto, Bull. Chem. Soc. Jpn. 59 (1986) 2044.

29. T. Yamamoto, Y. Hayashi and A. Yamamoto, Bull. Chem. Soc. Jpn. 51 (1978) 2091.

30. T. Hayashi, Y. Katsuro, Y. Okamoto and M. Kumada, Tetrahedron Lett. 22 (1981) 4449.

31. L. Wang and Z.-C. Chen, Synth. Commun. 30 (2000) 3607.

32. J. A. Miller, J. W. Dankwardt and J. M. Penney, Synthesis (2003) 1643.

33. H. E. Ramsden, A. E. Balint, W. R. Whitford, J. J. Walburn and R. Cserr, J. Org. Chem. 22 (1957) 1202.

34. U. Tilstam and H. Weinmann, Org. Proc. Res. & Dev. 6 (2002) 906.

35. V. Bonnet, F. Mongin, F. Trecourt, G. Breton, F. Marsais, P. Knochel and G. Queguiner, Synlett (2002) 1008.

36. K. Tamao, S. Kodama, I. Nakajima, M. Kumada, A. Minato and K. Suzuki, Tetrahedron 38 (1982) 3347.

37. G. D. Hartman, W. Halczenko and B. T. Phillips, J. Org. Chem. 51 (1986) 142.

38. H. Meng and W. Huang, J. Org. Chem. 65 (2000) 3894.

39. J. M. Kauffman, Synthesis (2001) 197.

40. T. Kamikawa and T. Hayashi, Synlett (1997) 163.

41. L. N. Pridgen, J. Org. Chem. 47 (1982) 4319.

42. H. Jendralla and L.-J. Chen, Synthesis (1990) 827.

43. T. Yamamoto, T. Nakamura, H. Fukumoto and K. Kubota, Chem. Lett. (2001) 502.

44. B. H. Lipshutz and P. A. Blomgren, J. Am. Chem. Soc. 121 (1999) 5819.

45. B. H. Lipshutz, P. A. Blomgren and S. K. Kim, Tetrahedron Lett. 40 (1999) 197.

46. A. O. King and E. Negishi, J. Org. Chem. 43 (1978) 358.

47. M. E. Angiolelli, A. L. Casalnuovo and T. P. Selby, Synlett (2000) 905.

48. E. Negishi, A. O. King and N. Okukado, J. Org. Chem. 42 (1977) 1821.

49. J. A. Miller and R. P. Farrell, Tetrahedron Lett. 39 (1998) 6441.

50. C. Dai and G. C. Fu, J. Am. Chem. Soc. 123 (2001) 2719.

51. S. A. Savage, A. P: Smith and C. L. Fraser, J. Org. Chem. 63 (1998) 10048.

52. M. Rottländer and P. Knochel, J. Org. Chem. 63 (1998) 203.

53. J. Wu, Y. Liao and Z. Yang, J. Org. Chem. 66 (2001) 3642.

54. T. Balle, K. Andersen and P. Vedsø, Synthesis (2002) 1509.

55. F. Effenberger, J. M. Endtner, B. Miehlich, J. S. R. Münter and M. S. Vollmer, Synthesis (2000) 1229.

56. D. R. Gauthier, R. H. Szumigala, P. G. Dormer, J. D. Armstrong, R. P. Volante and P. J. Reider, Org. Lett. 4 (2002) 375.

57. N. B. Mantlo, P. K. Chakravarty, D. L. Ondeyka, P. K. S. Siegl, R. S. Chang, V. J. Lotti, K. A. Faust, T.-B. Chen, T. W. Schorn, C. S. Sweet, S. E. Emmert, A. A. Patchett and W. J. Greenlee, J. Med. Chem. 34 (1991) 2919.

58. M. Amat, F. Seffar, N. Llor and J. Bosch, Synthesis (2001) 267.

59. G. Karig, J. A. Spencer and T. Gallagher, Org. Lett. 3 (2001) 835.

60. G. Karig, N. Thasana and T. Gallagher, Synlett. (2002) 808.

61. R. D. Rieke, S. J. Uhm and P. M. Hundall, J. Chem. Soc., Chem. Commun. (1973) 269.

62. K. Takagi, N. Hayama and S. Inokawa, Bull. Chem. Soc. Jpn. 53 (1980) 3691.

63. M. Amano, A. Saiga, R. Ikegami, T. Ogata and K. Takagi, Tetrahedron Lett. 39 (1998) 8667.

64. R. Ikegami, A. Koresawa, T. Shibata and K. Takagi, J. Org. Chem. 68 (2003) 2195.

65. T. N. Majid and P. Knochel, Tetrahedron Lett. 31 (1990) 4413.

66. A. S. B. Prasad, T. M. Stevenson, J. R. Citineni, V. Nyzam and P. Knochel, Tetrahedron 53 (1997) 7237.

67. J. Jensen, N. Skjærbæk and P. Vedsø, Synthesis (2001) 128.

68. C. Gosmini, Y. Rollin, J. Y. Nédélec and J. Périchon, J. Org. Chem. 65 (2000) 6024.

69. J. W. Tilley and S. Zawoiski, J. Org. Chem. 53 (1988) 386.

70. Y.-Q. Fang and G. S. Hanan, Synlett. (2003) 852.

71. M. Amat, S. Hadida, G. Pshenichnyi and J. Bosch, J. Org. Chem. 62 (1997) 3158.

72. B. A. Anderson, L. M. Becke, R. N. Booher, M. E. Flaugh, N. K. Harn, T. J. Kress, D. L. Varie and J. P. Wepsiec, J. Org. Chem. 62 (1997) 8634.

73. E. R. Larson and R. A. Raphael, J. Chem. Soc., Perkin Trans 1 (1982) 521.

74. H. Lam, S. M. Marcuccio, P. I. Svirskaya, S. Greenberg, A. B. P. Lever, C. C. Leznoff and R. L. Cerny, Can. J. Chem. 67 (1989) 1087.

75. A. M. Echavarren and J. K. Stille, J. Am. Chem. Soc. 109 (1987) 5478.

76. A. M. Echavarren and J. K. Stille, J. Am. Chem. Soc. 110 (1988) 1557.

77. T. R. Bailey, Tetrahedron Lett. 27 (1986) 4407.

78. A. F. Littke and G. C. Fu, Angew. Chem., Int. Ed. Engl. 38 (1999) 2411.

79. V. Farina, B. Krishnan, D. R. Marshall and G. P. Roth, J. Org. Chem. 58 (1993) 5434.

80. J. M. Saá, G. Martorell and A. G. Raso, J. Org. Chem. 57 (1992) 678.

81. G. Martorell, A. G. Raso and J. M. Saá, Tetrahedron Lett. 31 (1990) 2357.

82. W. J. Scott and J. K. Stille, J. Am. Chem. Soc. 108 (1986) 3033.

83. J. Mathieu and A. Marsura, Synth. Commun. 33 (2003) 409.

84. G. A. Grasa and S. P. Nolan, Org. Lett. 3 (2001) 119.

85. V. Farina, S. Kapadia, B. Krishnan, C. Wang and L. S. Liebeskind, J. Org. Chem. 59 (1994) 5905.

86. J. M. Saá and G. Martorell, J. Org. Chem. 58 (1993) 1963.

87. N. S. Nudelman and C. Carro, Synlett (1999) 1942.

88. S.-K. Kang, J.-S. Kim and S.-C. Choi, J. Org. Chem. 62 (1997) 4208.

89. W. J. Scott and J. E. McMurry, Acc. Chem. Res. 21 (1988) 47.

90. V. Farina and G. P. Roth, Tetrahedron Lett. 32 (1991) 4243.

91. A. L. Casado, P. Espinet, A. M. Gallego and J. M. Martinez-Ilarduya, Chem. Commun. (2001) 339.

92. M. Larhed and A. Hallberg, J. Org. Chem. 61 (1996) 9582.

93. K. Fugami, S. Ohnuma, M. Kameyama, T. Saotome and M. Kosugi, Synlett (1999) 63.

94. T. Bach and S. Heuser, J. Org. Chem. 67 (2002) 5789.

95. U. Asawapirom, R. Güntner, M. Forster, T. Farrell and U. Scherf, Synthesis (2002) 1136.

96. T. R. Kelly, A. Szabados and Y.-J. Lee, J. Org. Chem. 62 (1997) 428.

97. T. R. Hoye and M. Chen, J. Org. Chem. 61 (1996) 7940.

98. A. Arcadi, S. Cacchi, G. Fabrizi, F. Marinelli and L. Moro, Synlett (1999) 1432.

99. D. Li, B. Zhao and E. J. LaVoie, J. Org. Chem. 65 (2000) 2802.

100. G. Bringmann, R. Götz, P. A. Keller, R. Walter, M. R. Boyd, F. Lang, A. Garcia, J. J. Walsh, I. Tellitu, K. V. Bhaskar and T. R. Kelly, J. Org. Chem. 63 (1998) 1090.

101. M. Murata, S. Watanabe and Y. Masuda, Synlett (2000) 1043.

102. E. F. Córsico and R. A. Rossi, Synlett (2000) 227.

103. E. F. Córsico and R. A. Rossi, J. Org. Chem. 67 (2002) 3311.

104. P. F. H. Schwab, F. Fleischer and J. Michl, J. Org. Chem. 67 (2002) 443.

105. G. T. Crisp and S. Papadopoulos, Aust. J. Chem. 41 (1988) 1711.

106. N. Zhang, L. Thomas and B. Wu, J. Org. Chem. 66 (2001) 1500.

107. K. Kikukawa, K. Kono, F. Wanda and T. Matsuda, J. Org. Chem. 48 (1983) 1333.

108. S.-K. Kang, T.-G. Baik and S.-Y. Song, Synlett (1999) 327.

109. S.-Y. Lin, C.-L. Chen and Y.-J. Lee, J. Org. Chem. 68 (2003) 2968.

110. N. Tamayo, A. M. Echavarren and M. C. Paredes, J. Org. Chem. 56 (1991) 6488.

111. T. R. Kelly, Q. Li and V. Bhushan, Tetrahedron Lett. 31 (1990) 161.

112. R. Olivera, R. SanMartin and E. Dominiguez, Synlett (2000) 1028.

113. E. F. Córsico and R. A. Rossi, Synlett (2000) 230.

114. R.-A. Fallahpour, Synthesis (2000) 1665.

115. B. X. Colasson, C. Dietrich-Buchecker and J.-P. Sauvage, Synlett (2002) 271.

116. R.-A. Fallahpour, Synthesis (2003) 155.

117. R.-A. Fallahpour, Synthesis (2000) 1138.

118. S. Bedel, G. Ulrich, C. Picard and P. Tisnès, Synthesis (2002) 1564.

119. T. R. Kelly, Y.-J. Lee and R. J. Mears, J. Org. Chem. 62 (1997) 2774.

120. D. Garcia-Cuadrado, A. M. Cuadro, J. Alvarez-Builla and J. J. Vaquero, Synlett (2002) 1904.

121. D. J. Aldous, S. Bower, N. Moorcroft and M. Todd, Synlett (2001) 150.

122. G. Sotgiu, M. Zambianchi, G. Barbarella, F. Aruffo, F. Cipriani and A. Ventola, J. Org. Chem. 68 (2003) 1512.

123. A. Hucke and M. P. Cava, J. Org. Chem. 63 (1998) 7413.

124. B. Clapham and A. J. Sutherland, J. Org. Chem. 66 (2001) 9033.

125. S.-J. Luo, C.-M. Liu, Y.-M. Liang and Y.-X. Ma, Synth. Commun. 31 (2001) 3119.

126. C.-M. Liu, S.-J. Luo, Y.-M. Liang and Y.-X. Ma, Synth. Commun. 30 (2000) 2281.

127. C.-M. Liu, W.-Y. Liu, Y.-M. Liang and Y.-X. Ma, Synth. Commun. 30 (2000) 1755.

128. V. Farina, S. R. Baker, D. A. Benigni, S. I. Hauck and C. Sapino, J. Org. Chem. 55 (1990) 5833.

129. T. A. Rano, M. L. Greenlee and F. P. DiNinno, Tetrahedron Lett. 31 (1990) 2853.

130. M. E. Mowery and P. DeShong, J. Org. Chem. 64 (1999) 1684.

131. M. Murata, R. Shimazaki, S. Watanabe and Y. Masuda, Synthesis (2001) 2231.

132. K. Hirabayashi, A. Mori, J. Kawashima, M. Suguro, Y. Nishihara and T. Hiyama, J. Org. Chem. 65 (2000) 5342.

133. A. Mori and M. Suguro, Synlett (2001) 845.

134. Y. Hatanaka, K. Goda and Y. Okahara, Tetrahedron 50 (1994) 8301.

135. A. S. Manoso and P. DeShong, J. Org. Chem. 66 (2001) 7449.

136. L. J. Gooßen and A.-R. S. Ferwanah, Synlett (2000) 1801.

137. O. A. Reutov and O. A. Ptitsyna, Organometallic React. 4 (1972) 73.

138. N. A. Bumagin, P. G. More and I. P. Beletskaya, J. Organomet. Chem. 364 (1989) 231.

139. A. V. Tsvetkov, G. V. Latyshev, N. V. Lukashev and I. P. Beletskaya, Tetrahedron Lett. 41 (2000) 3987.

140. H. Malmberg and M. Nilsson, Tetrahedron 42 (1986) 3981.

141. E. Riguet, M. Alami and G. Cahiez, Tetrahedron Lett. 38 (1997) 4397.

142. R. D. Rieke, S.-H. Kim and X. Wu, J. Org. Chem. 62 (1997) 6921.

143. J. W. Han, N. Tokunaga and T. Hayashi, Synlett (2002) 871.

144. B. Weidmann, L. Widler, A. G. Olivero, C. D. Maycock and D. Seebach, Helv. Chim. Acta 64 (1981) 357.

145. I. Pérez, J. Pérez Sestelo and L. A. Sarandeses, J. Am. Chem. Soc. 123 (2001) 4155.

146. K. Takami, H. Yorimitsu, H. Shinokubo, S. Matsubara and K. Oshima, Org. Lett. 3 (2001) 1997.

147. I. Pérez, J. Pérez Sestelo and L. A. Sarandeses, Org. Lett. 1 (1999) 1267.

148. M. Kosugi, T. Tanji, Y. Tanaka, A. Yoshida, K. Fugami, M. Kameyama and T. Migita, J. Organometal. Chem. 508 (1996) 255.

5. THE SUZUKI-MIYAURA REACTION

5.1. General survey of the Suzuki-Miyaura reaction

The cross-coupling reactions of a wide variety of organoboron compounds with aryl halides and similar substrates such as aryl sulfonates or some other electrophilic aryl substrates to produce alkenes, arylalkenes, alkylarenes and biaryls etc. are well known as Suzuki-Miyaura (SM) reaction [1-9]. The latter, due to its general applicability and efficiency has became the most important C-C bond forming reaction, the "Grignard reaction of 21th century". Here, applications of the Suzuki-Miyaura cross-coupling reaction in the synthesis of biaryls are discussed. The SM synthesis of biaryls includes the reaction of arylboronic acids or esters with aryl halides (Ar-X, X = I, Br, Cl), sulfonates (ArOTf, OMs, etc.) or diazonium salts (ArN$_2^+$ BF$_4^-$) what proceeds smoothly to give corresponding biaryls in good to excellent yields [1-9], Scheme 1.

$$
\text{Ar—X} \quad + \quad \text{(RO)}_2\text{B—Ar'} \quad \longrightarrow \quad \left[
\begin{array}{c}
\text{base} \\
\text{Pd- or Ni-catalyst} \\
\text{r. t. or heat}
\end{array}
\right] \longrightarrow \quad \text{Ar—Ar'}
$$

Ar, Ar' = a wide variety of substituted aryls and heteroaryls

X = I, Br, Cl, OTf, OTs, OMs, N$_2^+$ BF$_4^-$

R = H or alkyls

Scheme 1

The boronic acids (or esters) are generally thermal stable and inert to water and oxygen, because the carbon-boron bond possesses a strong covalent character. However, under certain reaction conditions, the boronic acids or esters can act as carbanion-donors similar to other organometallics and, as the most importantly, without the practical difficulties connected with the latter. The coordination of a negatively charged base to the boron atom is an efficient method of increasing its carbanionic (nucleophilic) nature, to transfer the organic group from boron to the adjacent positive center, e.g. palladium. Thus activated organoboron compounds is readily undergoing transmetallation process to Mg(II), Zn(II), Al(III), Cu(I), Sn(II), Hg(II), Ag(I), and (here) more importantly to Pd(II) or Ni(II) as a key step of a wide range of selective carbon-carbon bond formations. An original SM method was based on palladium cataysts, but latter, a nickel-based catalysis was introduced as more valuable and cheaper alternative, at least for the certain substrates such as aryl chlorides and sulfonates. The SM reaction is strongly influenced by: employed catalyst, base, reaction solvent or generally the reaction medium, e.g. solvent-free or solid supported methodologies, an inert atmosphere, as well as reactivity of boronic acid (regarding to its stability and steric demands) and electrophilic reagent (aryl halides, triflate, tosylate or diazonium salt). The effect of all parameters have to be taken in account to understand the SM reaction properly. The first SM synthesis of biaryls was accomplished in two-phase system benzene (or toluene) / aqueous sodium carbonate solution in the presence of Pd(PPh$_3$)$_4$ as catalyst when several biaryls were obtained in good to excellent yield, Table 1.

Table 1. Synthesis of biaryls from various aryl bromides (1 eq.) and phenylboronic acid (PhB(OH)$_2$, 1.1 eq.) by the original SM method [1]

R	Yield (%)
2-CH$_3$	94
4-CH$_3$	88
2-CH$_3$O	99
4-CH$_3$O	40
4-Cl	74
4-CH$_3$O$_2$C	94

Excellent first results have shown the general efficiency in the cross-coupling reactions of arylboronic acids with aryl iodides and bromides, while the reactions with apparently less reactive chlorides failed under this conditions. Further, the SM reaction was significantly developed through a great number of contributions regarding improvements in all above mentioned factors which have strong influence on the reaction efficacy and wide applicability.

5.1.1. The base and solvent effect

Firstly recognized is the base effect since the SM reaction completely failed in its absence. In the early studies, sodium ethoxide in ethanol, and sodium acetate, hydroxide or carbonate in water were employed when the highest efficiency was proved for the latter one [1]. Since the solubility and basicity of the base strongly depend on the solvent used, these two parameters are closely connected. The SM reactions involving water-soluble bases, e.g. Na_2CO_3 or K_2CO_3, can be conducted in the two-phase systems where an organic halide is dissolved in an organic phase. As the latter serves benzene, toluene, cyclohexane, 1,2-dichloroethane or similar water inmiscible solvents [1]. Alternatively, the SM reaction can be accomplished in homogeneous medium by using dipolar aprotic or other water miscible solvents such as DMF, acetonitrile, DMAc, NMP, DMSO, tetrahydrofuran, 1,4-dioxane, DME, *t*-BuOH, ethanol, methanol or their mixtures [1-9]. Furthermore, the following bases were introduced as more effective alternatives in the SM cross-coupling reactions:

1. $Ba(OH)_2$ in 95% aqueous ethanol [10]
2. K_3PO_4 in toluene [11,12] or 1,4-dioxane [13]
3. Cs_2CO_3 in DMF [14] or 1,4-dioxane
4. KO*t*-butoxide in DME / *t*-BuOH [15]
5. TlOH in DMAc / water mixture [16]
6. Tl_2CO_3 in benzene [17], toluene [19], 1,2-dichloroethane [19], DMF [19], DME [19], 1,4-dioxane [19], cyclohexane [19], MeCN [19], or the most efficiently in THF [18,19]
7. Ag_2CO_3 in THF [18]
8. PhOM, where M = Na, K in THF [18] or benzene [19]
9. Fluoride bases: CsF in toluene [11], 1,4-dioxane [11], MeCN [20], DME [20], or MeOH / DME [20]; KF in MeOH / DME or toluene / water [20]; Et_4NF in MeCN or MeOH / DME [20].

Although the use of triethylamine and other trialkylamines has been reported, these compounds are generally far less efficient bases in the SM reactions [21]. The main point concerning the base effect is not only the basicity of an anion, but also the cation contribution is very important. To understand this, briefly discussion on the SM reaction mechanism is essential. Once again, the oxidative addition / reductive elimination mechanism, as fundamentally common processes, are operating, while only the transmetallation from boron to palladium (or nickel) is quite different than those described in previous Chapters since it is highly dependent on organometallics and reaction conditions. First, the aryl halide (**I**) (or other electrophilic aryl reactant) reacts with the palladium(0) catalysts to give an arylpalladium(II) halide (**XIII**) by oxidative addition. Secondly, the negatively charged base, e.g. alkoxide, reacts with the arylboronic acid (**XXI**) to generate the corresponding aryl(dihydroxy)boron alkoxide, an *ate*-complex **XXII** [2-6]. The latter reacts with the arylpalladium(II) alkoxide **XIIIa**, readily formed by the ligand-exchange reaction, to furnish a diarylpalladium(II) complex **XIV**, *via* the transmetallation process [2-6], presumably assisted with the metal cation (originated from the base) [15]. Whereas the function of the base is formation of the corresponding *ate*-complex **XXII** from the arylboronic acid or ester as well as conversion of the arylpalladium(II) halide into the arylpalladium(II) alkoxide (or hydroxide, acetate, carbonate etc.), the metal cation accelerates the reaction with the arylpalladium(II) species, through a complex **XXIII** [15,22]. The transmetallation process from boron to palladium(II) is probably realized through the four-membered (containing Pd-O-B-C(aryl)) transitional state [23]. Finaly, the *cis*-diarylpalladium(II) intermediate **XIVa** is affecting the fast reductive elimination to yield the biaryl (**II**) with regeneration of the catalytically active palladium(0) complex, closing the catalytic cycle [2], Scheme 2.

The nickel-catalysed SM reactions follow closely related reaction pathway. Although the SM reaction mechanism is not completely clear, several experimental evidences indicate the following facts:

1. The palladium(0) or nickel(0) complexes are truly catalytically responsible species. When metal complexes in higher oxidation states were used, e.g. $Pd(PPh_3)_2Cl_2$, previous reduction was essential to start the SM coupling reaction. This can be performed either by an external reducing agent or *in situ* with an excess of boronic acid. Since the oxidative addition of aryl chlorides to palladium(0) complexes (except for activated ones) is rather slow process, the latter are effective catalysts only for the cross-coupling reactions of aryl bromides, iodides and, of course, the most reactive diazonium salts. However, aryl chlorides, as well as various aryl sulfonates undergo oxidative addition to more reactive nickel(0) complexes.

2. The negatively charged base reacts with the arylpalladium(II) halide to give the arylpalladium hydroxide or alkoxide complex, which is able to form the dimeric palladium-boron complex **XXIII** what is crucial for the transmetallation process [2-6]. It is apparent that the metal cation (from the base) accelerates the formation of the latter, as clearly showed by Zhang and coworkers [15]. They have developed the SM coupling procedure for sterically bulky arylboronic acids when the clear influence of the anion basicity and the cation effect were discovered. The cationic radius is presumably an important parameter which influences the formation of dimeric

complex **XXIII**. Thus, too small cation destabilizes an optimal structure of the latter. For example, when the base containing the smaller cation, e.g. Na^+, was employed, the dramatical decreasing of the reaction rate and yield occured. Generally, larger cations, e.g. Ba^{2+}, Tl^+, Ag^+, etc., strongly accelerate the SM reactions. When the cation was complexed into the 18-crown-6 (18-C-6) and thus became unavailable, the reaction yield and the rate were sharply decreased, what strongly suggested the assistance of the metal cation in the SM reaction at the stage of complex **XXIII** formation. The similar results were obtained by using the weaker bases indicating that, at least in reactions with the sterically hindered arylboronic acids, the rate-determining step may be the formation of the complex **XIIIa**, from **XIII** [15]. For example, the cross-coupling reaction of sterically hindered boronic acid **234** with 3-methyl-2-bromopyridine (**235**) in the presence of $Pd(PPh_3)_4$ as the catalyst and various bases afforded the respective biaryl **236** in fair yield only when the base of appropriate basicity in connection with sufficiently large cation were employed, Table 2.

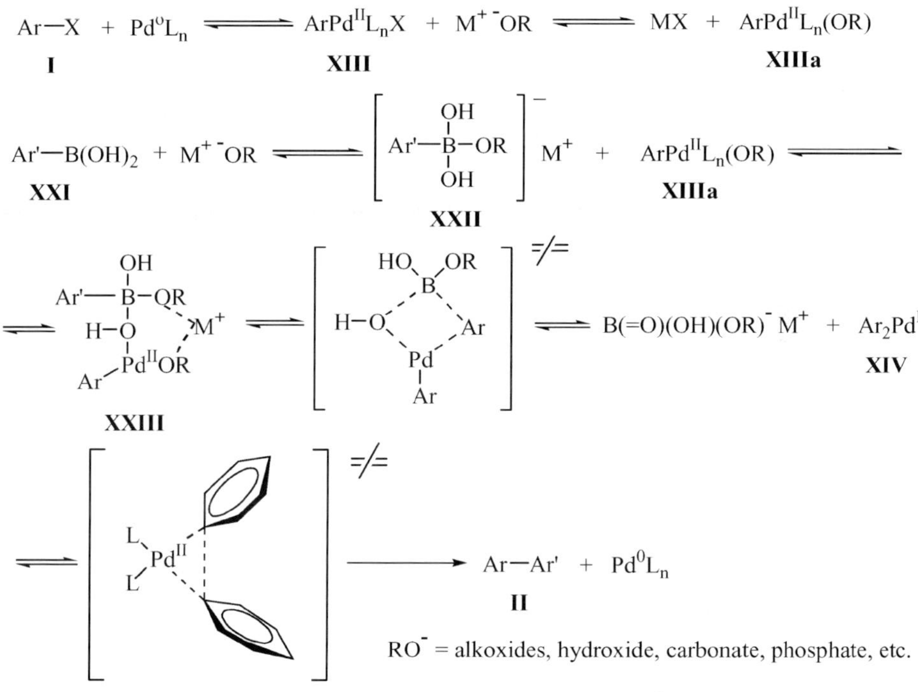

Scheme 2

RO^- = alkoxides, hydroxide, carbonate, phosphate, etc.

M^+ = Na^+, K^+, Ag^+, Tl^+, Ba^{2+}

L = ligands for stabilization of Pd^0 complexes

Table 2. The base and cation effect on the yield of **236** in the cross-coupling reaction of sterically hindered boronic acid **234** with 3-methyl-2-bromopyridine (**235**) [15]

Base	Time	Yield (%)
KO*t*-Bu	15 min	61
NaO*t*-Bu	5.5 h	60
KO*t*-Bu + 18-C-6	45 min	30
Cs₂CO₃	116 h	20
Ba(OH)₂	5 d	0

Practically, in the case of very bulky arylboronic acids, the stronger base is used to achieve reasonable conversions. The fluoride acts as very efficient base by forming an aryl(trifluoro)borate anion ($ArBF_3^-$) which further proceeds to biaryls in the similar manner. Separately prepared arylpalladium(II) alkoxides readily react with the arylboronic acids to give biaryls in the absence of base, what strongly supports the ligand-exchange reaction between the arylpalladium(II) halide and alkoxide (**XIII →　XIIIa**), hydroxide, or other anion originating from the base. Additional effects, caused by the steric demands and electronic properties of the ligands which stabilize the catalytically active palladium(0) or nickel(0) species, are discussed latter. Finally, to reach the successful cross-coupling reaction, all these factors must be taken in account: an optimaly large metal cation with sufficiently strong basic anion, efficient palladium or nickel complex with suitable stabilizing ligand, reaction medium, etc.

3. The nature of the transmetallation reaction is less known. Soderquist and coworkers [23] have introduced valuable mechanistic points regarding this process in the reactions with alkylboranes. Extending this results to arylboronic acids, the formation of the dimeric complex **XXIII** from **XIIIa** and **XXII**, which undergoes transmetallation reaction *via* the four-membered transitional state (Scheme 2), is reasonable explanation for this step [23]. The reductive elimination of resulting *cis*-diarylpalladium(II) intermediate **XIVa** proceeds through a three-membered transitional state with extrusion of the biaryl and regeneration of Pd(0) complex, what is generally accepted, see Chapter 3 [2-6].

5.1.2. *The catalyst effect*

The catalyst plays the most important role in the SM reaction as the reaction actually proceeds within the palladium or nickel coordinative sphere. Since only Pd(0) and Ni(0) complexes possess the catalytic activity, they can be used by:
- adding the stable Pd(0) or Ni(0) complexes,
- *in situ* reduction of common Pd(II) or Ni(II) complexes
- *in situ* reduction of simple Pd(II) or Ni(II) salts in the presence of zero-valent metal stabilizing ligands.

The SM reaction catalysts can be divided into three groups regarding their properties, what does not mean that some of the catalysts from the lower generation are less valuable than the any from the upper generation.

1. The SM-catalysts of the first generation are mainly consisted of simple palladium and nickel salts and complexes with the most common Pd(0)- and Ni(0)-stabilizing: mono-, or bidentate phosphine ligands. In this group, the palladium and nickel on charcoal as heterogeneous catalysts were put, despite quite recent use of these catalysts.

2. The second generation contains highly active palladium-catalysts, including the Bedford's and some other catalysts.

3. The third generation SM-catalysts are highly active and reusable heterogeneous catalysts.

Here, the first generation cataysts are presented whereas the latter two groups will be described in the following discussion. The first SM reactions were performed with tetrakis(triphenylphosphino)palladium(0), Pd(PPh$_3$)$_4$, which is still widely used, although relatively expensive catalyst for the labo-scale cross-coupling reactions with aryl iodides and bromides. As an alternative palladium catalysts: 5% Pd-C with [24] or without PPh$_3$ [25-27], Pd(OAc)$_2$ with [2-6] or without PPh$_3$ [10,28,29], Pd(SEt$_2$)$_2$Cl$_2$ [29], Pd(MeCN)$_2$Cl$_2$, Pd(PhCN)Cl$_2$ or Pd(PPh$_3$)$_2$Cl$_2$ with PPh$_3$ [2-6], Pd$_2$(dba)$_3$ with P(t-Bu)$_3$ or PCy$_3$ [31,32] have been most commonly employed. Among the other, less accessible phosphine based palladium-catalysts, the complexes *in situ* generated from Pd$_2$(dba)$_3$ and the ligands **237** [33], **238** [34], **239** [35], and **240** [36], or Pd(OAc)$_2$ and **238** [34] have been used. Substantially different are the phosphine ligands such as **241** [37], **242** [37], **243** [38], or several others, bearing a strong polar group(s) providing the water-solubility of both ligand and the resulting palladium-complexes. Beside the phosphine based complexes, the palladium-complexes *in situ* obtained by reaction of Pd$_2$(dba)$_3$ with *N*-heterocyclic carbene precursor **244** (in the presence of the base) [39], certain hydroxy-phosphines such as **245** [40], or Pd(OAc)$_2$ with readily accessible *N,N'*-dicyclohexyl-1,4-diazabutadiene **246** [41], or glyoxal bis(methylphenyl hydrazone) **247** [42], have been proved as efficient alternatives to the palladium-phosphine complexes.

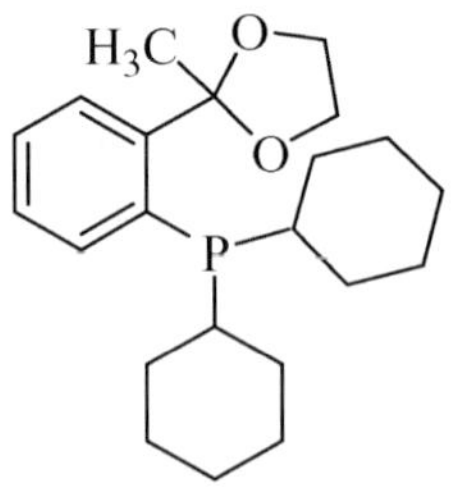

237   238   239

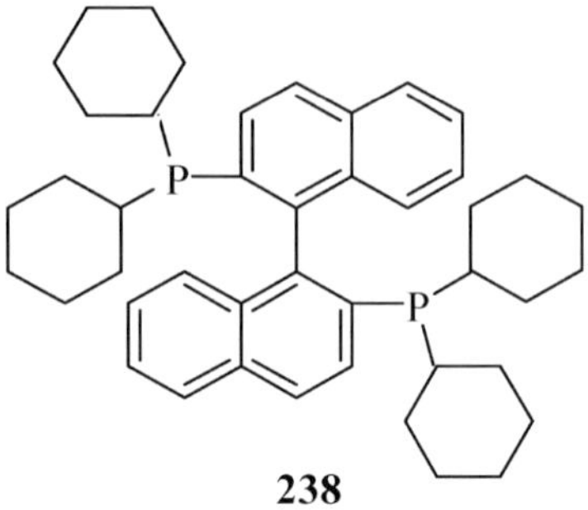

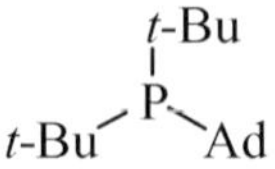

Ad = 1-adamantyl

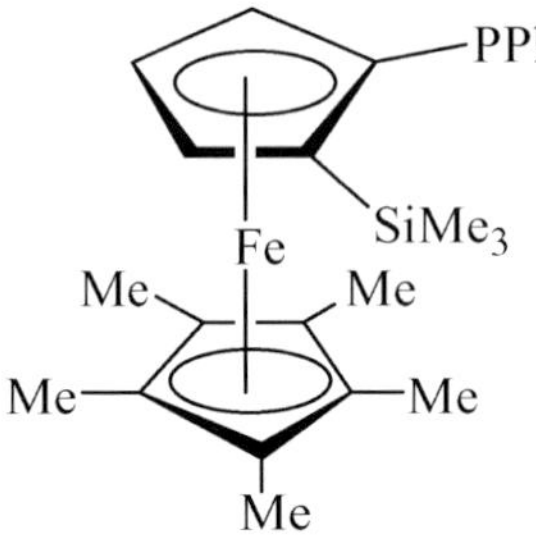

240    241    242

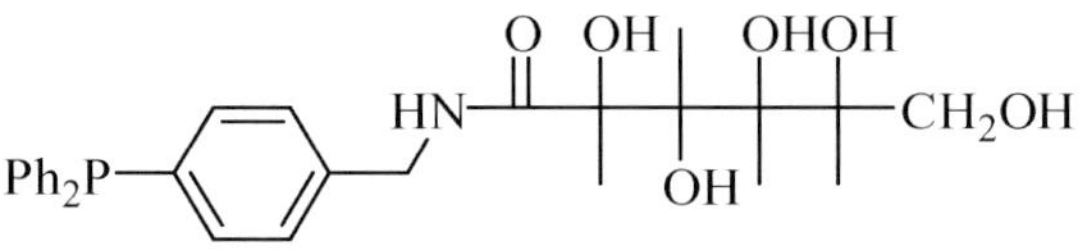

243

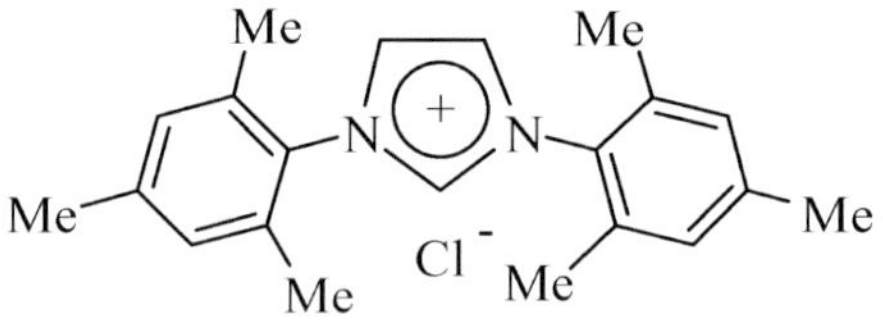

244 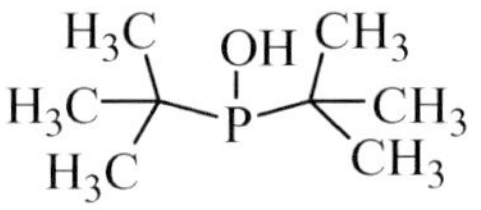   245

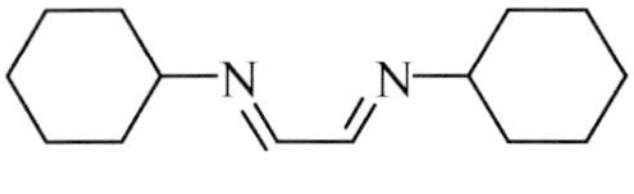

246 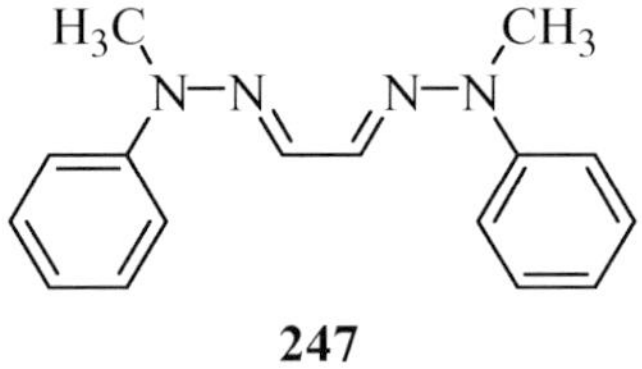   247

Among the nickel-based catalysts, beside the nickel on charcoal [43], mainly the various nickel(0)-phosphine complexes were employed [12,13,44,45]. Several common bidentate phosphine ligands such as dppf, dppe, dppp, dppb and tricyclohexylphosphine (PCy$_3$) are the most popular in the nickel-based SM cross-coupling reactions of less reactive aryl chlorides with arylboronic acids [44]. While the nickel complexes of electron-rich trialkylphosphines react readily with aryl chlorides, the simple triphenylphosphine, as far less efficient ligand, can be used in some instances [12]. As the nickel(0) complexes are rather oxygen- and light-sensitive, the corresponding stable nickel(II)-phosphine complexes are employed as precatalysts. The latter, e.g. Ni(dppf)Cl$_2$, is previously reduced with Zn, *n*-BuLi or DIBAlH [44]. Alternatively, a slight excess of the arylboronic acid can serve as the reducing agent for Ni(II), e.g. Ni(PCy)$_2$Cl$_2$, to the catalytically active Ni(0) complex, Ni(PCy$_3$)$_2$ [44,45].

dppe: n = 2

dppp: n = 3

dppb: n = 4

dppf

Generally, all palladium- and nickel-based catalysts of the first generation are usually loaded in amounts of 1-5 mol% [1-6], rarely under 1 mol% and sometimes 5-10 mol%. The palladium-based SM reactions can be performed in the presence of water, moreover in two-phase toluene / water system as the original procedure.

Contrarily, the nickel-catalysed SM reactions are always carried out in dry solvents. Since the arylnickel complexes have more ionic character than the corresponding arylpalladiums, the former are generally more reactive and sensitive to water (and other protic compounds) and furnish the protonated (reductive dehalogenation) arene from the parent aryl halide. However, the most common base in the nickel-catalysed SM reactions is K$_3$PO$_4$·nH$_2$O. The fact, that a small amount of dehalogenation products was isolated, supports the arylnickel protonation side-reaction, either with water (from solvent(s) or K$_3$PO$_4$·nH$_2$O) or any other protic source.

In contrast to palladium(0) complexes, the nickel(0) analogues are [11,12,43,44]:

- more reactive - applicable with less reactive substrates such as aryl chlorides and sulfonates,
- more sensitive to steric hindrances, caused either by an arylboronic acid or expecially an aryl halide structure and
- sensitive to the presence of a ligand-capable atom or group in the *ortho*-position to the halogen [43,44].

5.1.3. Reactivity of the substrates in the SM cross-coupling reaction

The relative reactivity of aryl halides, sulfonates and diazonium salts [46,47] follows an order:

$$\text{ArN}_2{}^+ \text{ X}^- \gg \text{ Ar-I } > \text{ Ar-Br } > \text{ Ar-Cl } \geq \text{ Ar-OTf } \geq \text{ Ar-OTs [45], Ar-OMs [48]}$$

The reactivity of aryl triflates [24,32,36,49] and other sulfonates [45,48] can be slightly different than is presented here, what depends on the catalyst and the reaction conditions used. For example, the SM reactions of aryl chlorides in the presence of aryl triflates catalysed by $Pd_2(dba)_3$ / $2Pt$-Bu_3 have given exclusively the biaryl derived from the chloride, whereas the use of $Pd(OAc)_2$ / $2PCy_3$ catalytic system leads to predominant SM reaction of aryl triflate. The latter results indicate the fact, that the (phosphine) ligand(s), not the palladium (nickel) source, controls chloride / triflate selectivity, as well as other closely reactive substrates.

The aryldiazonium salts are definitely the most reactive SM electrophilic reactants [46,47]. They are able to cross-couple with arylboronic acids in lower alcohols as solvents employing palladium(II) acetate, with [47] or without [46] any palladium(0)-stabilizing ligand in good to excellent yields, Table 3.

Table 3. An efficiency of the aryldiazonium salts in the SM reactions [46]

R_1	R_2	Yield (%)
4-Cl	H	80
2-CH$_3$	H	90
2-CH$_3$O	H	75
2-CH$_3$OOC	H	70 [a]
2-CH$_3$O	CH$_3$	40
2-CH$_3$OOC	CH$_3$	45 [a]

[a] The reactions were performed at room temperature.

Under certain reaction conditions, even activated arylmethylthioethers can be used as electrophiles in the SM reactions. Thus copper(I) 3-methylsalicylate (CuMeSal) does catalyse the cross-coupling reaction of 3-methylthio-1,2,4-triazine (**248**) with 2-thienyl boronic acid (**249**) to afford biaryl **250** in 70% yield, respectively [50], Scheme 3.

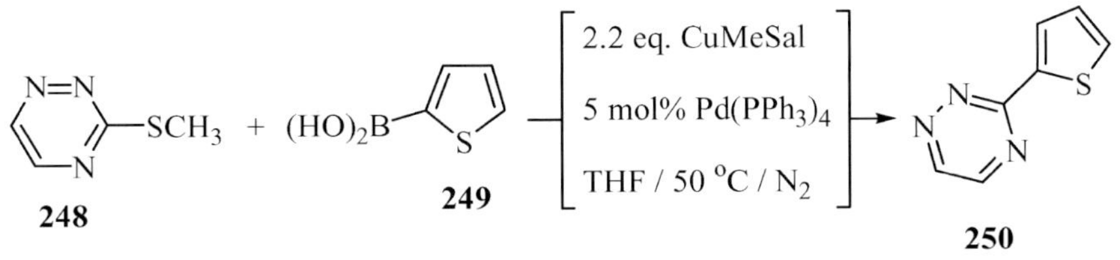

Scheme 3

Here, the copper(I) salt reacts with **248** to give an arylcopper(III) thiomethoxide intermediate, which readily affects the transmetallation process to palladium [49].

The reactivity of aryl chlorides can be increased by adding lithium bromide, and at aryl sulfonates with lithium chloride or bromide [43,49]. This observation is based on the well known palladium- or nickel-catalysed displacement reaction of chloride (or aryl sulfonates) to more reactive bromide, what increases the reaction rate [51,52].

The reactivity of arylboronic acids is mainly effected by the steric factors. Bulky substituents in *ortho*-positions significantly retard the formation of the *ate*-complex **XXII**, common arylboronic species which undergoes the transmetallation process to arylpalladium- or arylnickel complexes. However, as mentioned earlier, the use of much stronger bases such as potassium *t*-butoxide in dry solvents (DME, DME / *t*-BuOH) [15] or the use of thallium-bases [19] led to the successful SM reaction with reasonable bulky boronic acids.

5.1.4. The effect of an inert atmosphere

The SM reactions are commonly conducted under an argon or nitrogen atmosphere. Although some very fast SM reactions (under 1 h) can be carried out in air, the absence of oxygen is essential in many cases. The oxygen destroys the arylpalladium- and arylnickel species as well as palladium(0) or nickel(0) complexes. Additionally, arylboronic acids, under the thermal and basic SM conditions in the presence of oxygen, are affording the homo-coupling products in considerable amounts [10]. Generally, in the nickel-catalysed, and under ligandless palladium-catalysed SM reactions the deoxygenation of reaction mixtures is absolutely essential [10].

5.1.5. The side-reactions

The SM reactions are usually chemoselective and clean. However, in some instances the following side-reactions occur:

a) the reductive dehalogenation (RD) of aryl halides as a common side-reaction mainly in nickel-catalysed, and negligible in palladium-catalysed SM reactions as shown in the Scheme 4.

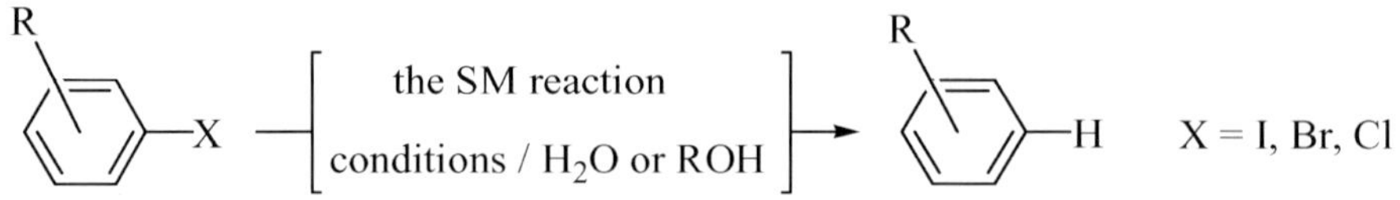

Scheme 4

Once again, the RD side-reaction proceeds by the protonation of arylnickel (or palladium) intermediates with water or other protic sources (see Chapter 3). The reaction is related to the quenching of Grignard reagent with water or alcohols to furnish a parent hydrocarbon.

b) The aryl-migration from the triarylphosphine ligands is well established side-reaction in the coupling reactions of aryl halides with nickel(0) complexes (see Chapter 3) and in the Karasch and Negishi cross-coupling reactions. The result of this side-reaction is the formation of biaryl, Ar-Ar', where one aryl-side is originated from aryl halide, whereas the other is from triarylphosphine ligand, Scheme 5.

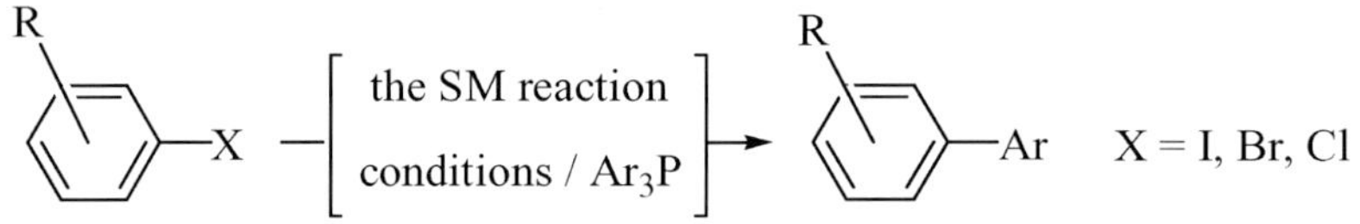

Scheme 5

The use of catalysts containing trialkylphosphines (or imines) as nickel(0)-stabilizing ligands avoids this side-reaction since alkyls are transfered less readily.

c) The hydrolytic deboronation (HD, or the protodeboronation) is another common side-reaction which converts arylboronic acids to the parent arene [15], Scheme 6.

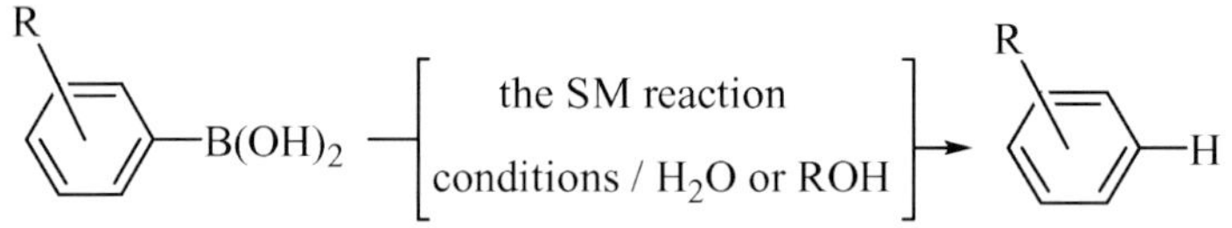

Scheme 6

In the case of bulky arylboronic acids, where the SM reaction is relatively slower, the yields of the HD side-reaction by-products may exceed 20-30% [15,19].

d) The homo-coupling reaction of arylboronic acids (AB-HC) is very important, and practically unavoidable unwanted side-event that occurs in almost all palladium- and nickel-catalysed SM reactions to afford a symmetrical biaryl, Scheme 7.

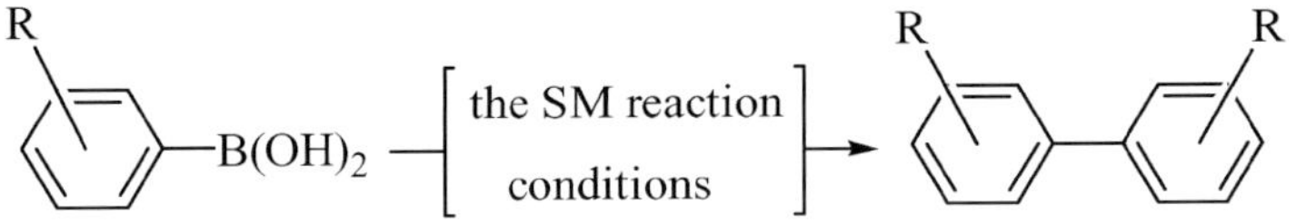

Scheme 7

The significance of the AB-HC side-reaction is enhanced in relatively slower, two-phase or other heterogeneous SM systems. Although the AB-HC reaction usually proceeds under oxidative conditions [10], the AB-HC reaction mechanism was somewhat clarified through the results described by Moreno-Mañas and coworkers [53]. This transformation includes two oxidative addition steps of arylboronic acid (**XXI**) to palladium(0) complex with formation of diarylpalladium(II) complex (**XXIV**), which, upon reductive elimination of biaryl (**II**), generates the parent palladium(0) complex, however, with apparent influence of some oxidants (oxygen) or by simple reductive elimination of hydrogen, at the stage of oxidative conversion of PdH$_2$ back to the parent palladium(0)-complex [53], Scheme 8.

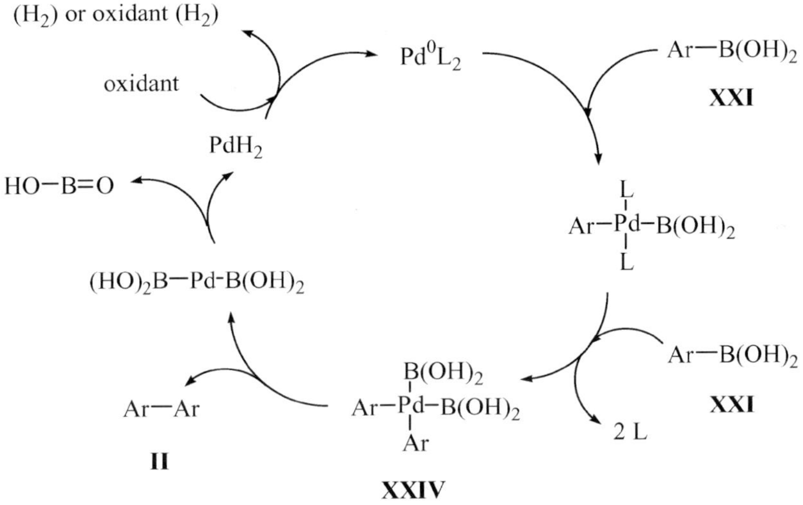

Scheme 8

Aryl halides containing amino group(s) under the SM reaction conditions have given deaminated biaryls in 40-70% yields, together with 20-50% of expected biaryls [54]. For example, the SM reaction of boronic acid **251** with 4-bromo-2-fluoroaniline (**252**) afforded only 12% of expected biaryl **253** and 72% of deaminated product **254** [54], Scheme 9.

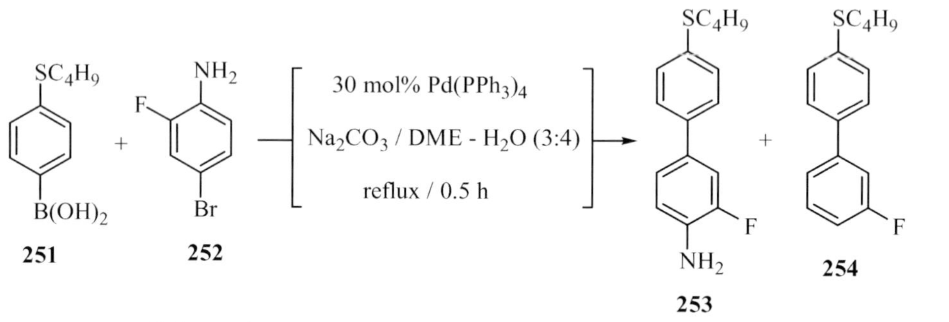

Scheme 9

Although the catayst was used in rather unusually high amount, the deamination side-reaction has to be taken in account. There is no evidence of deamination occuring when amino-group is in arylboronic acid counterpart. Finally there is a possibility for certain side-reaction to occur because of required basic reaction conditions. Some base-sensitive functional groups, e.g. esters, can be saponified, however, to avoid these difficulties, several successful alternative synthetic methods conducted in non-aqueous medium were developed.

5.2. Methods for performing the Suzuki-Miyaura synthesis of biaryls

Ever since an original Suzuki-Miyaura report, many different methods have been developed solving several basic problems concerning unreactive aryl halides or other electrophilic counterparts, inefficient catalysis, avoiding high (expensive) palladium or (toxic) nickel loading, inabillity to reuse homogeneous catalysts etc. Recently, the green-chemistry requirements such as environmental unfriendly organic solvents, toxic and cancer suspect chemicals and overall production costs have become important issues in the SM synthesis, as in all common metal-based catalytic processes. Here, the SM reaction methods regarding electrophilic counterparts are presented in order of their reactivity, because they are the most limiting factors in each particular reaction.

5.2.1. Methods of SM reactions involving aryldiazonium salts

Beside the already presented method based on Pd(OAc)$_2$-catalysis (Table 3), the SM reactions with aryldiazonium salts can be alternatively accomplished in the presence of palladium(0)-stabilizing ligand, *in situ* generated from **255**, to afford a wide variety of unsymmetrical biaryls in excellent yields [47]. As an illustrative example, 2-methoxy phenyldiazonium tetrafluoroborate (256) was cross-coupled with 4-*tert*-butylphenyl boronic acid (257) to give the biaryl 258 with a 86% yield [47], respectively, Scheme 10.

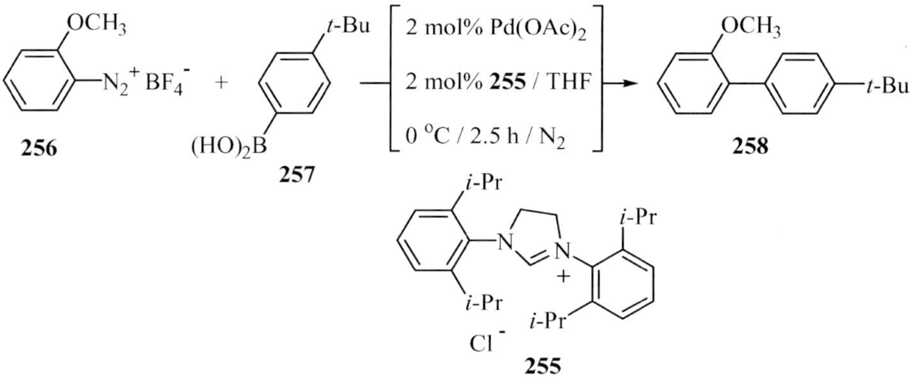

Scheme 10

The method was additionaly simplified by *in situ* aprotic diazotation with an organic nitrite in the presence of BF$_3$-etherate followed by cross-coupling with arylboronic acid in slightly lower yields. The latter variant gave 2-methoxybiphenyl (91) from 2-methoxyaniline (259) and phenylboronic acid (260) in 53% yield [47], Scheme 11.

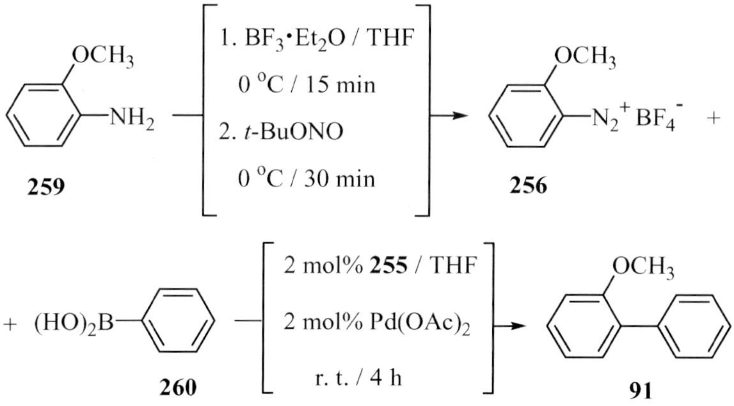

Scheme 11

The SM reaction involving the aryldiazonium salts must include a rapid formation of an arylpalladium(II) intermediate from diazonium salt (evolution of nitrogen occurs) and palladium(0) species *in situ* generated from palladium(II) acetate in ligandless method [46], or its complex with **255** [47], whereas the further mechanism follows the standard SM reaction pathway. These reactions are somewhat more sensitive to the steric hindrances furnishing with decreased yields of biaryls, but the final conclusion is still waiting for further probes of this very perspective approach. Although these reactions are far from wide-spread investigated SM processes, they are obviously very efficient, simple and clean methods.

5.2.2. Methods of the SM reactions with aryl iodides and bromides

Aryl iodides and bromides are the most commonly employed Suzuki-Miyaura electrophilic counterparts [1-9]. The SM methodology for aryl iodides and bromides is presented in the Tables 4 and 5.

Table 4. Methods for performing the SM reactions of arylboronic acids (and esters) with aryl iodides or bromides at elevated temperatures

Method	Catalyst (mol%)	Base (eq.)	Reaction conditions [a]	Lit. ref.
A	Pd(PPh$_3$)$_4$ [b] (3)	Na$_2$CO$_3$ (2)	PhMe / H$_2$O	1
B	Pd(PPh$_3$)$_4$ (5)	KOt-Bu (2)	DME / t-BuOH (10:1)	15
C	Pd(PPh$_3$)$_4$ [b] (3)	CsF, KF [c] (2)	DME, MeCN [d]	20
D [e]	Pd(PPh$_3$)$_3$ (3-6)	Tl$_2$CO$_3$ (1-1.2)	benzene, THF	19
E [e]	Pd(PPh$_3$)$_3$ (5)	Ag$_2$CO$_3$ (2)	THF	18
F	Pd(OAc)$_2$ (0.2)	K$_2$CO$_3$ (2.5)	acetone / H$_2$O (1:1)	28
G	Pd(OAc)$_2$ [f] (0.5)	K$_3$PO$_4$ (2)	DMF / r. t. - 130 °C	29
H	Pd(OAc)$_2$ / **243** (0.1)	K$_3$PO$_4$ (2)	H$_2$O / r. t. - 80 °C	38
I	Pd(OAc)$_2$ / **246** (3)	Cs$_2$CO$_3$ (2)	dioxane / 80 °C	41
J	10% Pd-C (5) / 4PPh$_3$	Na$_2$CO$_3$ (2)	DME / H$_2$O	24

[a] All methods were conducted at reflux temperature and under an inert atmosphere, unless otherwise noted.
[b] Beside PPh$_3$, other bidentate diphosphines, e.g. dppf, can be employed.
[c] n-Et$_4$NF is equally effective.
[d] Alternatively in DME / H$_2$O (2:1), DME / MeOH (2:1) or PhMe / H$_2$O (1:1)
[e] This methods work with aryldibutylboronic or arylpinacolboronic esters.
[f] In the presence of 20 mol% n-Bu$_4$NBr. Pd(SEt$_2$)Cl$_2$ has also been used.

The *Method A* is original SM procedure suitable for a wide variety of aryl iodides and bromides [1]. Alternatively, it can be performed either in water miscible organic solvents: acetone, DME, DMA, DMF, NMP, THF in the presence of aqueous Na$_2$CO$_3$ or K$_2$CO$_3$ as the base, or in dry dioxane, THF, toluene, DME, DMF in the presence of

K_2CO_3, Cs_2CO_3 or K_3PO_4 [1-9]. Beside triphenylphosphine, other diphosphine ligands have been used [1-9].

Table 5. Methods for performing the SM reactions of arylboronic acids with aryl iodides or bromides at room temperature

Method	Catalyst (mol%)	Base (eq.)	Reaction conditions [a]	Lit. ref.
K	$Pd(OAc)_2$ (5)	$Ba(OH)_2$ (1.5)	96% EtOH	10
L	$Pd(PPh_3)_4$ (2)	TlOH (1.5)	DMAc / H_2O (2:1)	16
M	$Pd_2(dba)_3$ (0.5) Pt-Bu_3 (1.2)	KF (3.3)	THF	31, 32
N	$Pd_2[P(1-Ad)_3]_2Br_2$ [b] (0.5)	KOH (3)	THF	35

[a] All methods were carried out at room temperature and under an inert atmosphere, unless otherwise noted.
[b] 1-Ad = 1-adamantyl

The two-phase, e.g. toluene / water, protocol has been described (Table 1), while here a non-aqueous procedure is illustrated. Sterically hindered biaryl **261** was prepared in absence of water in 75% yield from sterically hindered boronic acid **262** and bromide **263**, respectively [14], Scheme 12.

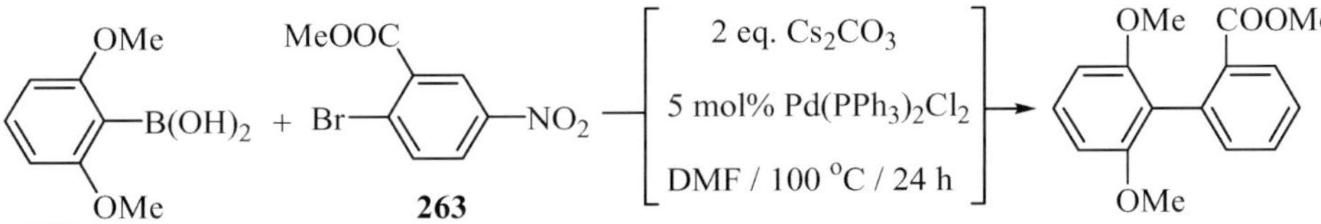

Scheme 12

Activated aryl chlorides have also been successfully coupled employing this original SM method. For example, 2-chloropyridine (**149**) reacts with phenylboronic acid (**260**) to give 2-phenylpyridine (**150**) in 71% isolated yield [30], Scheme 13, what represents an alternative protocol, performed in mixtures of water and water-miscible solvents. *Method B*, conducted with the stronger base, e.g. KOt-Bu, is developed for the bulky arylboronic acids, as already shown, Table 2 [15]. *Method C* (in several variants) is based on the fluoride bases, what is prefered for aqueous-base sensitive compounds [20]. Thus phenylboronic acid (**260**) was reacted with methyl 4-bromophenylacetate (**264**) applying fluorides with large cations to afford methyl 4-biphenylacetate (**265**) in the following yields: M = K (91%), Cs (95%), *n*-Bu$_4$N (80%), *n*-Et$_4$N (86%), Scheme 14.

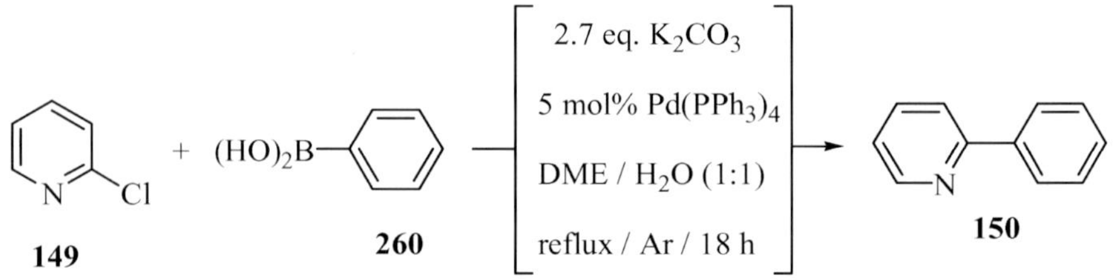

Scheme 13

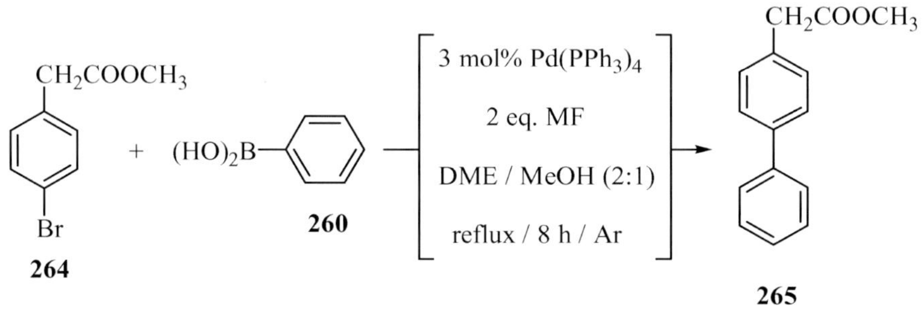

Scheme 14

The fluoride-mediated SM reactions have been carried out with aryl triflates in good yields. Several sensitive functional groups including tosylates have survived these reaction conditions [20].

Methods D and *E* are very efficient non-aqueous, rather neutral procedures applying silver(I) and thallium(I) carbonates, but useful only for small-scale labo-synthesis started from boronic esters [17-19]. An illustrative example is the reaction of 1,4-bispinacolboronic ester **266** with iodobenzene affording *p*-terphenyl (**267**) in essentially quantitative yield [18], respectively, Scheme 15.

Acceptable substitutes for extremely toxic Tl$_2$CO$_3$ or expensive Ag$_2$CO$_3$, are almost equally efficient sodium or potassium phenoxides [18,19].

Methods F and *G* are ligandless palladium-catalysed and therefore dramatically sensitive to the presence of an oxygen [28]. Beside aryl bromides, the Pd(OAc)$_2$ or Pd(SEt$_2$)Cl$_2$-catalysed SM reactions in *N,N*-dimethylformamide were effectively conducted with aryl chlorides [29]. When appropriate precautions were taken, the SM cross-coupling reaction of phenylboronic acid (**260**) with 4-nitro-bromobenzene (**168**) or 4-nitro-iodobenzene (**225**) furnished 4-nitrobiphenyl (**224**) with a quantitative yield within 0.75 or 2.5 h, Scheme 16.

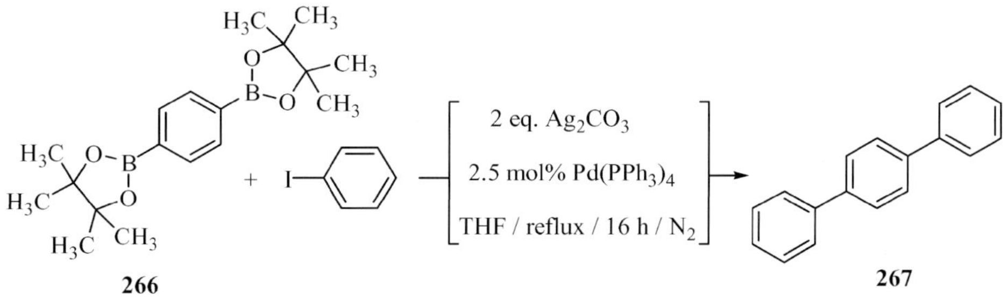

Scheme 15

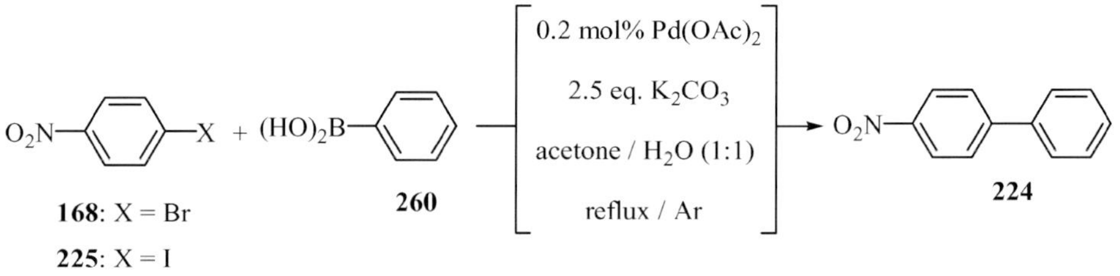

168: X = Br
225: X = I

260

224

Scheme 16

Other suitable reaction solvents are aqueous tetrahydrofuran, 1,2-dimethoxyethane or acetonitrile [28]. The same result was obtained by the *method G*, whereas 4-nitro-chlorobenzene was also coupled, in almost quantitative yield, within 2 h at 100 °C, or 87 h at room temperature. However, in the presence of tetra-*n*-butylammonium bromide (5 mol%), a soluble source of bromide anions, the SM reactions of aryl bromides have been effected in ethanol at room temperature in the presence of palladium(II) acetate or chloride (2 mol%) and potassium phosphate (2 eq.) as the base, even under exposure to air [55]. Palladium salts are reduced *in situ* with arylboronic acids to form catalytically active nano-sized palladium clusters (2-5 nm). The latter are stabilized by adsorbtion of one-layer bromide ions at the surface of each palladium-particle. Otherwise, the unstable nano-sized palladium-clusters are aggregated to the micro-sized catalytically inactive palladium black. In this manner, 2-bromonaphthalene (**268**) was reacted with 2-methoxyphenylboronic acid (**269**) to furnish the biaryl **270** in 98% yield [55], respectively, Scheme 17.

Method H includes the green-chemistry approach by using water as the only reaction solvent. The water-soluble palladium-catalyst from phosphine **243**, derived from D-glucono-1,5-lactone, in only 0.1 mol%, affects clean cross-coupling reactions of aryl iodides, bromides, as well as chlorides and triflates [38]. Thus 2-bromo benzonitrile (**116**) was coupled with 4-tolylboronic acid (**271**) in the presence of complex PdCl$_2$ and **243** to give biaryl **112** in 85% yield [38], Scheme 18.

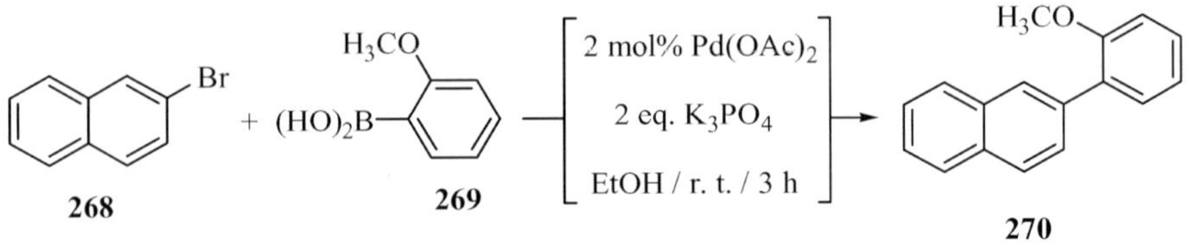

Scheme 17

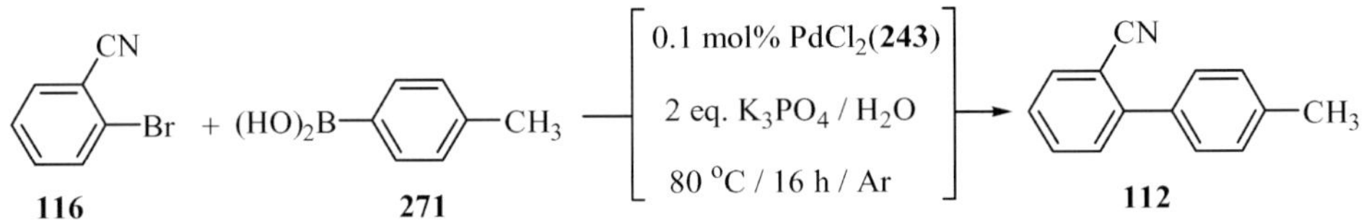

Scheme 18

However, at such low catalyst loading, the reactions are somewhat sensitive to the steric hindrances and with electron-rich aryl halides proceed more slowly. Indeed, the yield of a biaryl formed from 2-bromophenol was only 44%. Similarly, the catalysts *in situ* prepared from palladium(II) acetate and water-soluble phosphine ligands **241** and **242** have catalysed the SM reactions with aryl bromides (presumably iodides) at room temperature with excellent efficiency at Pd-loading in a range of 10^{-4} mol%, and thus became modern alternatives to the second (and third) generation SM catalysts (see below) [37]. *Method I* was investigated in the SM reactions with aryl bromides and chlorides [41]. Despite the use of palladium(II) acetate in relatively high amount, 3 mol%, this method is based on the readily available *N,N'*-dicyclohexyl-1,4-diazabutadiene (**246**), thus avoiding expensive (except PPh$_3$) phosphine ligands. An illustrative example is the coupling reaction of phenylboronic acid (**260**) and mesityl bromide **188** which afforded biaryl **189** in 97% yield [41], respectively, Scheme 19.

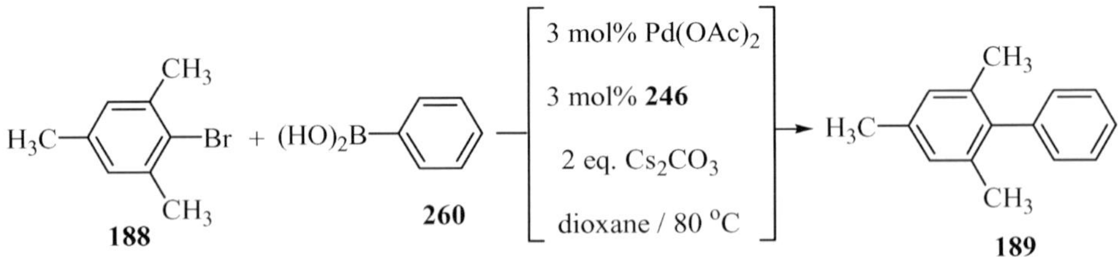

Scheme 19

Closely related method for SM reactions of aryl bromides using the complex of palladium(II) acetate with glyoxal bis(methylphenylhydrazone) (247) was shown to be similarly effective [42]. *Method J* is apparently the most economic approach employing easily regenerable and reusable Pd-C catalysts (5 mol%) [24-27]. The older variant has been conducted with triphenylphosphine (4 eq. to Pd, what means 20 mol%) as stabilizing ligand for catalytically active palladium(0)-species [24]. However, this complicates the purification procedure of less polar biaryls. The triphenylphosphine can be conveniently removed by reaction with methyl iodide, see Chapter 3, followed by filtration of resulting Wittig salt. The Pd-C catalysed SM reactions have been achieved also with aryl chlorides and triflates [24] bearing an activating electronegative group in *para*-position at about 80 °C in refluxing DME / H$_2$O (20:1) [25], or even unactivated ones at slightly higher reaction temperatures, 100 °C, in NMP / H$_2$O (10:3) [27]. Efficacy of this method is presented with an example where 4-bromobenzonitrile (272) is reacted with phenylboronic acid (260) to give biphenyl-4-carbonitrile (273) in essentially quantitative yield at very low catalyst loading [27], respectively, Scheme 20.

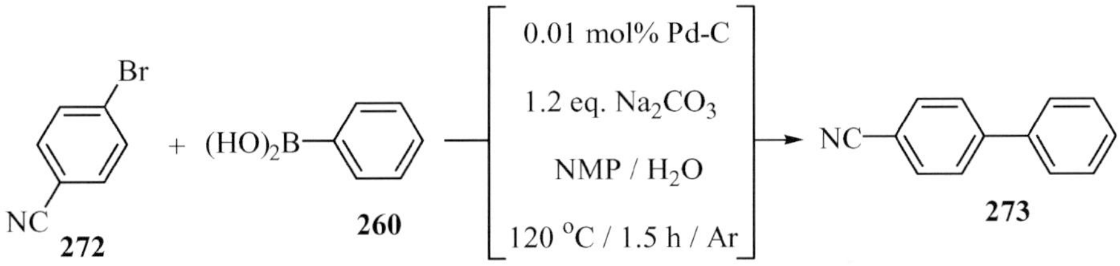

Scheme 20

Additionally, the Pd-C catalysed SM reactions with aryl iodides can be accomplished at room temperature. The reactions were alternatively carried out in water if aryl halides, e.g. iodophenols 274-276, are soluble in aqueous potassium carbonate to afford the biaryls 277-279 in good to excellent yields [26], Scheme 21.

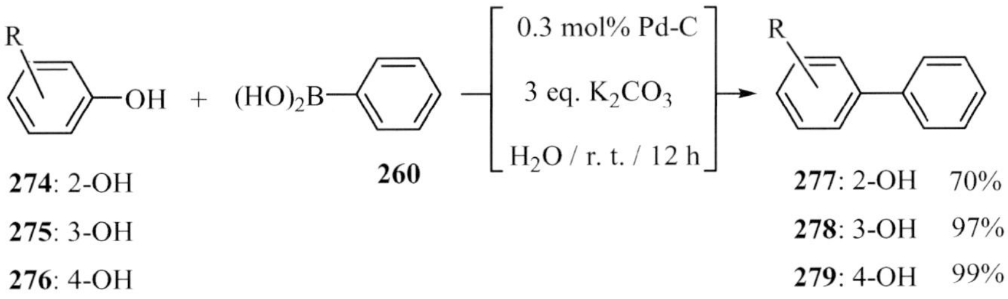

Scheme 21

Investigations have shown that the SM reactions catalysed by Pd-C actually proceed in homogeneous medium, since the certain amount of palladium is solubilized through the formation of arylpalladium(II) halides, ArPdX, and not at the surface of palladium particles [56]. The transmetallation is the rate-determining step, and not the oxidative addition, as one could predict.

Method K is a modification of the *method F* where barium hydroxide was used as the base in 96% ethanol to affect efficient SM reactions of aryl bromides at room temperature with excellent yields. In this manner, boronic acid **280** and 3,4-methylene dioxybromobenzene (**281**) were converted under very mild reaction conditions to the highly substituted biaryl **282** in 96% yield [10], respectively, Scheme 22.

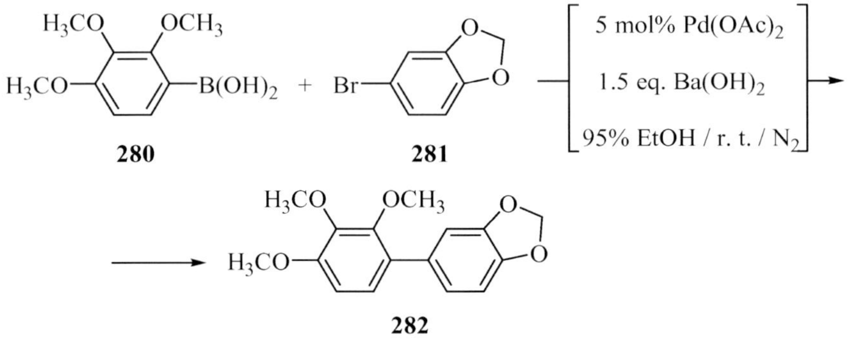

Scheme 22

Once again, absolute absence of oxygen is essential to obtain the high conversions and reproducible results in all ligandless (without any palladium(0)-clusters stabilizers) palladium (or nickel) catalysed methods [10]. Moreover, barium hydroxide, as quite strong base containing a large Ba^{2+} cation, additionally accelerates the coupling reaction which is able to proceed even at room temperature.

Method L is conducted with thallium hydroxide as the base and Pd(PPh$_3$)$_4$ as catalyst under very mild reaction conditions to afford a wide variety of biaryls, including highly encumbered, in fair yields [16]. The biaryl **189** was prepared from mesitylboronic acid (**283**) and iodobenzene in almost quantitative yield, Scheme 23.

Whereas tri-*ortho*-substituted biphenyls were obtained by this method in respective yields, synthesis of tetra-*ortho*-substituted analogues failed [16]. Among base-sensitive groups, only an ester function has survived thallium hydroxide-mediated reactions at room temperature. Concerning the extreme toxicity of thallium compounds, the method is applicable only in the small-scale labo-synthesis.

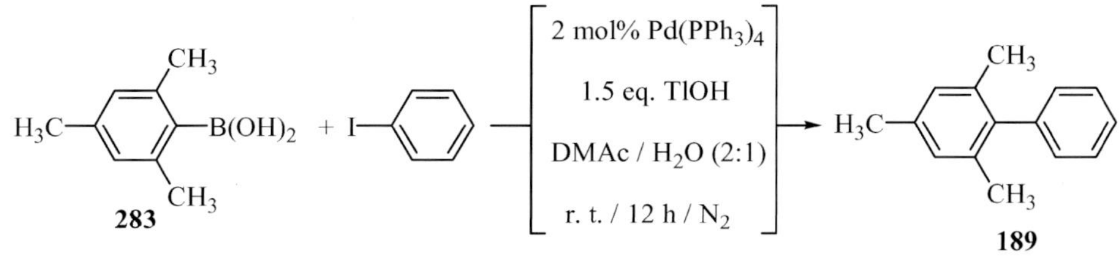

Schema 23

Methods M and N are based on different phosphine ligands which activate the resulting palladium-complex providing higher electron-density at palladium-metallic centre, more reactive catalyst, and very mild reaction conditions [32,35]. The excellent-yielding SM reactions of aryl iodides, bromides, as well as chlorides catalysed with $Pd_2(dba)_3$ / Pt-Bu_3 (1:2), in the presence of potassium fluoride as the base, smoothly proceed at room temperature [35], Scheme 24. All three isomeric 4-halo-acetophenones, **73**, **284**, **285**, in reaction with 2-tolylboronic acid (**286**) gave appropriate biaryl **287** in almost quantitative yields.

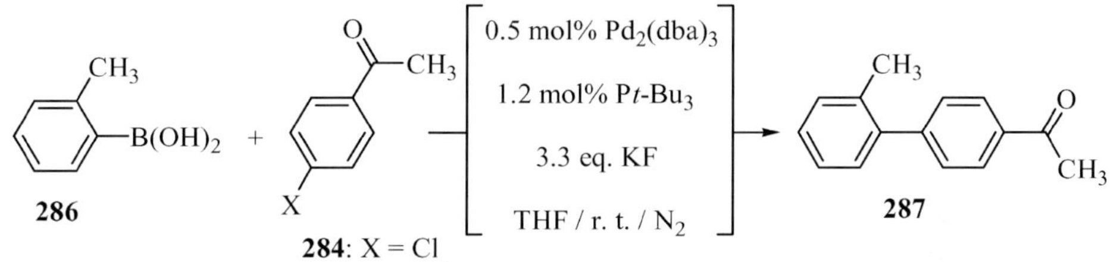

Scheme 24

However, unactivated aryl chlorides, with no electronegative substituents in the *ortho*- or *para*-position, require somewhat higher reaction temperatures, refluxing THF (65 °C) or dioxane (100 °C) and three-fold higher catalyst loading with KF, Cs_2CO_3 or K_3PO_4 as the bases [31,32]. An aromatic amino-group has remained intact.

5.2.3. Methods of the SM reactions with aryl chlorides

Aryl chlorides, as the least reactive halides, can be cross-coupled with arylboronic acids by different synthetic methods. Each particular method must include an apropriate catalyst, capable to react with rather unreactive C-Cl function to give a corresponding arylpalladium(II) (or nickel) chloride which further undergoes the standard SM reaction pathway. The original SM, $Pd(PPh_3)_4$ (3-5 mol%)-catalysed method performed in toluene / water (K_2CO_3), DMF (K_3PO_4) or in DME / water (K_2CO_3) mixture can be successfully applied in reactions with certain activated aryl chlorides, e.g. 2- and 4-chloropyridines [30]. As already mentioned, some methods for aryl iodides and bromides have been applied with aryl chlorides at similar or higher reaction temperatures. Here, additional, rather specific SM cross-coupling procedures for aryl chlorides are presented in the Table 6.

Method O uses variuos nickel-phosphine precatalysts, e.g. $Ni(PPh_3)_2Cl_2$, or $Ni(dppf)Cl_2$, which are sometimes previously reduced with zinc powder [12,13,44], DIBAlH [12,44], *n*-BuLi [44] or used directly [44]. Among the phosphine ligands, the efficiency follows the order [44]:

$$dppf > dppp > dppe > dppb, PPh_3$$

However, even inexpensive triphenylphosphine as $Ni(PPh_3)_2Cl_2$ can be successfully employed as precatalysts which is, under the reaction conditions, reduced in the presence of PPh_3 (2 eq.) to the catalytically active $Ni^0(PPh_3)_4$. For example, pharmaceutically important biaryl **112** was prepared in 97% yield from 2-chloro-benzonitrile (**114**) and 4-tolylboronic acid (**271**) [12], Scheme 25.

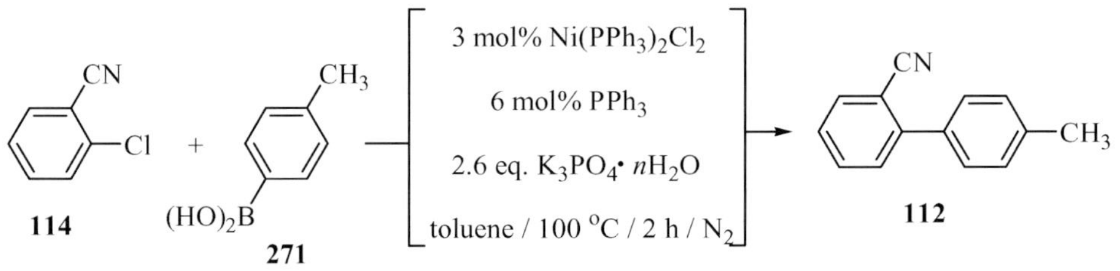

3 mol% $Ni(PPh_3)_2Cl_2$

6 mol% PPh_3

2.6 eq. $K_3PO_4 \cdot nH_2O$

toluene / 100 °C / 2 h / N_2

Scheme 25

Generally, a 10 mol%-excess of an arylboronic acid is applied in the SM reactions. Certain nickel-catalysed SM reactions, performed at higher temperatures, e.g. 100 °C, require increased excess of arylboronic acid. This usually gives respective homo-coupling side-product, often with similar properties as the cross-coupling product. This side-product causes difficulties during the purification procedure, either by chromatography (closed R_f values), or by crystalization (co-precipitation often occurs). Therefore, the methods which require higher excesses of arylboronic acids, e.g. 1.3-2 eq., are disfavourable. The $Ni(dppf)Cl_2$-catalysed SM reactions with aryl triflates and mesylates were efficiently used [44]. An aromatic amino-group is well tolerating, e.g. 4-chloroaniline was coupled with 4-tolylboronic acid in 81% yield (100 °C), respectively, without deamination [12].

Table 6. Methods for performing the SM reactions of arylboronic acids with aryl chlorides

Method	Catalyst (mol%)	Base (eq.)	Reaction conditions [a]	Lit. ref.
O	Ni(PPh$_3$)$_2$Cl$_2$ / 2 PPh$_3$ (3) [b]	K$_3$PO$_4$· H$_2$O (2.5-3)	toluene, dioxane [c]	12, 13, 43
P	Ni-C / 4 PPh$_3$ (10)	K$_3$PO$_4$ (3.6) LiBr [d] (2.5)	dioxane	42
Q	Pd(dba)$_2$ (0.5-1) **237** (1.5-3)	CsF (2)	dioxane or toluene	11, 33
R	Pd(OAc)$_2$ (0.5-2) **238** (0.75-3)	CsF (3) or K$_3$PO$_4$ [e]	dioxane r. t. - 100 °C	34
S	Pd$_2$(dba)$_3$ (1.5) **240** (6)	K$_3$PO$_4$· H$_2$O (2.4)	toluene r. t. - 70 °C	36
T	Pd$_2$(dba)$_3$ (1.5) **244** (3)	Cs$_2$CO$_3$ (2)	dioxane 80 °C	39
U	PdCl$_2$ (1.5-3) **245** (3-6)	CsF, K$_2$CO$_3$, K$_3$PO$_4$ (2)	dioxane, DME, THF	40

[a] The reactions were carried out at the reflux temperature and under an inert atmosphere, unless otherwise noted.
[b] The active catalyst was prepared by the previous reduction with Zn, *n*-BuLi, DIBAlH. In this method, several different mono- and bidentate, e.g. dppf, phosphine ligands can be employed.
[c] The lower catalyst loading (0.005-0.25 mol%) can be used at higher reaction temperatures (120 °C) in NMP / H$_2$O (10:3) with NaOH as the base.
[e] Slightly less efficient than CsF

Method P is applying the nickel on charcoal as heterogeneous catalyst in the presence of triphenylphosphine as stabilizing ligand for nickel(0)-species [43]. Several phosphine ligands were tested, but PPh$_3$ surprisingly proved to be the most effective. The Ni-C catalyst is usually prepared by impregnating the nickel(II) nitrate on commercially available charcoal (>200 mesh). This material is conveniently reduced in anhydrous dioxane suspension with *n*-BuLi (4 eq.) in the presence of PPh$_3$ (4 eq.). Anhydrous potassium phosphate in refluxing dioxane provides the best conditions for thus catalysed SM reactions. For example, 4-chlorobenzaldehyde (**107**) was reacted with phenylboronic acid (**260**) to afford 4-formylbiphenyl (**288**) in 85% yield [43], Scheme 26. This method gives good to excellent yields of biaryls, although it is quite sensitive to the *ortho*-steric encumbrances in aryl chlorides, as well as in boronic acids. In addition, amino- and acetamido-substituted aryl chlorides led to unapplicable reaction extent due to very slow reaction. The presence of nitro group is not tolerating since it strongly retards all nickel-catalysed reactions, see Chapter 3.

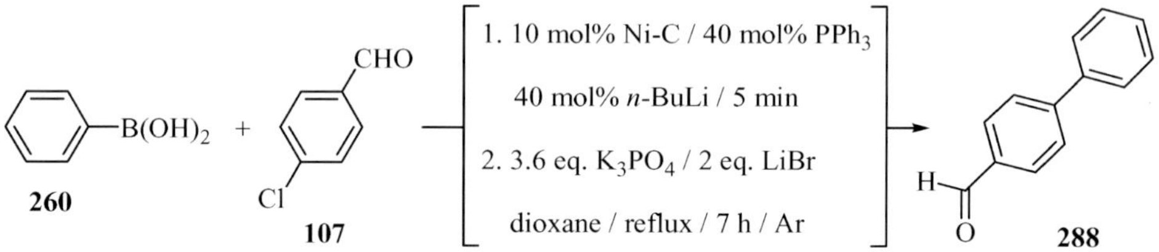

Scheme 26

Method Q is based on Pd$_2$(dba)$_3$ / ArPCy$_2$-type phosphine **237** catalysis and provides a convenient high-yielding SM procedure with aryl chlorides under reasonably mild reaction conditions (refluxing toluene, 105-110 °C) [11,33]. For instance, 2-chloro acetophenone (**289**) was coupled with 4-tolylboronic acid (**271**) to give **290** in 83% yield [33], respectively, Scheme 27. The method is not sensitive to the steric hindrances in both counterparts.

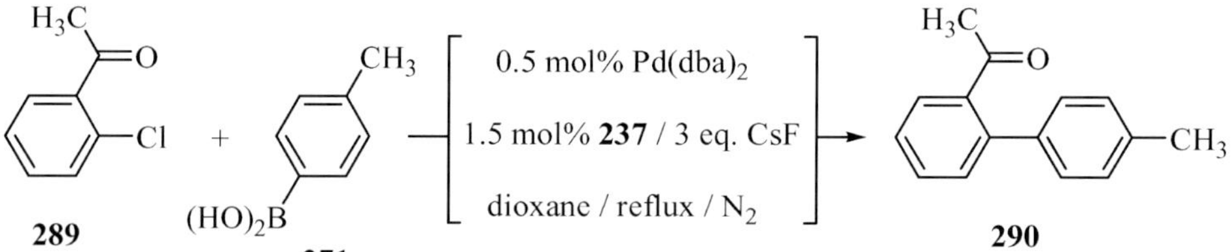

Scheme 27

In the *method R*, a bidentate ArPCy$_2$-type phosphine ligand **238** forms highly reactive palladium-catalyst capable to affect the SM reactions with aryl chlorides even at room temperature [34]. At higher reaction temperatures, lower catalyst loading can be employed with the same range of conversions. *Method U* includes palladium-complexes with air-stable phosphine *t*-Bu$_2$P(OH) (**245**) which efficiently catalysed the SM reactions with aryl chlorides and bromides to afford biaryls in good to high yields [40]. For example, both *methods R* and *U* gave excellent yields of 4-methoxybiphenyl (**78**) in the cross-coupling reaction of phenylboronic acid (**260**) with electron-rich (deactivated) 4-chloroanisole (**76**) [34,40], Scheme 28. However, all reactions were conducted with 50 mol% excess of arylboronic acids what may cause difficulties during the final purification. Additionally, arylboronic acids are still rather expensive materials. *Method S*, based on the palladium-complex of sterically congested ferrocene-phosphine **240**, and *method T*, which uses heterocyclic carbene **244** are versatile SM alternatives which led to efficient cross-couplings of unreactive aryl chlorides. Namely, 2-tolylboronic acid (**286**) and 4-chlorotoluene (**227**) produced biaryl **291** in excellent yields [36,39], Scheme 29.

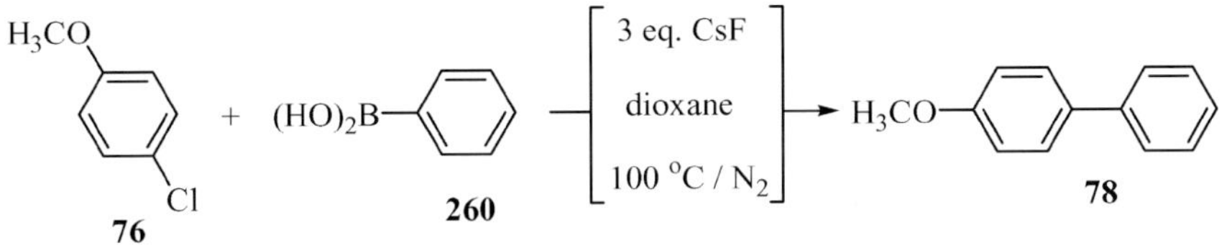

Method R : 0.5 mol% Pd(OAc)$_2$ / 0.75 mol% **238**, 93%

Method U : 1.5 mol% Pd$_2$(dba)$_3$ / 1.5 mol% *t*-Bu$_2$P(OH), 86%

Scheme 28

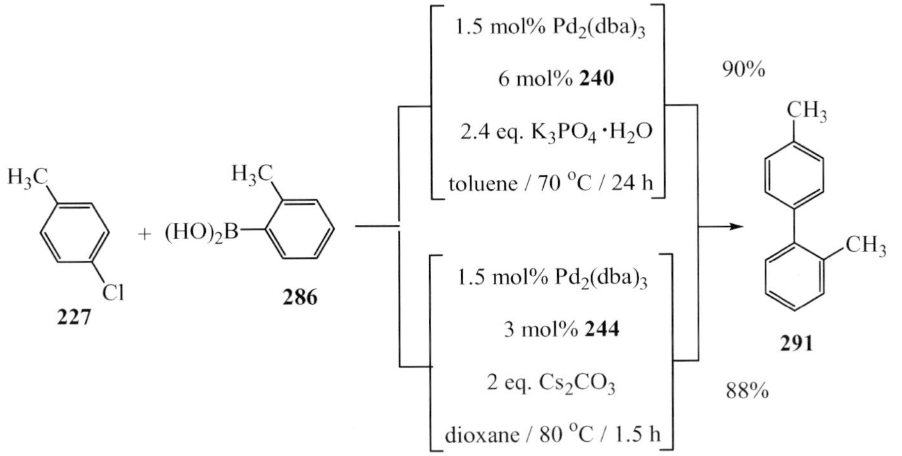

Scheme 29

The major drawback of these methods is relatively high amount of required ligands **240**, or **244**, and starting palladium complex.

5.2.4. Methods of the SM reactions with aryl sulfonates

Aryl sulfonates are the least reactive SM substrates since the C-O bond is less polarized and thus oxidative addition to palladium(0) or nickel(0) proceeds much slower. Furthermore, the steric demands of all common arylsulfonate groups are considerably higher than in any halide analogue. However, the SM reactions have been effectively carried out by some, above described methods. Further methods suitable for aryl sulfonates are presented in the Table 7. *Method V* is actually a modification of the original SM *method A* what clearly illustrates a wide applicability of Pd(PPh$_3$)$_4$-catalysed reactions [49]. However, an important improvement has been introduced by adding lithium chloride or bromide (2 eq.) which activates the aryl triflates for oxidative addition reaction [51,52]. This method can be accomplished in several similar variants, for example in refluxing dioxane with K$_2$CO$_3$ or K$_3$PO$_4$ as the base in the presence of LiCl or LiBr to furnish variuos substituted biaryls in moderate yields [49]. The coupling reaction of phenylboronic acid (**260**) with 4-nitrophenyl triflate (**292**) resulted in formation of 4-nitrobiphenyl (**224**) in 51% yield, whereas no reaction occurs in the absence of lithium chloride [49], Scheme 30.

Table 7. The Suzuky-Miyaura reactions of arylboronic acids with aryl triflates, mesylates and tosylates

Method	Catalyst (mol%)	Base (eq.)	Reaction conditions [a]	Lit. ref.
V	Pd(PPh$_3$)$_4$ (3)	Na$_2$CO$_3$ (2) LiCl [b] (2)	toluene / EtOH /H$_2$O	49
W	Pd(OAc)$_2$ (1) PCy$_3$ (1.2)	KF (3.3)	THF / r. t.	32
X [c]	Ni(PCy$_3$)$_2$Cl$_2$ + + 4 PCy$_3$ (1.5-3)	K$_3$PO$_4$ (2)	dioxane	44, 48

[a] All reactions were performed at reflux temperature in an inert atmosphere, unless otherwise noted.
[b] The role of LiCl is activation of an aryl triflate.
[c] Alternatively, Ni(dppf)Cl$_2$ in the presence of zinc powder can be used [48].

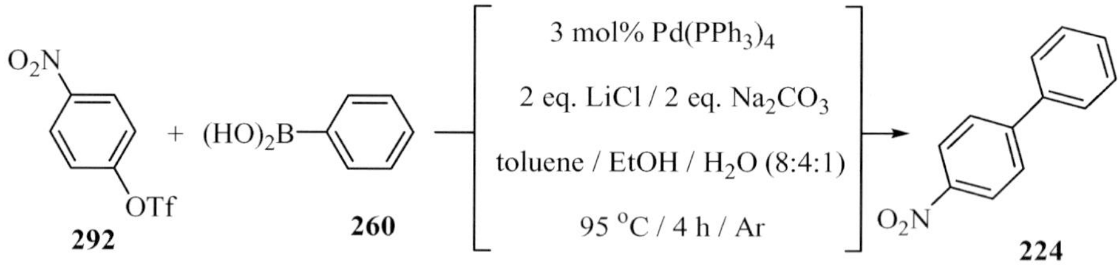

Scheme 30

The reactivity of palladium complexes can be enhanced by using more electron-rich aryl-dialkylphosphines, or even trialkylphosphines. The most general method for performing the SM reactions, employing Pd$_2$(dba)$_3$ and P*t*-Bu$_3$, with aryl iodides, bromides and chlorides, is *method M*, whereas in *method W* the complex of Pd(OAc)$_2$ and PCy$_3$ has given the best results with aryl triflates [32]. Both variants are conducted in tetrahydrofuran in the presence of potassium fluoride (3.3 eq.) as the base proceeding under very mild conditions (r. t.), to afford a wide variety of biaryls in very high to quantitative yields, Table 8.

Table 8. The yields of biaryls in the SM reactions of aryl triflates with arylboronic acids catalysed by Pd(OAc)$_2$ / PCy$_3$ [32] (*method W*)

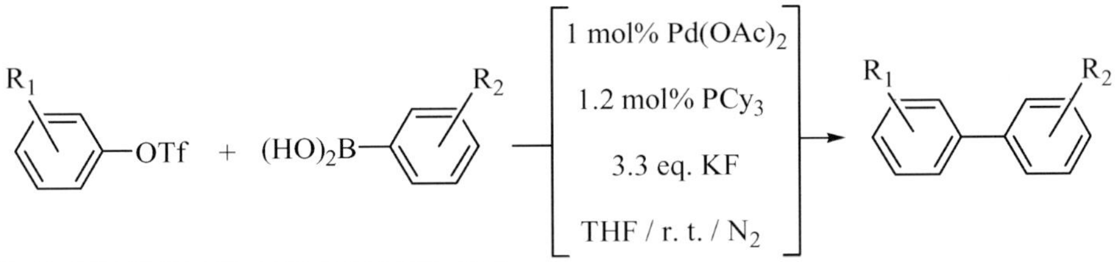

R$_1$	R$_2$	Yield (%)
4-CH$_3$	H	97
4-CH$_3$CO	2-CH$_3$	96
4-CH$_3$	2-CH$_3$	95
2-CH$_3$	2-CH$_3$	82 [a]

[a] 5 mol% Pd(OAc)$_2$ and 6 mol% PCy$_3$ were employed.

The *method X* affects the SM reactions catalysed by Ni^0(Cy$_3$)$_4$, which is *in situ* generated from Ni(PCy$_3$)$_2$Cl$_2$ and tricyclohexylphosphine (2 eq.), to give *para-* and *meta-*substituted biaryls in good to excellent yields [44]. Alternative phosphines such as dppf [48] can be used where previous reduction of the nickel(II)-precatalyst is not necessary since the arylboronic acid acts as an effective reductant. The method is sensitive to steric hindrances in both reactants, and resulted in significantly lower yields of biaryls, as can be seen from the reaction of phenylboronic acid (**260**) and tosylate **293**, derived from methyl salicylate, where methyl biphenyl-2-carboxylate (**294**) was isolated in 47% yield [44], Scheme 31.

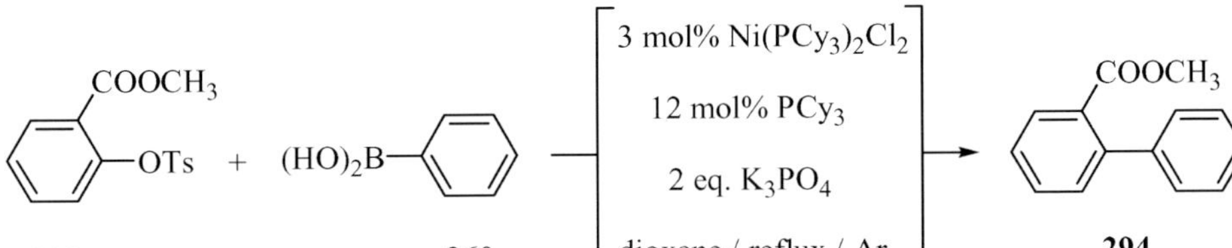

Scheme 31

For performing the SM reactions of aryl mesylates, triflates, tosylates, and besylates, Percec's method involving Ni(dppf)Cl$_2$ proved to be powerful and versatile alternative with moderate to good yields of biaryls. Thus the reaction of 4-acetylphenyl mesylate (**295**) with phenylboronic acid (**260**) resulted in 4-acetylbiphenyl (**68**) with a 51% yield [48], respectively, Scheme 32.

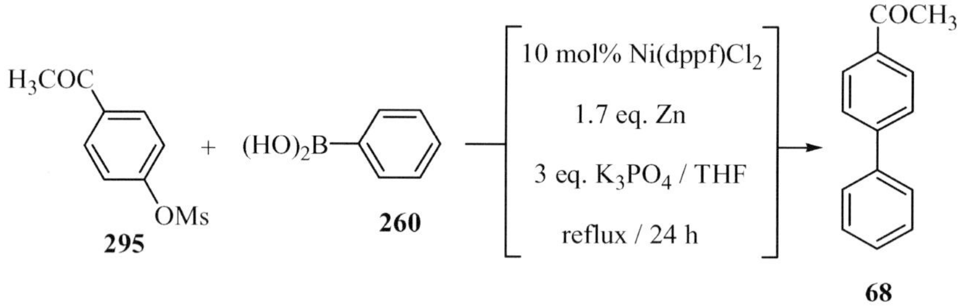

Scheme 32

Although the method is applicable to all sulfonates, and the arylboronic acid homo-coupling side-reaction is almost avoided (max. a few percent), required amount, 10 mol%, of Ni(dppf)Cl$_2$ is relatively high. Beside boronic acids, tetracoordinative boronate esters, such as **296**, as a result of faster transmetallation reaction, much readily react with aryl mesylates and tosylates in Ni(PPh$_3$)$_2$Cl$_2$-catalysed reactions to give biaryls in good to excellent yields. Thus compound **296**, easily *in situ* generated from the parent boronate ester **297** and *n*-BuLi, reacts with 4-cyanophenyl mesylate (**298**) to give 4-cyanobiphenyl (**273**) in 80% yield [57], respectively, Scheme 33. The main side-reaction involving tetracoordinative boronate esters is the formation of alkyl-transfer by-products, such as **299**, however, usually under 20%.

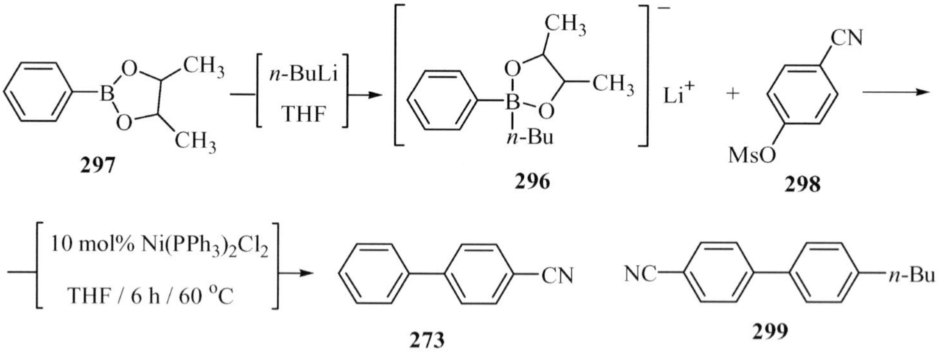

Scheme 33

The major advantages of this approach over the Percec's method are higher reaction yields and the use of inexpensive catalyst, Ni(PPh$_3$)$_2$Cl$_2$.

5.2.5. Applicability of the Suzuki-Miyaura synthesis of biaryls

The Suzuky-Miyaura cross-coupling reactions have been used in a great number of applications including an aryl-aryl bond formation in the variously substituted biphenyls [1-9,58-69] including highly substituted biphenyls [15,59-61], heterobiaryls [15,70-84] including phenylpyridines [15,30,70-72], phenylpyridine-oxides [70], phenylpyrimidines [70], phenylpyridazines [70,73], phenylpyrazines [70,74], phenylthiophenes [70,75], phenylfurans [70], phenylisatins [76], quinazolinylfurans [77], phenylbenzo[b]thiophenes [78], phenylpurines [79,80], thiophenylpurines [80], phenylimidazo[1,2a]pyridines [81], phenyl-7-azaindoles [82], thiophenyl-7-azaindoles [82], furyl-7-azaindoles [82], phenylisoxazoles [83], pyrimidylisoxazoles [83], sterically hindered naphthylpyridines [84], arylferrocenes [85], calix[4]arenyl phenylenes [86], mono-Cr(CO)$_3$-biphenyls [87], and many other classes of biaryls. Generally, the SM reaction conditions tolerate the presence of OH, OR, COOR, NH$_2$, NHR, CHO, C(=O)R, C(=O)Ar, COOH, NHCOOR, CH(OR)$_2$, C(OR)$_2$ and many heteroaryls containing ligand-capable heteroatoms: N, O, S. To illustrate wide applicability of the SM reactions, here an interesting example of the highly derivatized biaryl structure is given. Highly active antibacterial metabolite from the *Streptomyces griseorubiginosus* contains a biaryl building block which has been obtained by the SM reaction of boronic acid **300** with bromide **301** to afford the key biaryl intermediate **302** in 86% yield [58], respectively, Scheme 34.

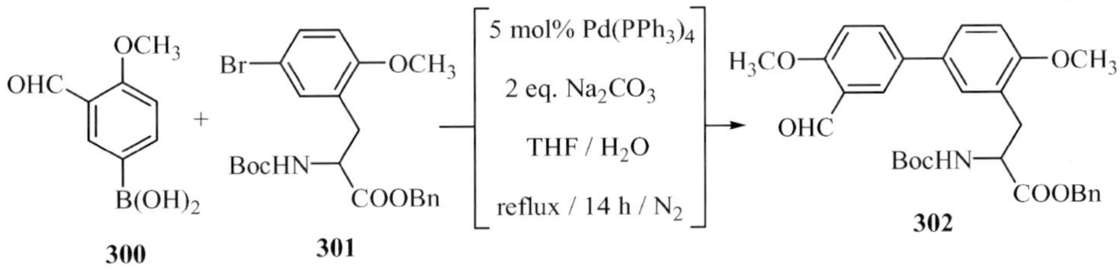

Scheme 34

Additional specific examples are from the ferrocene chemistry, where the huge progress has been achieved thanks to the Imrie's group results [85]. The ligandless palladium(II) acetate-based method in the presence of barium hydroxide as the base has been proved as a convenient procedure for producing arylferrocenes from the iodoferrocene as the starting material. For example, 4-formylphenyl ferrocene (**303**) was prepared in 63% yield from iodoferrocene (**304**) and 4-formylphenylboronic acid (**305**) [85], Scheme 35.

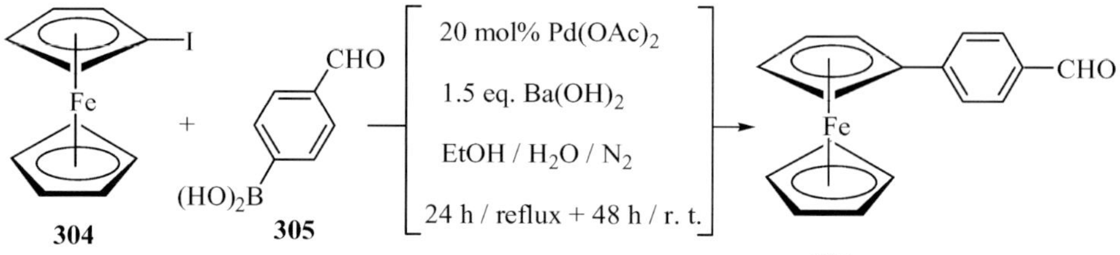

Scheme 35

Although the SM reactions of haloferrocenes proceed rather slow, they are clean and chemoselective, providing the production of arylferrocenes which would be hardly accessible by other strategies. Several papers regarding improved and alternative technics for performing the Suzuki-Miyaura reactions have been published. A useful procedure involves transmetallation of (half amount) simple aryl halides with only 0.5 eq. of *n*-BuLi, followed by quenching the resulted aryllithium reagent with trimethyl borate [88]. Thus obtained arylboronic ester is coupled under the standard SM conditions (*method A*) with an excess of parent aryl halide to furnish the symmetrical biaryl in fair yields. For example, 2-bromoanisole (**218**) was coupled by this way to afford 2,2'-dimethoxybiphenyl (**92**) in 56% yield [88], Scheme 36.

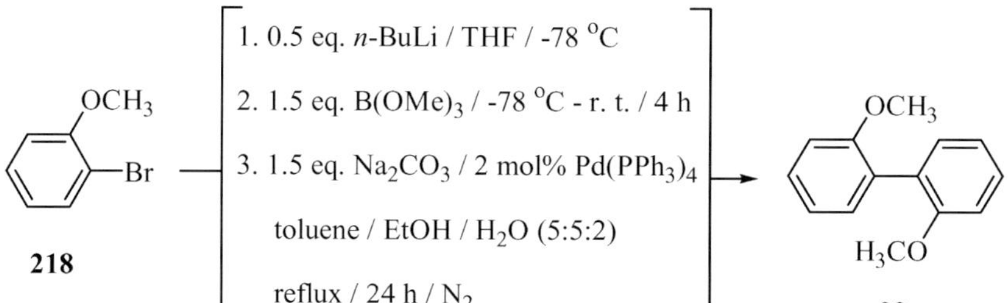

Scheme 36

The SM reactions in water [89-91] as the only reaction solvent or under solvent-free [92,93] conditions can be microwave-assisted with dramatical shortening of the reaction times from hours to minutes. For example, Leadbeater and coworkers [90] have described a preparation of biphenyl-4-carboxylic acid (**306**) in 97% yield from 4-bromobenzoic acid (**307**) and phenylboronic acid (**260**) by ligandless Pd(OAc)$_2$-based method applying the microwave-heating for only 10 minutes, Scheme 37.

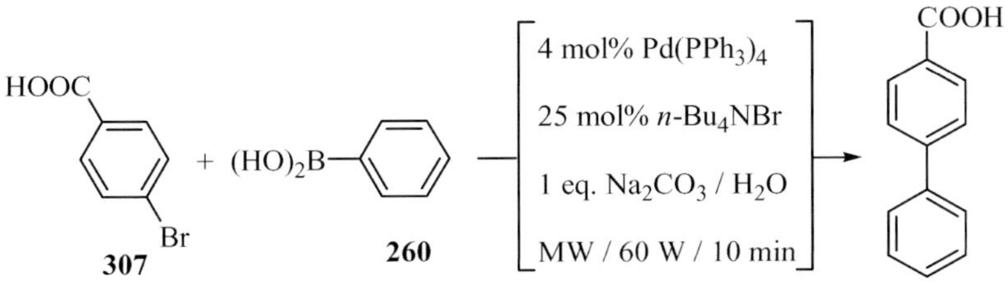

Scheme 37

Among the solvent-free methods, the SM reactions were conducted on Al_2O_3 as solid support in the presence of potassium fluoride as the base, premixed separately [92], or adsorbed directly on commercially available $KF-Al_2O_3$ [93]. The latter variant is applying palladium black (submicron powder) as catalyst, providing a simple regeneration (by filtration) and several possible reusages. Both solvent-free methods gave good yields in the SM reactions of aryl iodides and bromides [92,93]. Alternatively, the SM reactions can be carried out under solvent-free conditions without the microwave-heating. Nielsen and coworkers [94] have found that during the ball-milling of a mixture of aryl bromides (and iodides), arylboronic acid, Pd(PPh₃)₄, potassium carbonate, and sodium chloride as an inert solid diluent at room temperatures within 1 h, the SM reactions take place to produce biaryls in various yields. 3-Bromoaniline (**308**) was coupled with phenylboronic acid (**260**) to give 3-aminobiphenyl (**309**) in 89% yield [94], respectively, Scheme 38. Although the method is not generally applicable, it might be useful in certain instances.

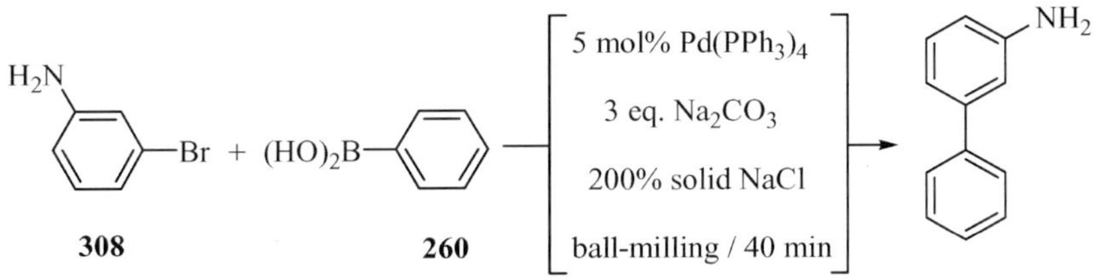

Scheme 38

The SM reactions are compatible with the solid-phase synthesis what is applied in the combinatorial chemistry [95,97]. The arylboronic acids can be immobilized on the polystyrene containing diethanolamine-moieties by forming boronate esters **XXV**. A wide variety of thus bounded arylboronic acids bearing a bromomethyl, carboxylic, amino or aldehyde-group have been converted by *N*- or *O*-alkylation, amidation, reductive alkylation and some other reactions, to amines, esters, amides, ureas, etc [96]. The resulting arylboronic acid can be regenerated from the resin by simple washing with THF / water (95:5) mixture, avoiding all purification difficulties connected with the classical organic synthesis.

PS—CH$_2$—N----B—Ar   PS = polystyrene

XXV

Generally, at least one Suzuki-Miyaura counterpart has to contain a connecting group which serves for immobilization on the suitable polymer (resin). The resin, often various polystyrenes (PS), is derivatized by introducing: NH$_2$, COOH or similar group. However, a number of different derivatized polymers are readily available. They are reacted with connecting group from one of the SM reactants to form immobilized arylboronic acid (or aryl halide). After the SM reaction of polymer-bounded arylboronic acid (aryl halide) with the aryl halide (arylboronic acid), and washing all impurities, the resulting biaryl is cleavaged from the resin in a very pure form, usually >95%. Interesting examples are the SM reactions from isoquinoline-chemistry that have been carried out on polystyrene bearing the carboxylic acid functions. The isoquinolines have been bounded by forming the corresponding Reissert-compound with the PS-COOH. For example, 6-bromo-5,8-dimethylisoquinoline (**310**), immobilized on the PS-COOH as Reissert intermediate **311**, affected the SM reaction with phenylboronic acid (**260**) to furnish the bounded biaryl **312**, which, upon hydrolysis, gave 6-phenyl-derivative **313** in 19% overall yield [97], Scheme 39.

Moreover, aryl halides bearing a carboxylic acid, bounded on polyethylene glycol (PEG 6000), readily affected the high-yielding SM reactions in liquid phase [95]. This approach works much efficiently than with solid-bounded reactants, as the reactions proceed in homogeneous phase, and allow the use of conventional spectroscopy for monitoring and characterization.

Aryltrifluoroborates, ArBF$_3^-$, are mostly used unconventional SM substrates [98]. Although the latter are transient intermediates in all SM reactions performed with fluoride-bases, the reaction can be affected with separately prepared potassium aryltrifluoroborates. Thus, by adding the methanolic solution of arylboronic acid to the concentrated aqueous solution of KHF$_2$, followed by evaporation and recrystallization, potassium aryltrifluoroborates were obtained in yields generally over 95% [98]. These reagents smoothly react with aryldiazonium salts in 1,4-dioxane at room temperature in the presence of palladium(II) acetate to give the respective biaryls in good to excellent yields. For instance, biphenyl **314** was produced in 96% yield from 4-methoxyphenyl boronic acid (**315**), *via* trifluoroborate **316**, and diazonium salt **317** [98], Scheme 40.

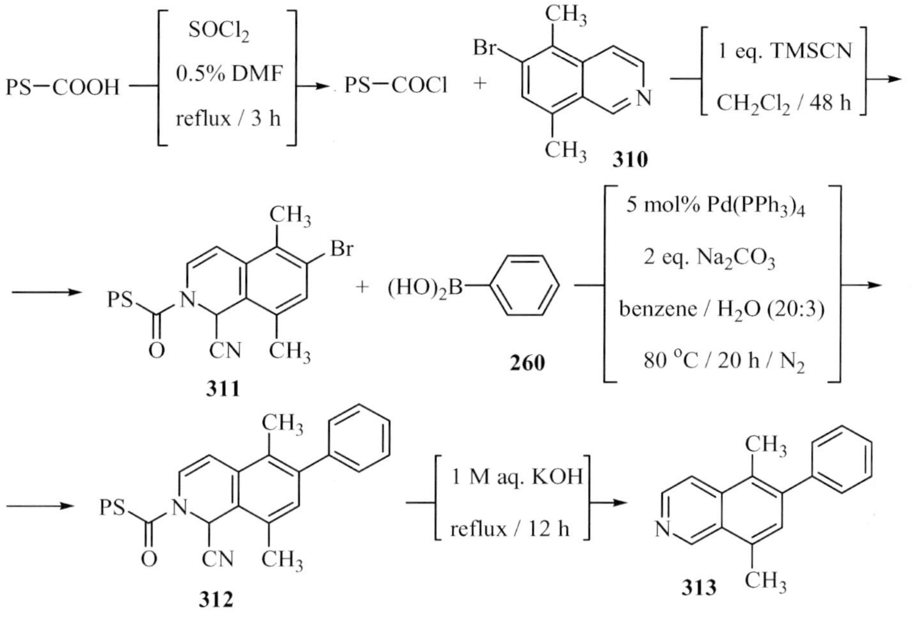

Scheme 39

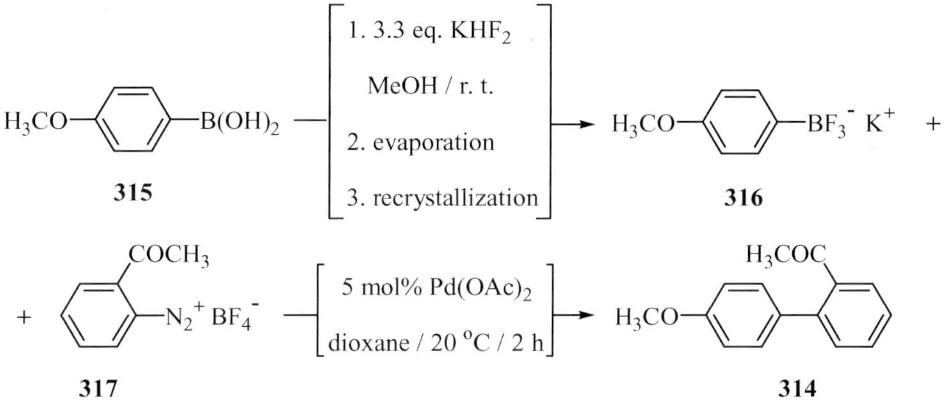

Scheme 40

The major drawback of this method, similarly to other SM reactions involving diazonium salts, is relatively high sensitiveness to the *ortho*-steric encumbrances. However, the yields in the reactions with sterically hindered substrates can be significantly improved by using $Pd_2[P(2\text{-tolyl})_3]_2(\mu\text{-OAc})_2$ as catalyst [98].

The preparation of arylboronic acids may sometimes have difficulties concerning either synthesis, or isolation procedure. To simplify the general strategy of SM synthesis of biaryls, Snieckus and his group [99-106] have published several excellent papers regarding *in situ* preparation of an arylboronic acid followed by cross-coupling reaction with aryl halides. Arenes bearing so-called directed metallation group (DMG): $CONR_2$ (R = Et, *i*-Pr) [99], $OCONR_2$ [100], OCH_2OCH_3 (MOM) [100], $NHCO_2t$-Bu [100], are *ortho*-metallated directly with the strong bases such as *n*-, *s*-, *t*-BuLi, or LDA to give the corresponding aryllithiums. Alternatively, aryl halides have been metallated with double amount of the strong base. Thus generated aryllithiums react with trimethyl-, tri-*n*-butyl- or tri-*i*-propyl borate to afford arylboronate esters. The latter are *in situ* hydrolyzed to obtain arylboronic acids, which are further used in the SM reactions without isolation. For example, *N,N'*-di-*i*-propylbenzamide (**318**), upon metallation with *s*-BuLi followed by reaction with trimethyl borate, and acid hydrolysis, furnished boronic acid **319**, which was then reacted by the original SM method with 2-bromothiazole (**320**) to produce **321** in 87% overall yield [99], respectively, Scheme 41.

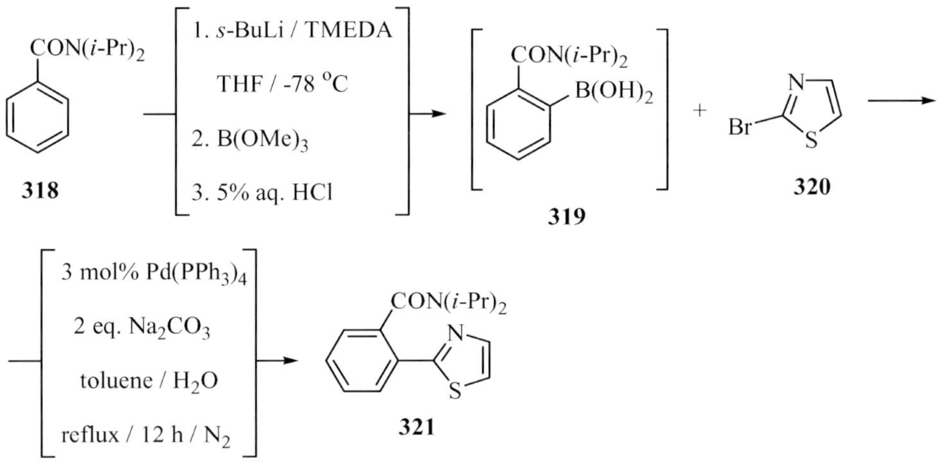

Scheme 41

The highly substituted biphenyls [99-106], phenylpyridines [99,100], phenylthiazoles [99], terphenyls [101], and some natural products [102] containing the biaryl structures have been successfully obtained using this strategy.

5.3. Modern Suzuki-Miyaura reaction catalysts

The most important factor playing a crucial role in the Suzuki-Miyaura reactions is the catalyst. The ligandless methods, involving Pd-C or various nano-sized palladium clusters formed from palladium(II) acetate in the presence of the bases such as $Ba(OH)_2$, K_2CO_3, or K_3PO_4 and reducing agent, despite their simplicity, in the absence of stabilizing agents, are very sensitive to oxygen, sometimes require high reaction temperatures, and are often unapplicable for several different substrates. The catalytically active zero-valent palladium or nickel-species have to be stabilized by adding an apropriate ligand to form palladium(0) or nickel(0)-complexes with enhanced activity and stability. Beside already described phosphine and imine ligands, several excellent papers concerning more active SM reaction catalysts have appeared in a last few years. They allow to affect the SM reactions at very low catalyst loading. The benefits of such approach are:

1. greatly decreased costs regarding much lower required quantities of (generally) expensive palladium chemicals,
2. avoiding the difficulties which arise from the catalyst, and catalyst-caused contaminations:
   - the level of residual heavy metals, what is one of the most important final analytical parameters in the pharmaceutical products,
   - the use of standard amount of SM catalysts, e.g. 2-3 mol%, led to contamination of the resulting crude product with approximately 10% (for example) triphenylphosphine,
   - the products may be significantly contaminated with the aryl-transfer side-products from triarylphosphines, whereas the use of greatly decreased amount of palladium-phosphine catalysts significantly avoids the formation of this compounds,
3. the use of third generation of the heterogeneous SM catalysts allows easily regeneration (by simple filtration), and reuse for several times with the same range of activity, thus avoiding all above mentioned problems with apparently great cost-benefits.

Herein, the second and the third generation SM catalysts are desribed. In the former group, all recently reported homogeneous SM catalysts with very high turnover number (TON) are put, while in the latter group are all readily regenerable and reusable heterogeneous catalysts, except Pd-C and Ni-C, which were described previously. Firstly, two important parameters regarding the catalyst activity and efficiency are defined. The *turnover number* (TON) shows the number of reactant moles converted to the desired product by one mole of the given catalyst. Since the arylboronic acids are usually employed in the SM reactions in a slight excess (10-50 mol%), the TON is expressed relative to aryl halides, which are actually the limiting reactants. The catalyst efficiency is shown by the *turnover frequency* (TOF, h^{-1}) value, which is defined as the number of reactant moles converted to the desired product per mole of the catalyst used per hour. For example, if the required amount of Pd(PPh$_3$)$_4$ in the standard optimized SM reaction is 1 mol% for 100% conversion of an aryl halide, then the TON is 100. Practically, with the use of Pd-C in only 0.0025 mol% (Pd to ArX), the TON can reach up to 36000 with the TOF up to 18000 h^{-1} [27].

Several authors have found that triphenylphosphine, in despite to its stabilizing ability for palladium(0) and nickel(0)-species, actually does deactivate the resulting complexes. However, palladium complexes with certain phosphines have shown very high activity, providing high TON's. Among these highly active phosphines and related phosphorus-based ligands are: various phosphite esters such as tri(2,4-di-*tert*-butylphenoxy)phosphite (**322**) [107], or triisopropylphosphite (P(O*i*-Pr)$_3$)[107], tetraphosphine-ligand tedicyp (**323**) [108,109], BuPAd$_2$ (Ad = 1-adamantyl, **324**) [110], certain Pd-1,6-diene complex **325** [111], 2-(di-*tert*-butylphosphino)biphenyl (**201**) [112,113], its cyclohexyl analogue **326** [112,113], or phenanthrenyl-9-Cy$_2$-phosphine **327** [114], diphospha-ferrocene **328** [115], and triferrocenyl phosphine **329** [116]. Some other phosphine and *N*-heterocyclic carbene ligands for palladium-catalysed SM reactions have also appeared, but with lower efficacy [117-119].

The corresponding palladium-complexes were *in situ* prepared by reaction with $Pd(OAc)_2$, or $Pd_2(dba)_3$, whereas the SM reactions were conducted in non-aqueous medium: refluxing toluene [107,110], xylene [108,109] or tetrahydrofuran [111] in the presence of K_2CO_3, CsF, K_3PO_4 or KF as the bases. On the other hand, at such low catalyst loading, the reactions have to be performed at elevated temperatures. This is connected with side-reactions arising from increased amounts of arylboronic acids, 1.5-2 eq. This is the serious disadvantage of all these reactions since the excess of the boronic acid gives mainly the homo-coupling product, causing difficulties during the purification procedure. These phosphines decrease the required amount of the palladium loading to a very low level, Table 9.

Table 9. The required amount of the palladium precatalysts with different phosphine ligands in the Suzuki-Miyaura reactions of various aryl halides with arylboronic acids

Ligand	Pd-salt (mol%)	TON	Lit. ref.
$P(Oi\text{-}Pr)_3$	0.0001	820 000	107
322	0.0001	up to 850 000	107
323	0.0001	21 - 6 800 000	108, 109
324	0.005	17400	110
325	0.05	1120-1940	111
326	0.001-0.05	up to 100 000 000	112, 113
328	0.005	up to 9 600 000	115

Although the complex of $Pd_2(dba)_3$ and ligand **329** has been employed at slightly higher loading, 0.036 mol%, it provided very efficient method for unactivated aryl chlorides, bearing electron-donating substituents [116]. The palladium complex of the ligand **327**, developed by Buchwald's group [114], is also somewhat outside of this discussion, however, it provides extremely efficient SM reactions with very bulky substrates at 1-5 mol% of palladium. Thus at the standard palladium-loadings (2-5 mol%), none of the standard SM methods will proceed with extraordinarily bulky substrates, at least not at this level of conversions. In contrast, the palladium-complex of ligand **327**, readily catalysed the reactions of both 9-chloro- and 9-bromoanthracene (**330**, **331**) with 2,6-dimethylphenyl boronic acid (**332**) to produce very bulky biaryl **333** in 80 and 82% yields [114], Scheme 42.

Several hindered *ortho*-di-, tri- and tetra-substituted biaryls were successfully obtained by this method in 58-98% yields [114]. Recently, the great breakthrough has been achieved since several very active palladacycle catalysts were developed [120-131]. The first report by Beller and coworkers [120] has shown very high activity of the palladacycle complex **89** (see Chapter 3) with TON's up to 74 000 in the SM reactions of aryl bromides and activated chlorides. Nájera's group [121,122] has developed several palladacycle catalysts derived from simple oximes, thus completely avoiding phosphine-based chemistry. Among them, complexes **334-336** were proved as effective catalysts for the SM reactions of aryl bromides and chlorides, even unactivated ones, providing excellent activity and efficiency at only 0.001-0.01 mol% loadings, accompanied with fair good yields at a reasonable excess (50 mol%) of the arylboronic acid. The TON's were in range 500-500 000, whereas the TOF values were within 100-198 000 h^{-1} [121,122].

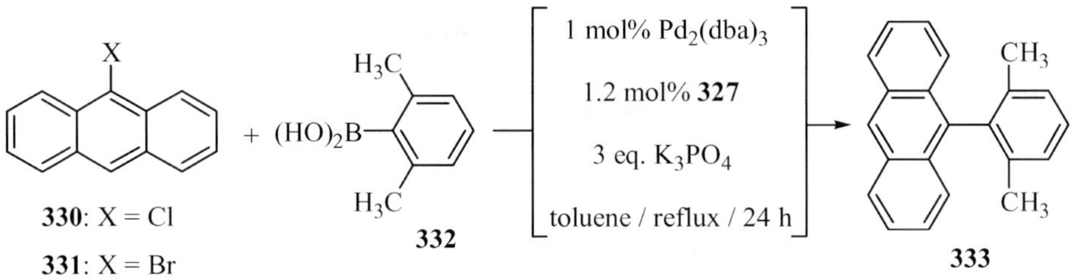

330: X = Cl
331: X = Br

332

333

Scheme 42

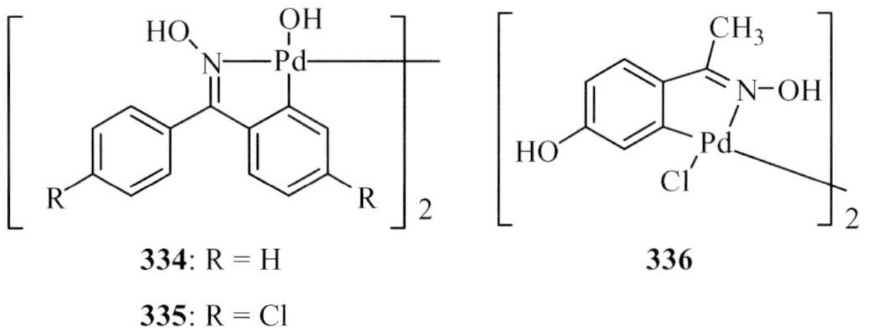

334: R = H
335: R = Cl

336

Several excellent papers reported by Bedford's group [123-130] dealing with very active palladium catalysts **337-340** show the great improvements in the SM reaction catalysis. The Bedford's catalysts are effective in the SM (and also Heck [123], and Stille [128]) reactions of aryl bromides (TON's up to 1 000 000 at 0.0001 mol%) [124,126], aryl chlorides (TON's up to 99 000 at 0.001 mol%) [125,127], both activated and unactivated, to afford the respective biaryls in good to high yields. For example, catalyst **339** was used in the SM reaction of (deactivated) 4-chloroanisole and phenylboronic acid to give 4-methoxybiphenyl with quantitative conversion at only 0.0015 mol% loading and TON of 33 000, respectively [127].

Among *N*-heterocyclic carbene oxime-palladacycles, complex **341** showed very high catalytic activity with aryl iodides, bromides and chlorides giving high TON's, 2000-92000, and TOF's, 200-4300 h^{-1} [131].

Still all these catalysts remain several difficulties connected with the homogeneous catalysis.

The third generation catalysts have to retain the high activity and efficiency, whereas their solid state allows easily separation from the reaction mixtures by simple filtration. Beside these properties, an ideal catalyst should be reused, at the same level of activity and efficiency, a number of times. The simplest catalysts with such properties are nano-sized palladium clusters. As already mentioned, they are formed by reduction of simple palladium salts, e.g. Pd(OAc)$_2$ or PdCl$_2$, in the presence of bases and arylboronic acid or any other reductant. Thus generated catalytically active nano-sized palladium clusters are unstable, and in the absence of stabilizer(s) are prone to rapid aggregation to higher aggregates well known as palladium black, accompanied with loss of catalytic activity. These heterogeneous catalytically active palladium-nanoparticles can be stabilized by:

- adding n-Bu$_4$NBr (TBAB) [129,132], tetra-n-octylammonium bromide [133], n-Bu$_4$NCl (TBAC) [134],
- using polar solvents and additives [134,135]: N,N-dimethylglycine (DMG), DMF, propylene carbonate, or acetone,
- using the block copolymers such as polystyrene-b-polyvinylpyridine [134,136],
- formation of palladium nano-sized particles on relatively stable copper clusters [133],
- employing well defined palladium hollow spheres [137], or by
- immobilization on the more suitable solid-supports (other than charcoal) [134,138].

The SM reaction performed in TBAB / water mixture is highly efficient ligandless method to afford biaryls in high yields with 0.1 (activated) - 1.0 (unactivated) mol% of palladium(II) acetate in the presence of potassium phosphate (or Cs$_2$CO$_3$) as the base [129]. However, this approach might be connected with difficulties during the work-up of reaction mixtures since the TBAB is a surfactant, and therefore complicates all extraction procedures. The use of polar solvents such as DMF or propylene carbonate is convenient way to stabilize the palladium nanoparticles as clearly shown by several successful "ligandless" SM methods described in previous discussion. Additionaly, the micelles of polystyrene-*b*-poly-4-vinylpyridine containing palladium clusters have been successfully used in analogous palladium-catalysed Heck-reactions [134-136]. The preparation of such stabilized palladium colloids generally involves the reduction of suitable palladium salt in the presence of colloide-stabilizer. This can be performed either by chemical, e.g. $LiBH(Et)_3$, or electrochemical reduction [135,136]. Copper forms quite stable nano-sized clusters obtained by reduction of copper(II) chloride with sodium borohydride in DMF as solvent and tetra-*n*-butylammonium bromide as stabilizer. Copper-clusters generated that way are quite active SM catalysts, however, exhibit a strong synergistic effect to palladium-catalysed process. In this manner, by simultaneous reduction of $CuCl_2$ and $PdCl_2$ under the same reaction conditions, the bimetallic Pd-Cu-clusters are formed showing a significant catalytic activity in the SM reactions with aryl iodides and bromides [133].

An interesting approach has been introduced with uniform palladium hollow spheres, produced from the palladium complexes immobilized on mercaptopropyl-derived silica spheres. The silica template is then removed by treating with HF to produce hollow palladium spheres. Thus obtained palladium has shown very high efficiency with no loss of activity after 7 reuses [137].

Finally, the palladium nano-particles can be stabilized by using more effective solid-supports. Shimizu reported that Pd^{2+}-sepiolite, prepared by simple ion-exchange procedure with palladium salts followed by drying, does catalyse the SM reactions with high activity and efficiency [138]. The sepiolite is a fibrous natural clay mineral, talc-type structured, with ribbons arranged in such way that tetrahedral sheet is continuous, but inverts in apical directions in adjacent ribbons. This generates the uniform size parallel-piped intracrystalline tunnels along the fibre (size 10.8 x 4.0 Å).

The second strategy includes the polymer bonded phosphine ligands [139-141]. A convenient polymeric phosphine, $PS-CH_2PPh_2$ (PS = polystyrene), was described by Miyaura [139]. This ligand has been produced by the chloromethylation of 1% cross-linked polystyrene with chloromethyl methylether in the presence of tin(IV) chloride. Thus obtained chloromethylated polystyrene was reacted with $LiPPh_2$ to give $PS-CH_2PPh_2$. The reaction of $Pd(COD)Cl_2$ and $PS-CH_2PPh_3$ proceeds smoothly to produce quantitatively polymer-supported palladium catalyst as an air-stable, bright yellow resin **342** [139], Scheme 43. The catalyst **342** (with 2.32 mmol/g of P and 0.256 mmol/g of Pd) has been employed in the SM reactions of activated aryl chlorides. For example, 2-chlorobenzonitrile (**114**) was reacted with 4-tolylboronic acid (**271**) to give **112** in 91% isolated yield [139], respectively, Scheme 44.

However, the reactions with more electron-rich chloroarenes were unsuccessful. The catalyst has been easily regenerated by simple filtration and reused several times with no significant loss of activity. Miyaura's polymer-supported palladium catalyst **342**, as diaryl-alkylphosphine, is less reactive than the corresponding aryl-dialkylphosphines.

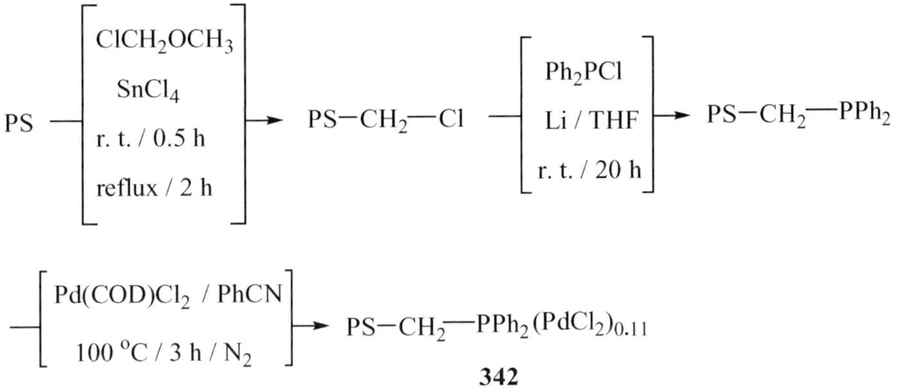

Scheme 43

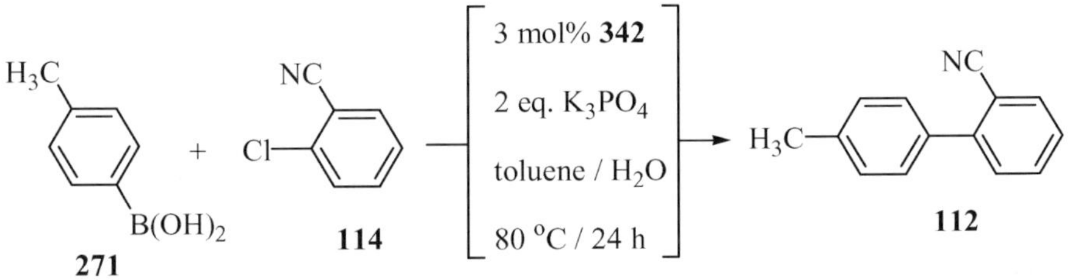

Scheme 44

Further improvements in the latter class of ligands have been introduced by appropriate immobilization of well established aryl-dialkylphosphines on the suitable polymers. Buchwald and coworkers [140] have developed ligand **326** supported on the Merrifield's peptide resin **343**. This forms the highly active palladium complex with Pd(OAc)$_2$ or Pd$_2$(dba)$_3$, wide applicable in several palladium-catalysed reactions including the SM reactions with unreactive aryl iodides, bromides and chlorides. Thus sterically demanded and deactivated bromo-*p*-xylene (**344**) and 2-tolylboronic acid (**286**) were reacted at 0.5 mol% of palladium-loading to produce 2,5,2'-trimethyl biphenyl (**345**) in essentialy quantitative yield [140], respectively, Scheme 45.

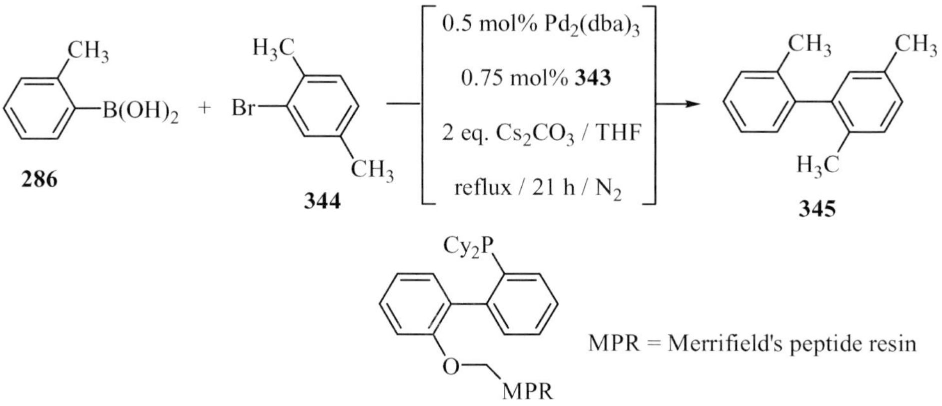

Scheme 45

Further efficient polymeric palladium complexes containing Pd(OAc)$_2$, Pd(dba), Pd(MeCN)Cl$_2$, Pd(PhCN)Cl$_2$ and (π-allyl)PdCl anchored onto the polyethylene-bounded triphenylphosphine are called FibreCat^{TM} catalysts [141]:

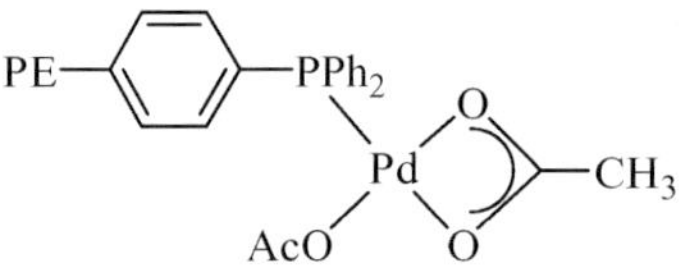

FC-Pd(OAc) 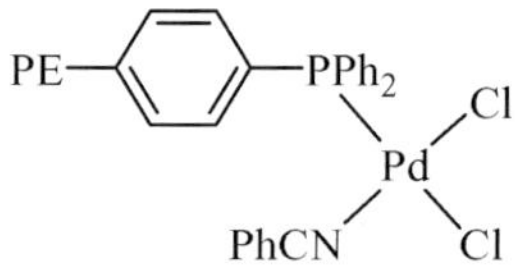   **FC-Pd(PhCN)** 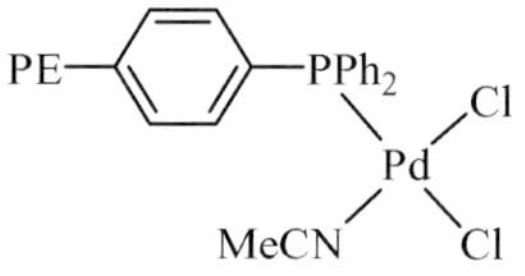   **FC-Pd(MeCN)**

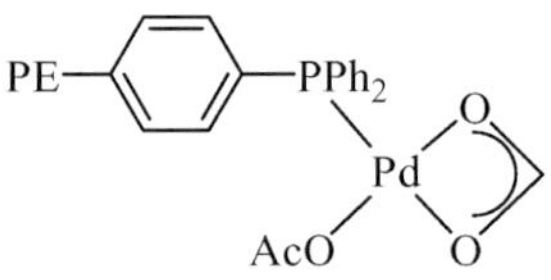

FC-Pd(π-allyl) 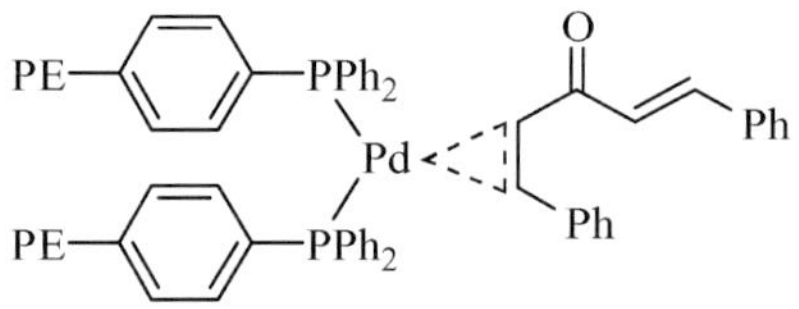   **FC-Pd(dba)**   PE = polyethylene

For instance, the **FC-Pd(OAc)** at 1 mol% palladium-loading catalysed the reaction of phenylboronic acid (**260**) and 2-bromotoluene (**346**) in ethanol / water mixture to give 2-methylbiphenyl (**347**) with quantitative conversion [141], Scheme 46.

However, the leaching of palladium, at least partially, occurs in some alternative reaction solvents, e.g. MIBK, and the catalyst turned black when the reaction was performed at higher temperatures and could not be reused.

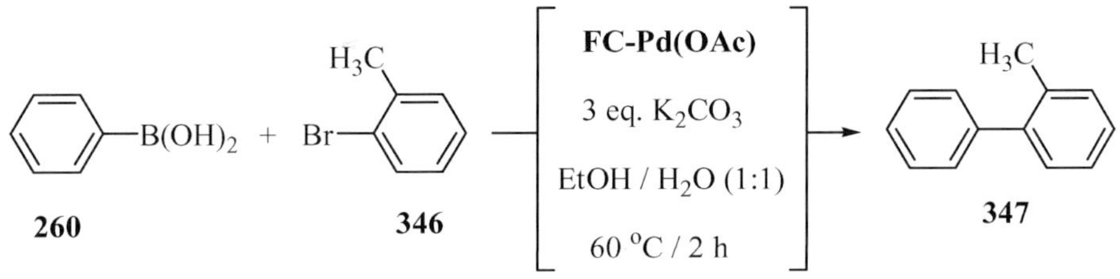

Scheme 46

The palladium-leaching has occured in reactions with aryl halides containing a ligand-capable heteroatoms, e.g. halopyridines. This phenomena is one of the most limiting factors of all modern heterogeneous catalysts' use since the product, contamined with the metal-residuals, can not be used for the certain purposes, e.g. active pharmaceutical ingredients (APIs). Darkening of the heterogeneous palladium-catalysts at elevated temperature reactions or after a few reuses indicates the slow decomposition of the original polymeric complex to the agglomerated (unactive) palladium-black. Since all SM reactions at low palladium-loadings require elevated temperatures and the high catalytic activity, stability to elevated temperatures even for several reuses as well as stability to leaching are the most important characteristics of the modern heterogeneous catalysts. Although the use of the latter catalysts avoids several difficulties in the SM, Heck and similar palladium-catalysed reactions, their further developments are still great multidisciplinary challenge [134,142].

5.4. Synthesis of arylboronic acids

Arylboronic acids, Ar-B(OH)$_2$, and esters, Ar-B(OR)$_2$, are the most important reactants in the Suzuki-Miyaura cross-coupling reactions. Although the carbon atom bearing the boron exhibits the carbanionic character, the arylboronic acids and esters are relatively air- and water-stable organometalloids. The latter are not common nucleophilic reagents as for instance, the Grignard reagents, even so, under the SM reaction conditions they are efficient formal aryl-carbanion donors. This enhanced stability allows to effect the SM reactions in the presence of a wide variety of reactive functional groups and conditions, including an aqueous, which is impossible for the most other organometallics. The classical approach to the arylboronic acid is the reaction of the aryl Grignard reagents or aryllithiums with trialkyl borates to afford arylboronic esters, which, upon mild acid hydrolysis, furnished the arylboronic acids [143,144], Scheme 47.
Organolithiums have rather limited applications due to their high reactivity with a number of functional groups, e.g. ester, nitrile, amide, etc.

$$Ar-M \quad + \quad B(OR)_3 \longrightarrow Ar-B(OR)_2 \quad + \quad MOR$$

$$\xrightarrow{\quad H_3O^+ \quad} Ar-B(OH)_2 \qquad M = Li, MgX \ (X = I, Br, Cl)$$

Scheme 47

Even though the use of less reactive Grignard reagents is the most popular, they are still no compatible with several common functional groups. The esters of lower alcohols: $B(OMe)_3$, $B(OEt)_3$, $B(Oi\text{-}Pr)_3$ and $B(On\text{-}Bu)_3$ serve as the boric acid reagents, and can be most easily prepared in fair yields by heating the mixture of boric acid or its anhydride, B_2O_3, and the desired alcohol in the presence of toluene or other suitable cosolvent which removes the water liberated in the reaction azeotropically. Adding the appropriate Grignard reagent to the solution of trimethyl borate in tetrahydrofuran furnished a dimethyl arylboronate at temperatures below approximately -50 °C. At higher temperatures, or by opposite mixing of reactants, the corresponding methyl diarylboronate, $Ar_2B(OCH_3)$, or even triarylboron, Ar_3B, are formed as the main products [145], Scheme 48.

$$Ar-B(OCH_3)_2 \quad + \quad M-Ar \longrightarrow Ar_2B(OCH_3) \quad + \quad MOCH_3$$

$$Ar_2B(OCH_3) \quad + \quad M-Ar \longrightarrow Ar_3B \quad + \quad MOCH_3$$

$$M = Li, MgX \ (X = I, Br, Cl)$$

Scheme 48

These side-reactions are avoided in the $B(OCH_3)_3$-based protocols with the use of very low reaction temperatures, e.g. -78 °C, by the use of an excess of $B(OCH_3)_3$ and by the order of mixing the reactants: the organometallic into the $B(OCH_3)_3$ and not *vice versa*. The latter, rather cryogenic technic, is common in the arylboronic acid and esters synthesis. As the reaction medium, all usual Grignard-performing dry solvents: THF, Et_2O, DME, various glymes, etc., have been employed. For instance, the sterically hindered mesitylboronic acid (**283**) was prepared by reaction of the Grignard reagent **125**, derived from mesityl bromide (**188**), and trimethyl borate followed by mild hydrolysis of resulting boronate ester **348** in 72% overall yield [143], respectively, Scheme 49.

The use of triisopropyl borate is the most popular nowadays since it generally allows to carry out the reactions at common reaction temperatures, between -25 and 0 °C. The acid catalysed hydrolysis of the arylboronic esters is affected by adding dilute aqueous hydrochloric, acetic acid or even water, at 0 °C to room temperature, to the reaction mixture containing the crude arylboronic ester, to produce the corresponding arylboronic acid. Apart from the standard procedure, the arylboronic esters can be obtained by direct Barbier-type reaction between an aryl halide, trialkyl borate and magnesium turnings [146].

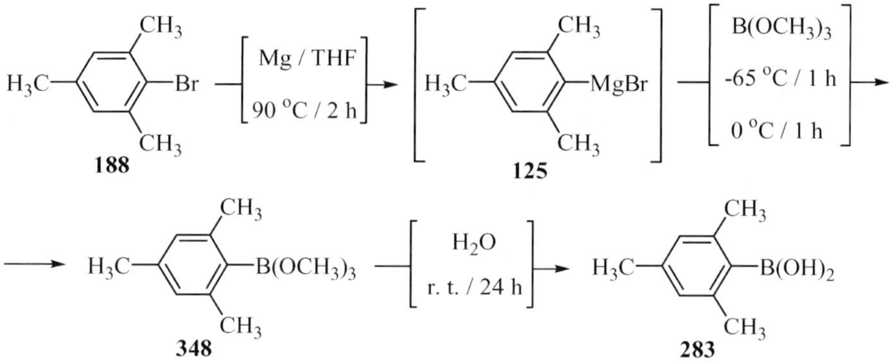

Scheme 49

Despite the fact that the arylboronic acids are quite weak acids, e.g. pK_a is around 12 in aqueous ethanol, in the reactions with sodium hydroxide solution they form the respective boronate salts, and the organic material could be extracted with dichloromethane from such solutions. Additionaly, the arylboronic acids under the influence of heating, long-term desiccation over drying agents or in the acidic media, undergo dehydration to form anhydrides such as **XXVI** and **XXVII**. The anhydrides can be converted back to the acids by boiling in water. Since the readily dehydration to the anhydrides, the acids often do not have simple melting point (m.p.), but rather two or more. The lowest m.p. corresponds to the primary dehydration point when the anhydride is formed, whereas the second one is acctualy its m.p. In certain more complicated instances, further m.p.'s are corresponded to the formation of different types of anhydrides. Among alternative boronates, cyclic 2-aryl[1,3,2]dioxaboronates (**XXVIII**) have been used as more suitable derivatives for high-vacuum distillation.

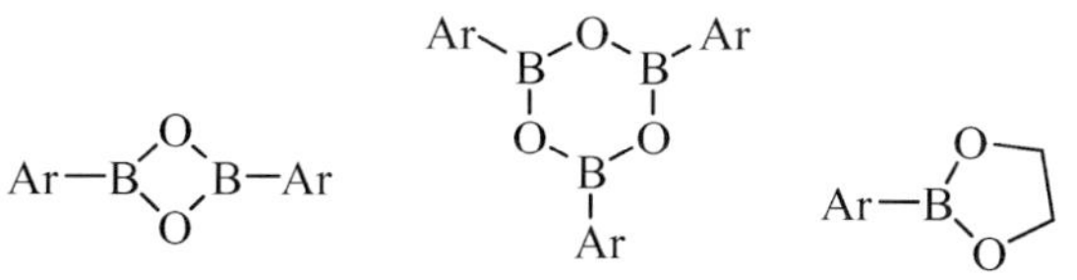

XXVI   **XXVII**   **XXVIII**

The latter have been prepared in high yields by the transesterification of the crude dialkyl arylboronates in refluxing two-phase ethyleneglycol / toluene mixture [147].

Except the metallation of aryl halides with the lithium metal and the Grignard reaction, the arylmetallics can be alternatively prepared by:

- reaction of aryl halides with the strong bases, e.g. *n*-BuLi [149,150], or
- metallation (deprotonation) reaction of relatively acidic arenes with the strong bases, e.g. LDA [151,152], or LTMP [153].

Thus formed, organometallics are trapped with trialkyl borates to produce whether the desired arylboronic [149-151,153] or diarylboronic esters [152], depending on the reaction conditions discussed above. The diarylboronic acids have also been successfully used in the $Pd(PPh_3)_4$-catalysed SM reactions with aryl triflates, under the standard conditions [152]. Beside trialkyl borates, *N,N*-dimethylaminoboron bromide has been trapped with organolithiums to give an unstable bis-aminoboronate **349**, which, upon esterification with pinacol, gave the 1,4-phenylene-bis-pinacolato boronate (**266**) in 21% overall yield [154], Scheme 50.

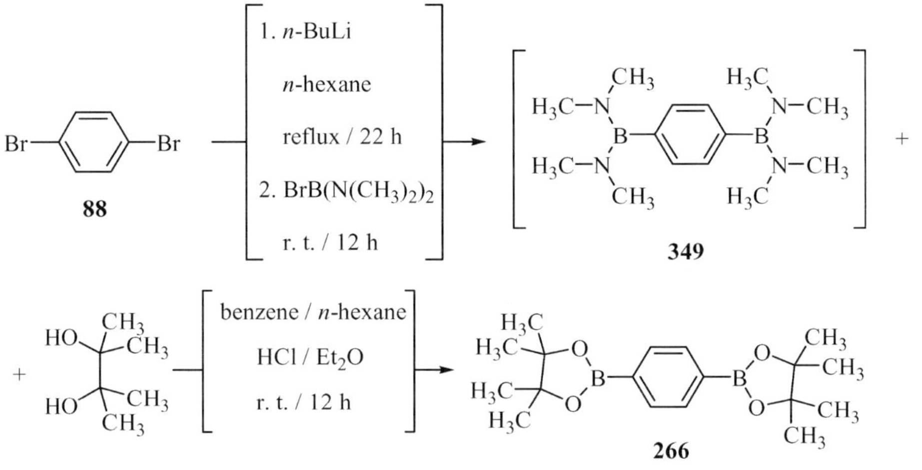

Scheme 50

The arylboronic esters and acids bearing a number of common functional groups, e.g. CHO, COOR, CN, etc., can not be obtained by any presented method since the highly reactive organometallics react with these functionalities. This is very serious limit of the SM reaction, because one of the aryl-counterparts is originated from the arylboronic acid or ester. However, the crucial break-through has been achieved when the direct syntheses of arylboronic esters, avoiding the organometallics, were developed by the following alternatives:

- cross-coupling reactions of aryl halides or triflates with the pinacol ester of diboron (**350**) [155-160],

- cross-coupling reactions of aryl halides or triflates with dialkoxyborane (HB(OR)$_2$) [161-165], or by
- borylation of arenes with the pinacolborane (**351**) [166-170].

The palladium-catalysed cross-coupling reaction of aryl halide with the pinacol ester of diboron (**350**) has been developed by the Miyaura's group [155]. Aryl iodides and bromides [155] in DMSO, DMF [157], or DME [160], triflates [156] in dioxane or DMF [157], or chlorides in THF [159] react with **350** in the presence of palladium catalyst and potassium acetate as the base to afford the corresponding arylboronic-pinacol esters in good to high yields. As catalyst, Pd(dppf)Cl$_2$ has been used in the reactions with aryl iodides, bromides and triflates, whereas Pd(dba)$_2$ / PCy$_3$ and Pd(OAc)$_2$ / imidazolium ligand **352** [159] have been employed in the reactions with less reactive aryl chlorides. Potassium acetate, as a weak base, was proved to be the most suitable since the stronger bases, e.g. K$_2$CO$_3$, K$_3$PO$_4$, or NaOEt, caused the formation of the respective symmetrical biaryls in considerable amount by the SM reaction. However, the strong evidences show that the acetate anion is not coordinated to the boron atom, thus avoiding the SM reaction pathway [155]. For instance, the reaction of methyl 4-bromobenzoate (**264**) and **350** afforded pinacolboronate **353** in 86% yield [155], respectively, Scheme 51.

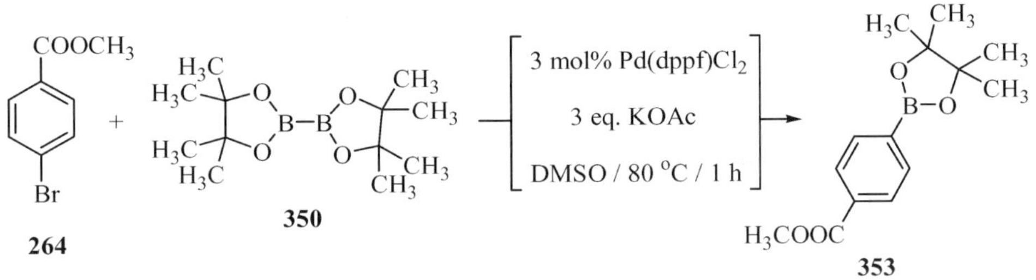

Scheme 51

The reactions involving aryl triflates have been equally effective, yet, as the reaction solvent dioxane must be used to avoid the difficulties associated with the competitive saponification of triflate, or the catalyst decomposition (DMSO is relatively strong oxidant). The aryl chlorides have been converted to the pinacolboronates by using Pd(OAc)$_2$ / imidazolinium chloride **352** in fair yields. For example, 6-chloroquinoline (**354**) was reacted with **350** to give the boronate **355** in 53% yield [159], Scheme 52.

Palladium-catalysed reaction of aryl halides or triflates with dialkoxyborane, HB(OR)$_2$, developed by Masuda's group [161,162], is based on the use of metalloid hydrides as metallating reagents to furnish the arylboronate ester, ArB(OR)$_2$, Scheme 53. The pinacolborane (**351**) [143-145], as the most common reagent, can be easily prepared by addition of BH$_3$·SMe$_2$ to a dichloromethane solution of pinacol at 0 °C [164]. Furthermore, catechol borane [162], and cedranediol borane [165] have also been employed. For instance, 2-iodothiophene (**97**) was converted to the boronate ester **356** in 80% yield [161], respectively, Scheme 54.

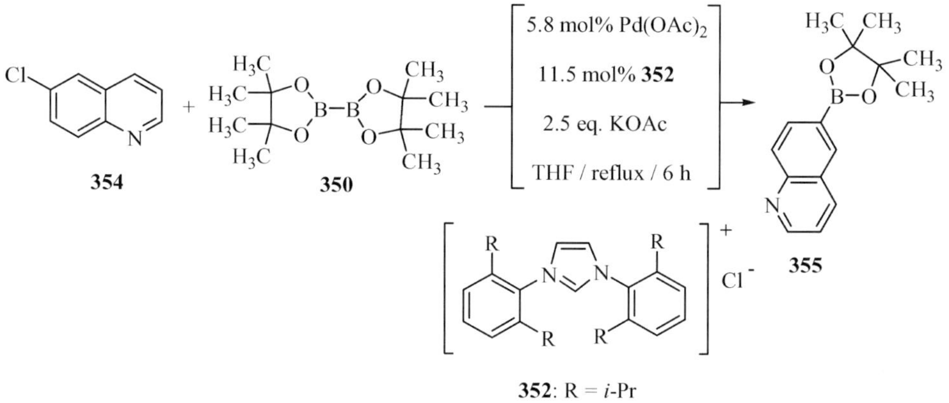

Scheme 52

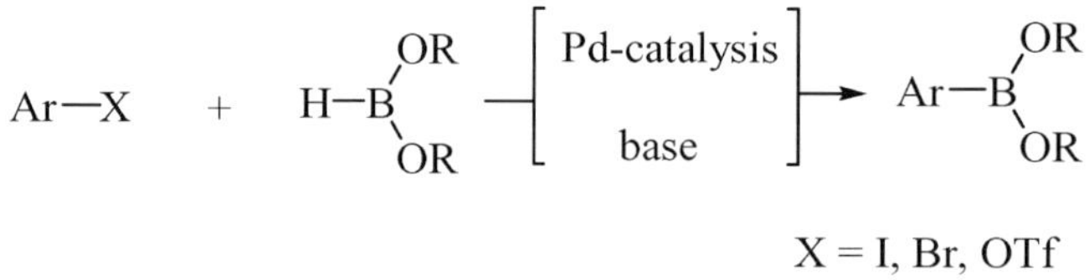

X = I, Br, OTf

Scheme 53

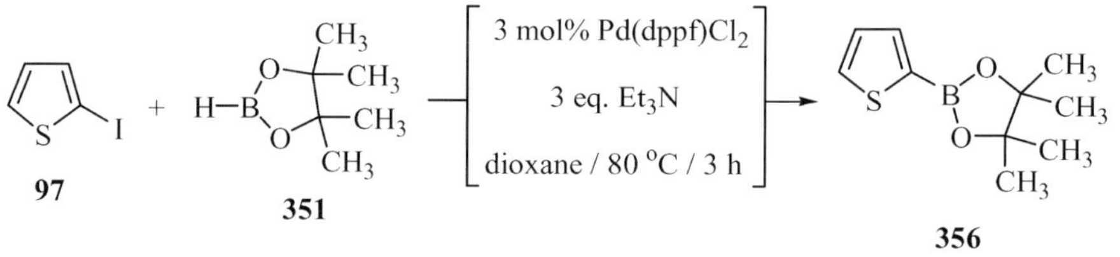

Scheme 54

The reaction is catalysed by the palladium phosphine complexes such as Pd(dppf)Cl$_2$ [161,162], Pd(OAc)$_2$ / **326** [163], or more inexpensive Pd(PPh$_3$)$_4$ [161,162] as slightly less efficient substitute in the presence of triethylamine as the most suitable base in dioxane [161-165]. The Pd(PPh$_3$)$_4$-catalysed reaction has been co-catalysed with copper(I) iodide [165] and it tolerates a wide variety of functional groups, even though the electron-withdrawing substituents significantly decrease the reaction rate. The main side-reaction is the reductive dehalogenation which may destroy up to approximately 30% of the starting aryl halide [161,163]. The Masuda reaction, as well as complementary Miyaura reaction with pinacol ester of diboron (**350**) are extremely valuable approaches to the arylboronic esters and acids with a great number of reactive functional groups. The third approach to the arylboronic esters and acids is direct borylation of arenes with dialkoxyboranes, $HB(OR)_2$ [166-170], under the influence of the $Cp*Rh(\eta^4-C_6Me_6)$ [167,168], $Cp*M^I(PMe_3)$ or $Cp*Ir(PMe_3)(H)(Cy)$ where M = Ir, Rh, $Cp* = \eta^5-C_5Me_5$, and $Cy = c-C_6H_{11}$ [166,169], $RhCl[Pi-Pr_3]_2(N_2)$ [170], $(Cp*RhCl_2)_2$ and the most conveniently $[IrCl(COD)]_2$ / bpy [150] as the catalysts. The pinacol borane (**351**), or pinacol ester of diboron (**350**) have been employed as the borylating reagents. The reactions involving electron-rich substrates such as anisole or N,N-dialkylanilines, as well as electron-poor arenes, e.g. ethyl benzoate or N,N-dialkyl benzamides, with **351** proceed smoothly at elevated temperatures to afford arylboronate esters (**XXIX**) in good yields [167]. The unsymmetrically substituted benzenes have given the mixture of all three possible isomers, but the *meta*-isomer is predominant, Scheme 55.

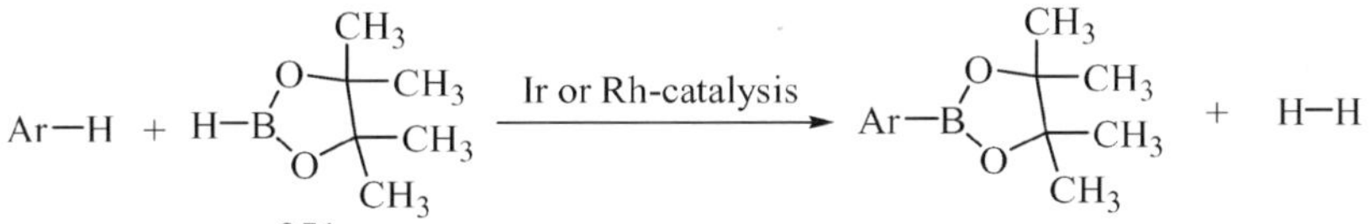

Scheme 55

The reactions have been conducted in an excess of the arene to be borylated as the reaction medium [169,170], or in cyclohexane as an inert solvent [168]. For example, Miyaura's group [170] described a rather convenient method for borylation of simple arenes with **350** under mild reaction conditions, Table 10.

Table 10. Iridium-catalysed borylation of arenes with pinacol ester of diboron (**350**) [170]

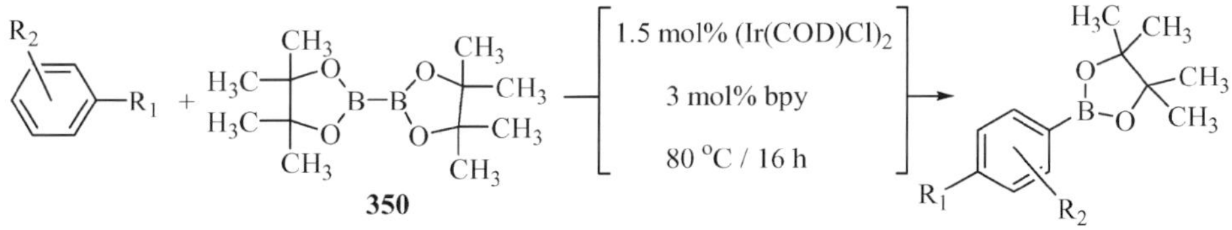

R_1, R_2	Yield, % ($o : m : p$)
H, H	95
H, OCH_3	95 (*1:74:25*)
1,2-di-OCH_3	86
1,3-Br, OCH_3	73
H, CF_3	80 (*0:70:30*)

5.4.1. Other properties of arylboronic acids

The arylboronic acids, as relatively stable organometalloids, can be homo-coupled to symmetrical biaryls [171]. The practical consideration of this interesting reaction for the synthesis of symmetrical biaryls is described in the Chapter 7. Aqueous hydrogen peroxide in the presence of copper(II) acetate and triethylamine as the base converts the arylboronic acids to phenols or symmetrical diarylethers [172]. Moreover, the arylboronic acids react smoothly with phenols, anilines or N-monosubstituted amides in the presence of Cu(OAc)$_2$ and Et$_3$N to give the respective diarylethers [173], diarylamines [174], or N-arylamides [174]. Apparently all these reactions proceed via free-radical mechanism. Additionaly, manganese(III) acetate, as one-electron oxidant, reacts with arylboronic acids to generate the free-aryl radicals which undergo homolytic arylation of a liquid arene, analogously to the Gomberg-Bachmann-Hey reaction, see Chapter 7. The arylboronic acids are relatively stable to dilute acids at room temperature. However, at elevated temperatures the hydrolytic deboronation become very fast process to produce the parent arene, analogously as the quenching of the Grignard reagents with water (or H$_3$O$^+$). Sterically unhindered arylboronic acids can survive the boiling in dilute hydrochloric acid for several hours, while the sterically demanding analogues rapidly decompose within a few minutes. Alkyl-chain oxidations to aldehyde [175] or carboxylic acid [176] are further useful reactions which are compatible with the unprotected Ar-B(OH)$_2$ function. For instance, the carefully controlled oxidation of 2-tolylboronic acid (**286**) with basic aqueous potassium permanganate solution afforded, otherwise difficultly accessible compound, 2-carboxyphenylboronic acid (**357**) in 76% yield [176], respectively, Scheme 56.

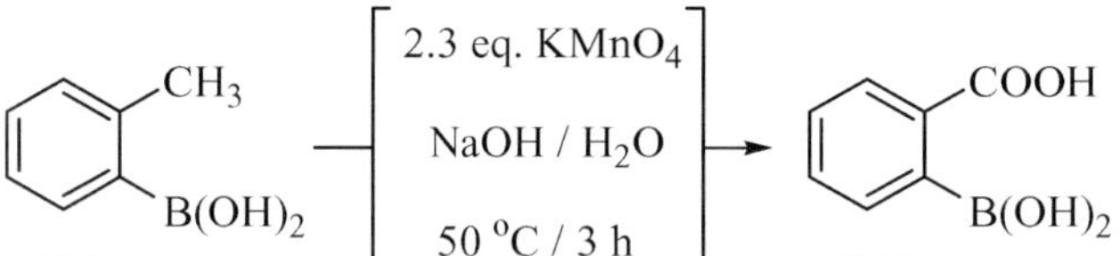

286   **357**

Scheme 56

The formation of the diphenic acid at higher reaction temperatures strongly indicates the generation of the free aryl-radicals under the influence of the strong oxidant.

An extensive review of arylboronic acid chemistry in older literature was written by Lappert [177], while an excellent and comprehensive review of heterocyclic boronic acids chemistry was summarized by Tyrrell and Brookes [178]. Since the arylboronic acids and esters became an important chemical intermediates, a number of these substances can be purchased. However, they are still quite expensive materials.

5.5. Selected synthetic procedures

5.5.1. The original Suzuki-Miyaura procedure: Preparation of 4-chlorobiphenyl (42) [1]

| 260 | 178 | 42 |

A 50 ml-flask was charged with Pd(PPh$_3$)$_4$ (0.35 g, 0.3 mmol, 3 mol%), benzene (20 ml), 4-chloro-bromobenzene (**178**, 1.91 g, 10 mmol), and aqueous solution of Na$_2$CO$_3$ (2.12 g, 20 mmol in 10 ml of water) under a nitrogen atmosphere, and then phenylboronic acid (**260**, 1.34 g, 11 mmol, 1.1 eq.) in ethanol (96%, 5 ml) was added. The mixture was refluxed for 6 h under vigorous stirring. After the reaction was complete, the residual phenylboronic acid was oxidized by 30%-H$_2$O$_2$ (0.5 ml) at room temperature for 1 h. The product was extracted with diethyl ether, washed with saturated aqueous sodium chloride, and dried (Na$_2$SO$_4$). The residue, after evaporation of the solvent, was distilled to afford 1.40 g (74%) 4-chlorobiphenyl (**42**), b.p. 156 °C / 15 mmHg, m.p. 77 °C.

Note: Instead benzene, toluene or xylene can be used as reaction solvent.

5.5.2. The ligandless palladium-catalysed SM procedure: Preparation of 3,4-methylenedioxy-2',3',4'-trimethoxybiphenyl (282) [10]

| 280 | 281 | 282 |

A suspension of 3,4-methylenedioxy-bromobenzene (**281**, 110 mg, 0.55 mmol), Ba(OH)$_2$·8H$_2$O (240 mg, 0.75 mmol, 1.4 eq.) in 95%-aqueous ethanol (6 ml) was degassed with nitrogen. Then, 2,3,4-trimethoxyphenylboronic acid (**280**, 110 mg, 0.5 mmol) was added, followed by Pd(OAc)$_2$ (6 mg, 0.025 mmol, 5 mol%) under N$_2$. The reaction mixture was stirred at room temperature for 2 h under N$_2$. Water (10 ml) was added and the mixture was extracted with dichloromethane (3x10 ml). Combined organic phases were dried (Na$_2$SO$_4$), filtered, and evaporated to dryness. The 138 mg (96%) of pure product **282** was isolated by preparative chromatography.

Note: The ligandless Pd(OAc)$_2$ or Pd(SEt)$_2$Cl$_2$-catalysed SM reactions can be accomplished in DMF (r. t. to 130 °C) in the presence of K$_3$PO$_4$ as the base and *n*-Bu₄NBr as palladium-black stabilizer [29] or in THF (acetone) / water (1:1) mixtures with K₂CO₃ [28]. The absolute absence of atmospheric oxygen is usually essential for successful SM reaction.

5.5.3. The SM reaction of sterically hindered arylboronic acids: Preparation of 2-(2'-methoxy-4',6'-di-tert-butylphenyl)-3-methylpyridine (236) [15]

A mixture of 2-bromo-3-methylpyridine (**235**, 170 mg, 1 mmol) and Pd(PPh₃)₄ (58 mg, 0.05 mmol, 5 mol%) in DME (3 ml) was degassed for three times by freeze-pump-thaw method and then gently heated and stirred in a nitrogen atmosphere until the suspension disappeared. Afterward, 2,4-di-*tert*-butyl-6-methoxyphenylboronic acid (**234**, 265 mg, 1 mmol) and potassium *tert*-butoxide (225 mg, 2 mmol, 2 eq.) in *tert*-butyl alcohol (0.5 ml) were added successively. The mixture was degassed for another three times and refluxed at 95 °C for 15 min under nitrogen. Then, the mixture was cooled to room temperature, filtered through Celite and then the Celite was washed with dichloromethane (20 ml). The filtrate was evaporated and the product, $R_f = 0.40$, was purified by preparative chromatography over silica gel with *n*-hexane / ethyl acetate (6:1) as an eluent to obtain 200 mg (64%) of pure biaryl **236**, m.p. 114-115 °C.

5.5.4. Thallium hydroxide-mediated room temperature SM reactions with sterically hindered substrates: Preparation of 2,4,6-trimethyl-2'-nitrobiphenyl (358) [16]

To a solution of Pd(PPh₃)₄ (23 mg, 0.02 mmol, 2 mol%) and mesitylboronic acid (**283**, 180 mg, 1.1 mmol) in DMAc (6 ml), an aqueous solution of TlOH (350 mg, 1.5 eq. in 3.5 ml of H₂O), and 2-nitro-bromobenzene (**94**, 202 mg, 1 mmol) were added under nitrogen. The heterogeneous reaction mixture was stirred for 12 h at room temperature.

The product was extracted with dichloromethane, washed with brine, dried over MgSO$_4$, filtered and filtrate evaporated. From the crude product, 200 mg (83%) of pure 2,4,6-trimethyl-2'-nitrobiphenyl **(358)** was isolated by preparative chromatography: yellowish crystals, m.p. 47 °C.

5.5.5. Silver carbonate as the base in the SM reaction with arylboronic ester: Preparation of p-terphenyl (267) [18]

In a flask equipped with a Dean-Stark apparatus, a stirred suspension of phenyl-1,4-diboronic acid bispinacol ester (**266**, 100 mg, 0.3 mmol), iodobenzene (104 mg, 0.51 mmol), silver carbonate (167 mg, 0.6 mmol) and Pd(PPh$_3$)$_4$ (9 mg, 0.025 mmol, 5 mol%) in anhydrous THF (20 ml) was refluxed in the dark for 16 h under nitrogen. The suspension was then filtered through Celite. The solvent was evaporated under reduced pressure. The residue was dissolved in dichloromethane (30 ml) and successively washed with 2 M aqueous hydrochloric acid (10 ml) and water (10 ml). The aqueous layers were extracted with dichloromethane (3x20 ml). The combined organic extracts were washed with aq. NaHCO$_3$ (40 ml) and brine (40 ml), dried (MgSO$_4$) and concentrated *in vacuo*. From the crude product, 58 mg (100%) of pure *p*-terphenyl (**267**), m.p. 212-214 °C, was isolated by preparative chromatography on silica gel column with petroleum ether / diethyl ether (95:5) as an eluent.

5.5.6. The SM reaction of sterically hindered arylboronic ester in the presence of sodium phenoxide as the base: Preparation of 5-methoxycarbonyl-2-methoxy-2',4',6'-trimethylbiphenyl (359) [19]

In a flask equipped with a Dean-Stark apparatus and a condenser, mesitylboronic acid-2,2-dimethyl-1,3-propanediol ester (**360**, 87 mg, 0.37 mmol, 1.1 eq.), methyl 3-iodo-4-methoxybenzoate (**361**, 100 mg, 0.34 mmol), Pd(PPh$_3$)$_4$ (26 mg, 0.022 mmol, 6 mol%), and sodium phenoxide (44 mg, 0.37 mmol, 1.1 eq.) were suspended in benzene (20 ml). The reaction mixture was stirred at reflux under nitrogen for 118 h. Then the mixture was cooled, filtered through Celite and 2 M aqueous solution of hydrochloric acid (20 ml) was added to the filtrate with stirring. The organic layer was washed with water, and the aqueous layers were extracted with dichloromethane (3x20 ml). The combined organic layers were washed with brine, dried over MgSO$_4$ and concentrated *in vacuo*. From the crude product, 77 mg (80%) of pure 5-methoxy carbonyl-2-methoxy-2',4',6'-trimethylbiphenyl (**359**) was obtained, m.p. 156-159 °C.

5.5.7. Fluoride-mediated SM reaction: Preparation of methyl biphenyl-4-acetate (265) [20]

To a stirred mixture of phenylboronic acid (**260**, 0.30 g, 2.5 mmol, 1.1 eq.), methyl 4-bromophenylacetate (**264**, 0.52 g, 2.25 mmol), and powdered cesium fluoride (0.76 g, 5.0 mmol, 2 eq.) in DME (8 ml) was added Pd(PPh$_3$)$_4$ (87 mg, 3 mol%). The reaction mixture was flushed with argon and maintained refluxed under argon for 24 h. The mixture was then cooled and diluted with ethyl acetate (20 ml) and water (20 ml), and the ethyl acetate layer was dried (Na$_2$SO$_4$), and concentrated. From the crude product, 0.50 g (98%) of pure methyl biphenyl-4-acetate (**265**), m.p. 19-21 °C, was isolated by preparative chromatography on silica gel using *n*-hexane / ethyl acetate (5:1) as an eluent.

Note: Instead of expensive CsF, cheap KF, Et$_4$NF or *n*-Bu$_4$NF have been employed, whereas the DME as reaction medium can be replaced with MeOH / DME (1:2), H$_2$O / DME (1:2), H$_2$O / toluene (1:1) or MeCN.

5.5.8. Heterogeneus Pd-C catalysed SM reaction of aryl iodides in water: Preparation of 3-phenylphenol (278) [26]

A mixture of 10% Pd-C (3 mg), potassium carbonate (415 mg, 3 mmol), phenylboronic acid (**260**, 122 mg, 1 mmol) and 3-iodophenol (**275**, 220 mg, 1 mmol) in water (10 ml, flushed with nitrogen) was vigorously stirred at room temperature for 12 h under nitrogen. The reaction mixture was quenched with 1.5 M aqueous hydrochlorid acid solution (10 ml) to precipitate a white solid. The latter was dried, extracted with ethyl acetate, filtered and evaporated. From the crude product, 165 mg (97%) of pure 3-phenylphenol (**278**) was isolated by preparative chromatography over silica gel (*n*-hexane / dichloromethane).

5.5.9. The SM reactions of aryl bromides, chlorides and triflates: Preparation of biphenyl-4-carbonitrile (273) [27]

A vigorously stirred mixture of 4-bromobenzonitrile (**272**, 910 mg, 5 mmol), phenylboronic acid (**260**, 730 mg, 6 mmol, 1.2 eq.), sodium carbonate (636 mg, 6 mmol) and 5% Pd-C (1 mg, 0.01mol % Pd) in NMP / water (5 ml, 10:4, v/v) was heated at 120 °C under nitrogen for 1.5 h. To the reaction mixture, cooled to room temperature, water (20 ml) was added, and extracted with dichloromethane (3x20 ml). Combined organic layers were successively washed with water, brine and dried (Na$_2$SO$_4$). Evaporation of the solvent, followed by preparative chromatography (SiO$_2$), afforded 896 mg (100%) of pure biphenyl-4-carbonitrile (**273**).

Note: The SM reactions of aryl bromides and chlorides catalysed by Pd-C can be performed at lower temperatures (80 °C) in DMAc / H$_2$O (20:1) as the reaction medium [25], or in the presence of PPh$_3$ (4 eq. to Pd) with Na$_2$CO$_3$ as the base in DME / H$_2$O (2:1) at approx. 80 °C [24].

5.5.10. The room temperature Pd$_2$(dba)$_3$ / Pt-Bu$_3$ or PCy$_3$-catalysed SM reactions with aryl iodides, bromides, chlorides and triflates: Preparation of 2-acetyl-5-(2-methylphenyl)thiophene (362) [32]

A solution of 2-acetyl-5-chlorothiophene (**363**, 1.61 g, 10 mmol), 2-tolylboronic acid (**286**, 1.50 g, 11 mmol, 1.1 eq.), potassium fluoride (1.92 g, 33 mmol, 3.3 eq.) in anhydrous THF (20 ml) was deggased with argon and charged with $Pd_2(dba)_3$ (45 mg, 0.05 mmol, 0.5 mol%) followed by Pt-Bu_3 (20 mg, 0.1 mmol, 1 mol%). The reaction mixture was stirred under argon at room temperature overnight. The solvent was evaporated and the product was isolated by preparative chromatography (silica gel) to afford 2.14 g (99%) of pure 2-acetyl-5-(2-methylphenyl)thiophene (**362**).

Note: The same procedure can be applied for the SM reactions with aryl iodides and bromides, whereas aryl triflates require 1 mol% $Pd(OAc)_2$ / 1.2 mol% PCy_3 to achieve high conversions to the desired biaryls at room temperature. For the electron-rich aryl chlorides, slightly higher amounts of the catalyst, 1.5 mol% $Pd_2(dba)_3$ / 4.5 mol% Pt-Bu_3 must be used in refluxing THF or dioxane.

5.5.11. Nickel-catalysed SM reactions with aryl tosylates: Preparation of methyl biphenyl-2-carboxylate (294) [45]

To a mixture of 2-methoxycarbonylphenyl tosylate (**293**, 3.04 g, 10 mmol), phenyl boronic acid (**260**, 1.83 g, 15 mmol,1.5 eq.) and potassium phosphate (4.25 g, 20 mmol, 2 eq.) in dioxane (20 ml), deggased with argon was added $Ni(PCy_3)_2Cl_2$ (210 mg, 0.3 mmol, 3 mol%) followed by tricyclohexylphosphine (340 mg, 0.12 mmol, 4 eq. to Ni). The reaction mixture was refluxed under argon for 60 h. Then the solvent was evaporated and the residue was chromatographed on silica gel column (n-hexane / dichloromethane) affording 1.0 g (47%) of pure methyl biphenyl-2-carboxylate (**294**).

5.5.12. Palladium-catalysed SM reactions of aryldiazonium salts: Preparation of 2-metoxybiphenyl (91) [46]

A solution of phenylboronic acid (**260**, 245mg, 2 mmol), 2-methoxyphenyldiazonium tetrafluoroborate (**256**, 490 mg, 2.2 mmol, 1.1 eq.) and $Pd(OAc)_2$ (45 mg, 0.2 mmol,

10 mol%) in methanol (5 ml) was heated under reflux for 1 h. After being cooled to room temperature, the reaction mixture was filtered through Celite, and the filtrate was diluted with water and extracted with diethyl ether (3x10 ml). Removal of ether followed by silica gel chromatography (5% EtOAc in petroleum ether) gave 276 mg (75%) of pure 2-methoxybiphenyl (**91**).

5.5.13. Synthesis of arylboronic acids via Grignard reaction of sluggishly reactive aryl halides: Preparation of mesitylboronic acid (283) [143]

To a dry magnesium turnings (10.00 g, 0.41 mol, 1.37 eq.) heated to 90 °C, a solution of mesityl bromide (**188**, 59.67 g, 0.30 mol) in dry THF (60 ml) was added slowly, with stirring under nitrogen. The reaction started within 2 min, and it was quenched briefly by external cooling. The rate of addition of the halide was then adjusted to maintain the reactants at about 90 °C. After the addition, the reaction mixture was kept at about 95 °C for 2 h. The cooled reaction mixture (65 °C) was diluted with dry THF (120 ml), transferred by nitrogen pressure into a graduated dropping funnel, whereas the analysis of an aliquot usually indicates 96% yield of the mesitylmagnesium bromide (**125**).

A THF solution of mesityl Grignard reagent (**125**, 0.275 mol) and a solution of trimethyl borate (34 ml, 31.11 g, 0.3 mol, 1.1 eq.) were added simultaneously within about 20 minutes to vigorously stirred dry diethyl ether (200 ml) maintained at -60 to -70 °C. After an additional half an hour at -70 °C, the reaction-mass was allowed to warm to   0 °C and was stirred at that temperature for 11 h. It was hydrolyzed by the addition of water (100 ml) and allowed to stand overnight. The organic layer was separated from the pasty aqueous phase and combined with three ether extracts of the latter. The ether phase was washed with water (2x), dilute hydrochloric acid (3x) and twice more with water. The solvent was evaporated, and water (560 ml) was added. The mixture was heated to 100 °C with stirring (15 minutes) and then allowed to stand at rooom temperature overnight. After additional cooling in an ice-salt bath for 2 h, the crystals were collected, washed with ice-water (3x20 ml), and dried at room temperature in high vacuum. A second crop was isolated from the combined filtrate and washings by concentration and cooling. The first and second crops afforded 33.50 g (72%) of mesitylboronic acid (**283**), which can be further purified from water (1:20, m / v; 70% recovery).

5.5.14. Synthesis of arylboronic acids by in situ quench procedure with n-BuLi and triisopropyl borate: Preparation of 3-thienylboronic acid (364) [148]

365   **364**

A 50 ml round-bottomed flask equipped with a temperature probe, a magnetic stirrer and a septum was charged with toluene (16 ml) and THF (4 ml) and put under a nitrogen atmosphere. The flask was further charged with triisopropyl borate (2.8 ml, 12 mmol,1.2 eq.) and 3-bromothiophene (**365**, 1.63 g, 10 mmol). The mixture was cooled to -70 °C using a dry ice / acetone bath. *n*-Butyllithium (2.5 M in *n*-hexane, 4.8 ml, 12 mmol, 1.2 eq.) was added dropwise *via* a syringe pump over 1 h, and the mixture was stirred for additional 0.5 h, while the temperature was held at -70 °C. The acetone / dry ice bath was removed, and the reaction mixture was then allowed to warm to -20 °C before 2 M aqueous sodium hydroxide solution (10 ml) was added. When the mixture reached room temperature, it was transferred to a 100 ml separatory funnel and the layers were separated. The organic layers were evaporated *in vacuo* to provide solids which were recrystallized from acetonitrile to yield 1.16 g (91%) of 3-thienylboronic acid (**364**).

5.5.15. Esterification of arylboronic acids: Preparation of phenyl-1,4- diboronic acid bis-pinacol ester (266) [18]

366   **266**

A suspension of phenyl-1,4-diboronic acid [18] (**366**, 1.00 g, 6.03 mmol), pinacol (1.71 g, 14 mmol, 2.32 eq.) and anhydrous magnesium sulfate (2.00 g, 16 mmol) in anhydrous methanol (10 ml) was stirred overnight (approx. 24 h) at 30 °C. The solvent was evaporated under reduced pressure and the residue was extracted with ethyl acetate (30 ml). The organic phase was washed with water (10 ml), dried (MgSO$_4$) and then concentrated *in vacuo*. The crude product was purified by silica gel chromatography (petroleum ether / dichloromethane, 7:3; to pure dichloromethane) to give 1.93 g (97%) of pure phenyl-1,4-diboronic acid bis-pinacol ester (**266**) as a colourless solid, m.p. 237-238 °C.

5.5.16. Palladium-catalysed reaction of aryl halides with pinacol borane (351): Preparation of 4-ethoxycarbonylphenylboric acid pinacol-ester (367) [161]

A solution of ethyl 4-iodobenzoate (**210**, 276 mg, 1 mmol), pinacol borane (**351**, 190 mg, 1.5 mmol, 1.5 eq.) and triethylamine (0.42 ml, 0.30 g, 3 mmol, 3 eq.) in dioxane (4 ml) was flushed with nitrogen, and Pd(dppf)Cl$_2$ (22 mg, 3 mol%) was added. The reaction mixture was stirred at 80 °C under a nitrogen atmosphere for 2 h. After being cooled, the reaction mixture was poored into water (20 ml) and extracted with dichloromethane (3x20 ml). Combined organic layers were washed with water and brine, dried (Na$_2$SO$_4$), filtered, and evaporated. From the crude product, 218 mg (79%) of pure 4-ethoxycarbonylphenyl pinacolboronate (**367**) was obtained.

5.5.17. Palladium-catalysed reaction of aryl halides with pinacol ester of diboron (350): Preparation of 4-acetylphenylboronic acid-pinacol ester (368) [155]

A mixture of pinacol ester of diboron (**350**, 280 mg, 1.1 mmol), potassium acetate (300 mg, 3 mmol) and Pd(dppf)Cl$_2$ (3 mol%) was flushed with nitrogen. DMSO (6 ml) and 4-bromoacetophenone (**73**, 200 mg, 1 mmol) were then added. After being stirred at 80 °C for 1 h under nitrogen, the product was extracted with toluene, washed with water and dried over Na$_2$SO$_4$. Kugelrohr distillation *in vacuo* gave 197 mg (80%) of pure 4-acetylphenyl-pinacolboronate (**368**).

5.5.18. The SM reactions catalysed by Miyaura's diphenylphosphine-bounded palladium heterogeneous catalyst (342): Preparation of the catalyst and the cross-coupling reaction of 4-tolylboronic acid (271) with 2-chloro pyridine (149) [139]

To a mixture of 1% cross-linked polystyrene (PS) beads (200-400 mesh, 45 g) and chloromethyl methylether (225 ml), tin(IV) chloride (5 ml) was added and the resulting suspension stirred at room temperature for 30 minutes, followed by 2 h at the reflux temperature. The reaction mixture was cooled to room temperature, and product was filtered and washed with dichloromethane (3x30 ml). After drying in high vacuum for 3 h, 64 g of chloromethylated-PS was obtained. To the suspension of thus prepared, chloromethylated-PS in THF (300 ml), Ph_2PCl (25 g) and lithium (1.80 g) sliced to small pieces were added, and the mixture was then stirred for 20 h at room temperature under argon. The beads were filtered and successively washed with methanol, chloroform / methanol (2:3, 3:1 and 9:1, v/v), and finally with pure chloroform. The beads were dried *in vacuo* under nitrogen at 100 °C for 6 h to give the (diphenylphosphinomethyl)polystyrene (PS-CH_2PPh_2, 16.84 g) with 2.86 mmol/g of P. A mixture of PS-CH_2PPh_2 (2.00 g, 5.72 mmol of PPh_2 group on the resin) and Pd(COD)Cl_2 (0.172 g, 0.6 mmol) in benzonitrile (30 ml) was heated at 100 °C for 3 h under nitrogen. The yellow color of solution faded completely to yield the bright yellow beads. The beads were filtered, washed with acetone (6x20 ml), dichloromethane (5x20 ml), and diethyl ether (3x20 ml), and finally dried of residual solvents for 5 h at 60 °C *in vacuo* to afford the polymer-supported palladium catalyst 342.

A flask equipped with a condenser, a septum inlet and a mechanical stirrer was charged with a polymer-supported palladium catalyst (342, 0.10 g, 2.47 mmol of $PdCl_2$ on the resin) and 4-tolylboronic acid (271, 0.177 g, 1.3 mmol), and was flushed with nitrogen. Toluene (3 ml), 2-chloropyridine (149, 0.114 g, 1.0 mmol) and 2 M $K_3PO_4 \cdot nH_2O$ (1.0 ml, in water, 2 mmol) were added. The resulting mixture was then stirred for 24 h at 80 °C under nitrogen. The reaction mixture was cooled to room temperature, diluted with dichloromethane (10 ml) and filtered. The filtrate was washed with water (20 ml), and the water layer was extracted with dichloromethane (3x10 ml). Combined organic phases were dried (Na_2SO_4), filtered, and evaporated to dryness. The residue was chromatographed over silica gel to give 154 mg (91%) of 2-(4-tolyl)pyridine (369) as a pale yellow liquid.

5.6. Conclusion

The Suzuki-Miyaura cross-coupling reaction of arylboronic acids or esters with aryl halides, aryl sulfonates or aryldiazonium salts is the most important aryl-aryl, C-C bond forming reaction. During last two decades the methodology of SM reaction has been greatly improved for each class of electrophilic reactants and boronic counterparts providing extremely versatile, general, and very efficient access to all kind of biaryls. Palladium-based catalysis is more general, while nickel complexes have been effective in the SM reactions with aryl sulfonates, chlorides and bromides. In this Chapter, also the basis of arylboronic acid-chemistry is described in order to popularize this valuable, although relatively less known intermediates.

5.7. References

1. N. Miyaura, T. Yanagi and A. Suzuki, Synth. Commun. 11 (1981) 513.
2. N. Miyaura and A. Suzuki, Chem. Rev. 95 (1995) 2457.
3. A. Suzuki, Acc. Chem. Res. 15 (1982) 178.
4. A. Suzuki, Pure Appl. Chem. 57 (1985) 1749.
5. A. Suzuki, Pure Appl. Chem. 63 (1991) 419.
6. A. Suzuki, Pure Appl. Chem. 66 (1994) 213.
7. I. P. Beletskaya, J. Organomet. Chem. 250 (1983) 551.
8. I. P. Beletskaya, Bull. Acad. Sci. USSR, Div. Chem. Sci. 39 (1990) 2013.
9. S. P. Stanford, Tetrahedron 57 (1998) 263.
10. E. M. Campi, W. R. Jackson, S. M. Marcuccio and C. G. M. Naeslund, J. Chem. Soc., Chem. Commun. (1994) 2395.
11. X. Bei, H. W. Turner, W. H. Weinberg and A. S. Guram, J. Org. Chem. 64 (1999) 6797.
12. K. Inada and N. Miyaura, Tetrahedron 56 (2000) 8657.
13. A. F. Indolese, Tetrahedron Lett. 38 (1997) 3513.
14. Y. Y. Ku, T. Grieme, P. Raje, P. Sharma, H. E. Morton, M. Rozema and S. A. King, J. Org. Chem. 68 (2003) 3238.
15. H. Zhang, F. Y. Kwong, Y. Tian and K. S. Chan, J. Org. Chem. 63 (1998) 6886.
16. J. C. Anderson, H. Namli and C. A. Roberts, Tetrahedron 53 (1997) 15123.
17. Y. Hoshino, N. Miyaura and A. Suzuki, Bull. Chem. Soc. Jpn. 61 (1988) 3008.
18. H. Chaumeil, C. Le Drian and A. Defoin, Synthesis (2002) 757.
19. H. Chaumeil, S. Signorella and C. Le Drian, Tetrahedron 56 (2000) 9655.
20. S. W. Wright, D. L. Hageman and L. D. McClure, J. Org. Chem. 59 (1994) 6095.
21. W. J. Thompson and J. Gaudino, J. Org. Chem. 49 (1984) 5237.

22. N. Miyaura, T. Ishiyama, H. Sasaki, M. Ishikawa, M. Satoh and A. Suzuki, J. Am. Chem. Soc. 111 (1989) 314.
23. K. Matos and J. A. Soderquist, J. Org. Chem. 63 (1998) 461.
24. G. Marck, A. Villiger and R. Buchecker, Tetrahedron Lett. 35 (1994) 3277.
25. C. R. LeBlond, A. T. Andrews, Y. Sun and J. R. Sowa, Org. Lett. 3 (2001) 1555.
26. H. Sakurai, T. Tsukuda and T. Hirao, J. Org. Chem. 67 (2002) 2721.
27. R. G. Heidenreich, K. Köhler, J. G. E. Krauter and J. Pietsch, Synlett (2002) 1118.
28. T. I. Wallow and B. M. Novak, J. Org. Chem. 59 (1994) 5034.
29. D. Zim, A. L. Monteiro and J. Dupont, Tetrahedron Lett. 41 (2000) 8199.
30. O. Lohse, P. Thevenin and E. Waldvogel, Synlett (1999) 45.
31. A. F. Littke and G. C. Fu, Angew. Chem. Int. Ed. 37 (1998) 3387.
32. A. F. Littke, C. Dai and G. C. Fu, J. Am. Chem. Soc. 122 (2000) 4020.
33. X. Bei, T. Crevier, A. S. Guram, B. Jandeleit, T. S. Powers, H. W. Turner, T. Uno and W. H. Weinberg, Tetrahedron Lett. 40 (1999) 3855.
34. D. W. Old, J. P. Wolfe and S. L. Buchwald, J. Am. Chem. Soc. 120 (1998) 9722-9723.
35. J. P. Stambuli, R. Kuwano and J. F. Hartwig, Angew. Chem. Int. Ed. 41 (2002) 4746.
36. S. Y. Liu, M. J. Choi and G. C. Fu, Chem. Commun. (2001) 2408.
37. K. H. Shaughnessy and R. S. Booth, Org. Lett. 3 (2001) 2757.
38. M. Ueda, M. Nishimura and N. Miyaura, Synlett (2000) 856.
39. C. Zhang, J. Huang, M. L. Trudell and S. P. Nolan, J. Org. Chem. 64 (1999) 3804.
40. G. Y. Li, J. Org. Chem. 67 (2002) 3643.
41. G. A. Grasa, A. C. Hillier and S. P. Nolan, Org. Lett. 3 (2001) 1077.
42. T. Mino, Y. Shirae, M. Sakamoto and T. Fujita, Synlett (2003) 882.
43. B. H. Lipshutz, J. A. Sclafani and P. A. Blomgren, Tetrahedron 56 (2000) 2139.
44. S. Saito, S. Oh-tani and N. Miyaura, J. Org. Chem. 62 (1997) 8024.
45. D. Zim, V. R. Lando, J. Dupont and A. L. Monteiro, Org. Lett. 3 (2001) 3049.
46. S. Sengupta and S. Bhattacharyya, J. Org. Chem. 62 (1997) 3405.
47. M. B. Andrus and C. Song, Org. Lett. 3 (2001) 3761.
48. V. Percec, J. Y. Bae and D. H. Hill, J. Org. Chem. 60 (1995) 1060.
49. A. Huth, I. Beetz and I. Schumann, Tetrahedron 45 (1989) 6679.
50. F. A. Alphonse, F. Suzenet, A. Keromnes, B. Lebret and G. Guillaumet, Synlett (2002) 447.
51. R. Cramer and D. R. Coulson, J. Org. Chem. 40 (1975) 2267.
52. V. V. Grushin and H. Alper, Chem. Rev. 94 (1994) 1047.
53. M. Moreno-Mañas, M. Pérez and R. Pleixats, J. Org. Chem. 61 (1996) 2346.
54. M. Hird, A. J. Seed and K. J. Toyne, Synlett (1999) 438.
55. Y. Deng, L. Gong, A. Mi, H. Liu and Y. Jiang, Synthesis (2003) 337.

56. D. A. Conlon, B. Pipik, S. Ferdinand, C. R. LeBlond, J. R. Sowa, Jr., B. Izzo, P. Collins, G.-J. Ho, J. M. Williams, Y.-J. Shi and Y. Sun, Adv. Synth. Catal. 345 (2003) 931.
57. Y. Kobayashi and R. Mizojiri, Tetrahedron Lett. 37 (1996) 8531.
58. U. Schmidt, R. Meyer, V. Leitenberger and A. Lieberknecht, Angew. Chem. Int. Ed. 28 (1989) 929.
59. M. G. Johnson and R. J. Foglesong, Tetrahedron Lett. 38 (1997) 7001.
60. T. R. Hoye and M. Chen, J. Org. Chem. 61 (1996) 7940.
61. J. W. Benbow and B. L. Martinez, Tetrahedron Lett. 37 (1996) 8829.
62. Q. S. Hu, D. Vitharana, X. F. Zheng, C. Wu, C. M. S. Kwan and L. Pu, J. Org. Chem. 61 (1996) 8370.
63. S. Kotha and A. K. Ghosh, Synlett (2002) 451.
64. K. Monde, Y. Tomita, M. L. Gilchrist, A. E. McDermott and K. Nakanishi, Israel J. Chem. 40 (2000) 301.
65. F. J. Zhang, C. Cortez and R. G. Harvey, J. Org. Chem. 65 (2000) 3952.
66. S. Kumar, Synthesis (2001) 841.
67. S. Kumar, J. Org. Chem. 62 (1997) 8535.
68. S. Kumar and T. Y. Kim, J. Org. Chem. 65 (2000) 3883.
69. S. Kotha, K. Lahiri and N. Sreenivasachary, Synthesis (2001) 1932.
70. Y. Gong and H. W. Pauls, Synlett (2000) 829.
71. A. Sutherland and T. Gallagher, J. Org. Chem. 68 (2003) 3352.
72. P. R. Parry, M. R. Bryce and B. Tarbit, Synthesis (2003) 1035.
73. E. Sotelo and E. Ravina, Synlett (2002) 223.
74. P. Jeanjot, F. Bruyneel, A. Arrault, S. Gharbi, J.-F. Cavalier, A. Abels, C. Marchand, R. Touillaux, J.-F. Rees and J. Marchand-Brynaert, Synthesis (2003) 513.
75. N.-X. Wang, Synth. Commun. 33 (2003) 2119.
76. V. Lisowski, M. Robb and S. Rault, J. Org. Chem. 65 (2000) 4193.
77. M. S. McClure, F. Roschangar, S. J. Hodson, A. Millar and M. H. Osterhout, Synthesis (2001) 1681.
78. A. Heynderickx, A. Samat and R. Guglielmetti, Synthesis (2002) 213.
79. M. Havelkova, M. Hocek, M. esnek and D. Dvorak, Synlett (1999) 1145.
80. M. Havelkova, D. Dvorak and M. Hocek, Synthesis (2001) 1704.
81. C. Enguehard, J. L. Renou, V. Collot, M. Hervet, S. Rault and A. Gueiffier, J. Org. Chem. 65 (2000) 6572.
82. M. Allegretti, A. Arcadi, F. Marinelli and L. Nicolini, Synlett (2001) 609.
83. J. S. D. Kumar, M. M. Ho, J. M. Leung and T. Toyokuni, Adv. Synth. Catal. (2002) 1146.
84. A. C. Spivey, T. Fekner, S. E. Spey and H. Adams, J. Org. Chem. 64 (1999) 9430.

85. C. Imrie, C. Loubster, P. Engelbrecht and C. W. McCleland, J. Chem. Soc., Perkin Trans 1 (1999) 2513.

86. A. Dondoni, C. Ghiglione, A. Marra and M. Scoponi, J. Org. Chem. 63 (1998) 9535.

87. K. Kamikawa, T. Watanabe and M. Uemura, J. Org. Chem. 61 (1996) 1375.

88. N. G. Andersen, S. P. Maddaford and B. A. Keay, J. Org. Chem. 61 (1996) 9556.

89. C. G. Blettner, W. A. König, W. Stenzel and T. Schotten, J. Org. Chem. 64 (1999) 3885.

90. N. E. Leadbeater and M. Marco, J. Org. Chem. 68 (2003) 888.

91. D. Badone, M. Baroni, R. Cardamone, A. Ielmini and U. Guzzi, J. Org. Chem. 62 (1997) 7170.

92. M. Melucci, G. Barbarella and G. Sotgiu, J. Org. Chem. 67 (2002) 8877.

93. G. W. Kabalka, L. Wang, R. M. Pagni, C. M. Hair and V. Namboodiri, Synthesis (2003) 217.

94. S. F. Nielsen, D. Peters and O. Axelsson, Synth. Commun. 30 (2000) 3501.

95. C. G. Blettner, W. A. König, G. Rühter, W. Stenzel and T. Schotten, Synlett (1999) 307.

96. M. Gravel, K. A. Thompson, M. Zak, C. Bérubé and D. G. Hall, J. Org. Chem. 67 (2002) 3.

97. B. A. Lorsbach, J. T. Bogdanoff, R. B. Miller and M. J. Kurth, J. Org. Chem. 63 (1998) 2244.

98. S. Darses and J.-P. Genêt, Tetrahedron Lett. 38 (1997) 4393.

99. M. J. Sharp and V. Snieckus, Tetrahedron Lett. (1985) 5997.

100. M. J. Sharp, W. Cheng and V. Snieckus, Tetrahedron Lett. 28 (1987) 5093.

101. W. Cheng and V. Snieckus, Tetrahedron Lett. 28 (1987) 5097.

102. B. I. Alo, A. Kandil, P. A. Patil, M. J. Sharp, M. A. Siddiqui, V. Snieckus and P. D. Josephy, J. Org. Chem. 56 (1991) 3763.

103. J. M. Fu, B. P. Zhao, M. J. Sharp and V. Snieckus, J. Org. Chem. 56 (1991) 1683.

104. C. M. Unrau, M. G. Campbell and V. Snieckus, Tetrahedron Lett. 33 (1992) 2773.

105. W. Wang and V. Snieckus, J. Org. Chem. 57 (1992) 424.

106. J. M. Fu, B. P. Zhao, M. J. Sharp and V. Snieckus, Can. J. Chem. 72 (1994) 227.

107. A. Zapf and M. Beller, Chem. Eur. J. 6 (2000) 1830.

108. M. Feuerstein, H. Doucet and M. Santelli, Synlett (2001) 1458.

109. M. Feuerstein, F. Berthiol, H. Doucet and M. Santelli, Synlett (2002) 1807.

110. A. Zapf, A. Ehrentraut and M. Beller, Angew. Chem. Int. Ed. 39 (2000) 4153.

111. M. G. Andreu, A. Zapf and M. Beller, Chem. Commun. (2000) 2475.

112. J. P. Wolfe, R. A. Singer, B. H. Yang and S. L. Buchwald, J. Am. Chem. Soc. 121 (1999) 9550.

113. J. P. Wolfe and S. L. Buchwald, Angew. Chem. Int. Ed. 38 (1999) 2413.

114. J. Yin, M. P. Rainka, X. X. Zhang and S. L. Buchwald, J. Am. Chem. Soc. 124 (2002) 1162.

115. X. Sava, L. Ricard, F. Mathey and P. Le Floch, Organometallics 19 (2000) 4899.

116. T. E. Pickett and C. J. Richards, Tetrahedron Lett. 42 (2001) 3767.

117. N. Kataoka, Q. Shelby, J. P. Stambuli and J. F. Hartwig, J. Org. Chem. 67 (2002) 5553.

118. A. Fürstner and A. Leitner, Synlett (2001) 290.

119. A. J. Arduengo, R. Krafczyk and R. Schmutzler, Tetrahedron 55 (1999) 14523.

120. M. Beller, H. Fischer, W. A. Herrmann, K. Öfele and C. Brossmer, Angew. Chem. Int. Ed. 34 (1995) 1848.

121. L. Botella and C. Najera, Angew. Chem. Int. Ed. 41 (2002) 179.

122. D. A. Alonso, C. Najera and M. C. Pacheco, J. Org. Chem. 67 (2002) 5588.

123. D. A. Albisson, R. B. Bedford and P. N. Scully, Tetrahedron Lett. 39 (1998) 9793.

124. D. A. Albisson, R. B. Bedford, S. E. Lawrence and P. N. Scully, Chem. Commun. (1998) 2095.

125. R. B. Bedford and C. S. J. Cazin, Chem. Commun. (2001) 1540.

126. R. B. Bedford and S. L. Welch, Chem. Commun. (2001) 129.

127. R. B. Bedford, C. S. J. Cazin and S. L. Hazelwood, Angew. Chem. Int. Ed. 41 (2002) 4120.

128. R. B. Bedford, C. S. J. Cazin and S. L. Hazelwood, Chem. Commun. (2002) 2608.

129. R. B. Bedford, M. E. Blake, C. P. Butts and D. Holder, Chem. Commun. (2003) 466.

130. R. B. Bedford, S. L. Hazelwood and M. E. Limmert, Chem. Commun. (2002) 2610.

131. S. Iyer and A. Jayanthi, Synlett (2003) 1125.

132. M. T. Reetz and E. Westermann, Angew. Chem. Int. Ed. 39 (2000) 165.

133. M. B. Thathagar, J. Beckers and G. Rothenberg, Adv. Synth. Catal. 345 (2003) 979.

134. B. M. Bhanage and M. Arai, Catalysis Rev. 43 (2001) 315.

135. M. T. Reetz and G. Lohmer, Chem. Commun. (1996) 1921.

136. S. Klingelhöfer, W. Heitz, A. Greiner, S. Oestreich, S. Förster and M. Antonietti, J. Am. Chem. Soc. 119 (1997) 10116.

137. S. W. Kim, M. Kim, W. Y. Lee and T. Hyeon, J. Am. Chem. Soc. 124 (2002) 7642.

138. K. I. Shimizu, T. Kan-no, T. Kodama, H. Hagiwara and Y. Kitayama, Tetrahedron Lett. 43 (2002) 5653.

139. K. Inada and N. Miyaura, Tetrahedron 56 (2000) 8661.

140. C. A. Parrish and S. L: Buchwald, J. Org. Chem. 66 (2001) 3820.

141. T. J. Colacot, E. S. Gore and A. Kuber, Organometallics 21 (2002) 3301.

142. I. P. Beletskaya and A. V. Cheprakov, Chem. Rev. 100 (2000) 3009.

143. R. T. Hawkins, W. J. Lennarz and H. R. Snyder, J. Am. Chem. Soc. 82 (1960) 3053.

144. M. F. Hawthorne, J. Am. Chem. Soc. 80 (1958) 4291.

145. M. F. Hawthorne, J. Am. Chem. Soc. 80 (1958) 4293.

146. B. Jiang, Q. F. Wang, C. G. Yang and M. Xu, Tetrahedron Lett. 42 (2001) 4083.

147. K: T. Wong, Y. Y. Chien, Y. L. Liao, C. C. Lin, M. Y. Chou and M. K. Leung, J. Org. Chem. 67 (2002) 1041.

148. W. Li, D. P. Nelson, M. S. Jensen, R. S. Hoerrner, D. Cai, R. D. Larsen and P. J. Reider, J. Org. Chem. 67 (2002) 5394.

149. S. Caron and J. M. Hawkins, J. Org. Chem. 63 (1998) 2054.

150. P. R. Parry, C. Wang, A. S. Batsanov, M. R. Bryce and B. Tarbit, J. Org. Chem. 67 (2002) 7541.

151. E. Vazquez, I. W. Davies and J. F. Payack, J. Org. Chem. 67 (2002) 7551.

152. D. D. Winkle and K. M. Schaab, Org. Proc. Res. & Dev. 5 (2001) 450.

153. J. Kristensen, M. Lysén, P. Vedsø and M. Begtrup, Org. Lett. 3 (2001) 1435.

154. M. Wasgindt and E. Klemm, Synth. Commun. 29 (1999) 103.

155. T. Ishiyama, M. Murata and N. Miyaura, J. Org. Chem. 60 (1995) 7508.

156. T. Ishiyama, Y. Itoh, T. Kitano and N. Miyaura, Tetrahedron Lett. 38 (1997) 3447.

157. A. Giroux, Y. Han and P. Prasit, Tetrahedron Lett. 38 (1997) 3841.

158. C. Malan and C. Morin, J. Org. Chem. 63 (1998) 8019.

159. A. Fürstner and G. Seidel, Org. Lett. 4 (2002) 541.

160. P. Appukkuttan, E. Van der Eycken and W. Dehaen, Synlett (2003) 1204.

161. M. Murata, S. Watanabe and Y. Masuda, J. Org. Chem. 62 (1997) 6458.

162. M. Murata, T. Oyama, S. Watanabe and Y. Masuda, J. Org. Chem. 65 (2000) 164.

163. O. Baudoin, D. Guenard and F. Gueritte, J. Org. Chem. 65 (2000) 9268.

164. C. E. Tucker, J. Davidson and P. Knochel, J. Org. Chem. 57 (1992) 3482.

165. Y. L. Song and C. Morin, Synlett (2001) 266.

166. C. N. Iverson and M. R. Smith, J. Am. Chem. Soc. 121 (1999) 7696.

167. J.-Y. Cho, C. N. Iverson and M. R. Smith, J. Am. Chem. Soc. 122 (2000) 12868.

168. M. K. Tse, J. Y. Cho and M. R. Smith, Org. Lett. 3 (2001) 2831.

169. J. Y. Cho, M. K. Tse, D. Holmes, R. E. Maleczka and M. R. Smith, Science 295 (2002) 305.

170. T. Ishiyama, J. Takagi, K. Ishida, N. Miyaura, N. R. Anastasi and J. F. Hartwig, J. Am. Chem. Soc. 124 (2002) 390.

171. D. J. Koza and E. Carita, Synthesis (2002) 2183.

172. J. Simon, S. Salzbrunn, G. K. S. Prakash, N. A. Petasis and G. A. Olah, J. Org. Chem. 66 (2001) 633.

173. D. A. Evans, J. L. Katz and T. R. West, Tetrahedron Lett. 39 (1998) 2937.

174. D. T. C. Chan, K. L. Monaco, R. P. Wang and M. P. Winters, Tetrahedron Lett. 39 (1998) 2933.

175. X. Zhou and K. S. Chan, J. Org. Chem. 63 (1998) 99.

176. B. Tao, S. C. Goel, J. Singh and D. W. Boykin, Synthesis (2002) 1043.

177. M. F. Lappert, Chem. Rev. 56 (1956) 959.

178. E. Tyrrell and P. Brookes, Synthesis (2003) 469.

6. SYNTHESIS OF BIARYLS AND POLYARYLS BY OXIDATIVE COUPLINGS OF ARENES

6.1. Introduction

The formation of aryl-aryl, C-C bond between two aromatic rings bearing no electrophilic and nucleophilic sites can be accomplished by variuos methods of oxidative couplings (OCA), Scheme 1.

OCA reagent: Pd(II), Tl(III), Fe(III), Cu(II), V(V), Mo(V), etc.

Scheme 1

Electron-rich arenes such as thiophenes, pyrroles, phenols, anilines, alkoxyarenes, etc. react smoothly with a number of oxidants to give the coupling biaryl products. Additionally, benzene and slightly less electron-rich aromatics like halobenzenes have been coupled to the corresponding biaryls and even polyaryls by certain methods for performing the oxidative coupling reactions. In overall transformation, the OCA reaction involves dehydro-coupling since two hydrogen radicals are formally eliminated from two molecules of arenes to form the new aryl-aryl bond. The latter reaction has been employed in the synthesis of a wide variety of naturally occuring products, mainly alkaloids, phenolic biaryls, as well as conducting polymers poly-*p*-phenylene, polythiophenes, and biphenyl itself [1].

6.2. Mechanisms of oxidative couplings of arenes to biaryls and polyaryls

The oxidative coupling of arenes (OCA) can be affected with numerous reagents capable to generate an aryl radical-cation from the parent arene. The reaction is actually related to the Scholl reaction where the same conversion is conducted in the presence of Lewis acid to form the new aryl-aryl bond. The reagents and conditions which produce the aryl cation-radicals from arenes are: Brönsted acids, e.g. H_2SO_4, CF_3COOH, CF_3SO_3H; Lewis acids, e.g. $AlCl_3$, $AlBr_3$, $FeCl_3$, $FeBr_3$, SO_3, PF_5, HgX_2, PbX_2, ZnX_2, MgX_2, SnX_4; metal salts of Ag(II), Ti(II), Ti(III), Cu(III), Cu(II), Cr(II), Co(III), Co(I), Mn(III), Pb(IV), Ce(IV), Ir(IV), Mo(V), V(IV), As(V) and Sb(V); halogens, e.g. Br_2, I_2, IBr, ICl; charge transfer complexes forming reagents, e.g. chloranil, tetracyanoethylene, tetracyanomuconitrile; irradiation, e.g. photoionization (UV), X-irradiation, etc; electrochemical anodic oxidation; as well as oxygen and sometimes catalyst surfaces, e.g. alumina, silica-alumina, zeolites, etc [1]. Thus generated, aryl radical-cation reacts with the second molecule of arene to give a bicyclohexadiene radical-cation, $BCD^{+\bullet}$, which undergoes the elimination of hydrogen (H^+ followed the $2e^-$ abstraction by oxidant) to form the biaryl. When an additional oxidant is absent, the liberation of molecular hydrogen occurs (the Scholl reaction), whereas in a number of OCA reaction methods involving various oxidants, the $BCD^{+\bullet}$ is oxidatively converted to the final biaryl structure. Once formed biaryl can further react with $BCD^{+\bullet}$, or an original aryl radical-cation, to finish at the stages of variuos oligo- to highly polymeric aryls. However, depending upon the substrate and conditions, the reaction may follow the classical Friedel-Crafts reaction pathway: generation of aryl cations, e.g. by the influence of a protic acid, followed by aryl-cationic arylation of the second molecule of arene to form a bicyclohexadienyl cation which, upon elimination of H^+, gives cyclohexadienyl-arene. The latter is converted to the biaryl by elimination of two hydrogen atoms. The reaction mechanism, specially for oxidative polymerization of arenes, is far more complicated, and was excellently rewieved by Kovacic's group [1]. However, for basic understanding of the oxidative coupling of arenes, the following pathway is reasonable (for example with $AlCl_3$), Scheme 2.

Once again, the impact of external oxidant is crucial at the stage of bicyclohexadienyl radical-cation, $BCD^{+\bullet}$, or cyclohexadienyl arene-radical, $CHD-Ar^\bullet$, to convert these to the expected biaryl **8**.

The oxidative coupling reactions of certain electron-rich arenes under suitable reaction conditions proceed, at least partially, *via* free-radical mechanism, Scheme 3. The phenolate anion is oxidized by suitable one-electron oxidant to the phenoxyl-radical whose tautomeric form is aryl-radical on the adjacent carbon atom. The symmetrical biaryl is formed by coupling of the latter, whereas the unsymmetrical one is produced by free-radical arylation of the second arene molecule, usually in an intramolecular process, Scheme 3. Typical examples are the oxidative couplings of phenols with $K_3[Fe(CN)_6]$ in basic media and with copper(II)-amine complexes.

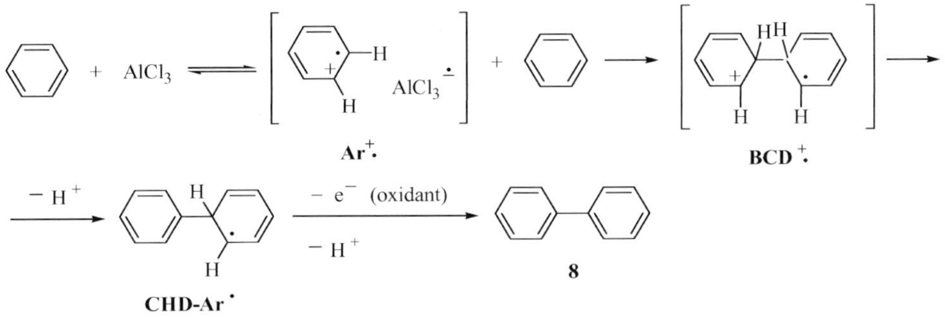

Scheme 2

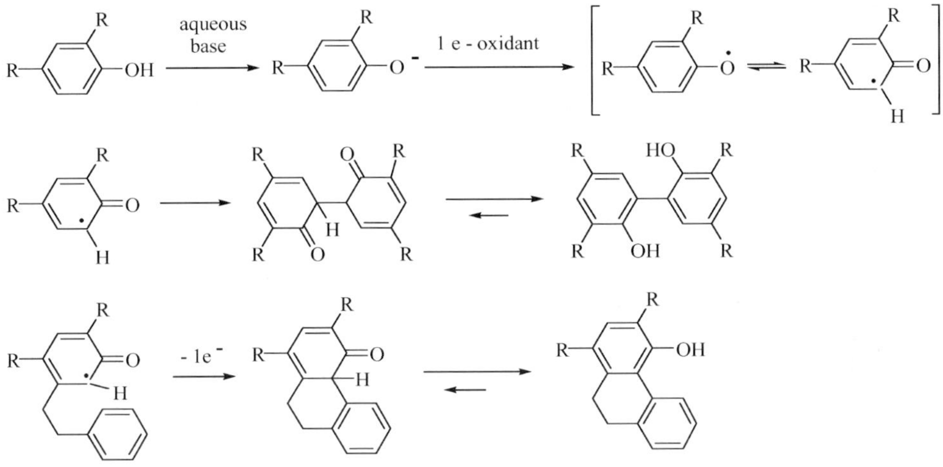

Scheme 3

6.3. Oxidative couplings of arenes to biaryls and polyaryls

Oxidative couplings of simple arenes, whose reactivity is comparable with benzene itself, to biaryls have been performed with stoichiometric palladium salts [2-5]. Biphenyl (**8**) is obtained in high yield from two molecules of benzene by the influence of palladium(II) acetate or mixture of palladium(II) chloride (1 eq.) and sodium acetate (2 eq.). The reaction apparently involves the electrophilic palladation of benzene to give phenylpalladium(II) acetate (**370**) which further reacts with the second benzene molecule to generate diphenylpalladium(II) (**371**). The latter is prone to rapid reductive elimination of biphenyl with liberation of palladium black [3,5]. However, in the presence of suitable external oxidant, the outcoming palladium(0) species is oxidized back to the palladium(II) acetate. In this manner, the oxidative coupling of arenes can be carried out catalytically. The preparation of biphenyl (**8**) from benzene is illustrative, Scheme 4.

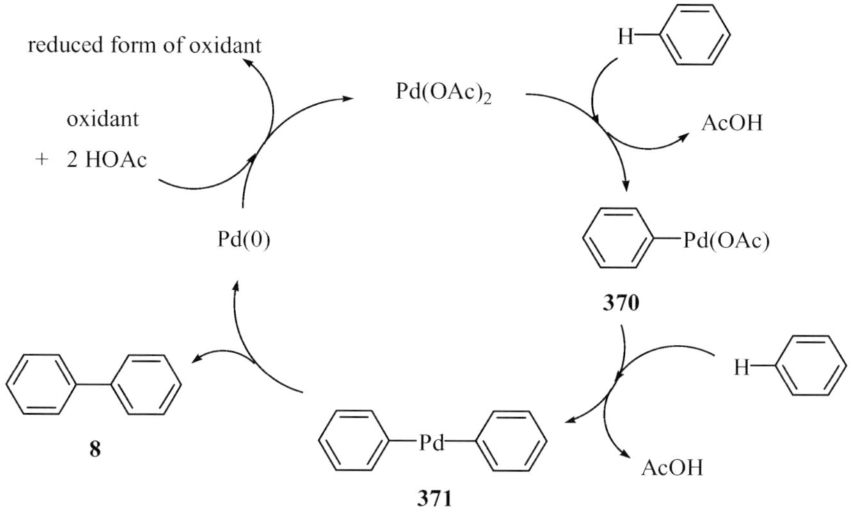

Scheme 4

This classical stoichiometric palladium couplings are usually conducted with palladium(II) acetate in glacial acetic or trifluoroacetic acid as solvent [2-5]. Alternatively, the reactions have been performed with a catalytic amount of Pd(OAc)$_2$ in the presence of acetic acid in 1,4-dioxane or acetonitrile under air [6]. On the other hand, an efficient catalytic method for synthesis of biphenyl from benzene is based on the catalytic amount of Pd(OAc)$_2$ and heteropoly acids such as H$_4$PMo$_{11}$VO$_{40}$·nH$_2$O as catalytic oxidant for palladium(0) to palladium(II) species, with oxygen as very cheap and ecologically acceptable terminal oxidant. Thus biphenyl (**8**) has been obtained in 14% yield, with terphenyls (3.3%) and phenol (0.5%) as by-products [7], Scheme 5.

Among other co-catalysts, MoO$_2$(acac)$_2$, TiO(acac)$_2$ (0.5 eq. to Pd) were proved as the most effective alternatives in the Pd(OAc)$_2$-catalysed oxidative dimerization of benzene to biphenyl [8]. Moreover, the preparation of biaryls from arenes can be connected with the reductive coupling of the corresponding aryl halides since both reactions are palladium-catalysed processes. Whereas the former converts the palladium(II) to palladium(0), the latter reaction turns this redox back to the palladium(II), closing the catalytic cycle. In this way, the mixture of benzene and chlorobenzene in the presence of palladium(II) chloride (2.8 mol%) and tetrahexylammonium chloride (2.3 mol%, stabilizer for Pd-nanoparticles) has been converted to biphenyl with relatively high chemoselectivity [9]. In overall process, chlorobenzene actually serves as stoichiometric reductant.

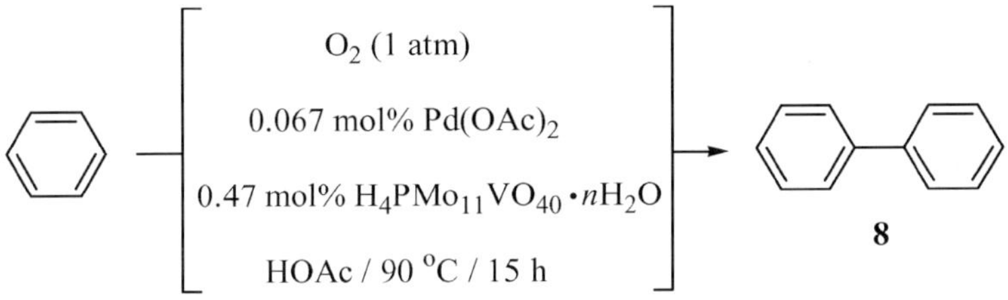

Scheme 5

Apart from palladium-catalysed processes, the oxidative coupling of arenes have been affected by thallium(III) salts [10-23]. Among them, thallium trifluoroacetate (TTFA), usually in the presence of trifluoroacetic acid, has been found to act as extremely versatile reagent. The thallium(III)-mediated OCA reactions proceed *via* radical-cation mechanism described above [10,11]. Sometimes, the reactions have been conducted in the presence of boron trifluoride-etherate as an additional Lewis acid. The reactions of simple electron-rich benzenes: anisole, alkylanisoles, dialkoxybenzenes, and various aryl, and alkoxynaphthalenes, as well as their halogenated derivatives, proceed smoothly to give the respective biaryls in moderate to excellent yields [12,13]. Thallium trifluoroacetate was prepared by refluxing thallium(III) oxide with trifluoroacetic acid (TFA) containing 10-20% water for 12 h. Such TFA solutions can be used directly, or after evaporation of water and TFA. However, TTFA is commercially available, but sensitive to moisture, what can be avoided by *in situ* preparation of the reagent, by reaction of stable Tl_2O_3 with trifluoroacetic anhydride in dichloromethane at room temperature. Moreover, the oxidative coupling reactions can be conducted by subsequent adding of the arene to be coupled, and trifluoroacetic anhydride to the thallium(III) oxide in dichloromethane [13], Table 1.

Since the oxidative coupling of two arene molecules actually involves the elimination of two electrons by Tl(III), the reaction stoichiometrically requires one molar equivalent of TTFA per two equivalents of arene, but, TTFA has been often employed in slight excess. The efficacy of thallium(III) trifluoroacetate-mediated oxidative coupling of relatively electron-rich benzenes is presented in the Table 2.

Table 1. Methodology of the OCA reactions involving thallium(III) trifluoroacetate
[12,13,15]

Reagent (eq. to Ar)	Solvent [a]	Lit. ref.
$Tl(CF_3COO)_3$ (0.5-1.4) + $BF_3 \cdot Et_2O$ (1.2, 2, 8)	A, B, C, D	12, 13, 15
Tl_2O_3 (0.52) + 6 CF_3COOH [b]	B, C, D	15, 16
Tl_2O_3 / 3 $(CF_3CO)_2O$	B, C, D	15, 16

[a] Solvent: A = CCl_4; B = CH_2Cl_2; C = MeCN; D = TFA;
[b] The OCA can be performed with or without $BF_3 \cdot Et_2O$.

Table 2. The yields of symmetrical biaryls obtained by oxidative coupling of arenes
with $Tl(CF_3COO)_3$ [12]

R_1	R_2	R_3	R_4	Conv. (%) [a]	Yield (%) [b]
H	H	H	H	100	16
Me	H	Me	H	63	60
Me	Me	Me	H	77	74
MeO	H	Me	H	85	74
MeO	H	Br	H	92	88
MeO	Me	Br	H	85	99

[a] Based on the amount of starting material which is recovered.
[b] Based on consumed starting material.

This reactions have been successfully used in the synthesis of various naturally occuring products including a number of isoquinoline alkaloids [19,23], lignans [15-18,20,21], colchinol derivatives [22], and in several other instances. An interesting example is oxidative coupling of compound **372** to *N*-acetylcolchinol (**373**) in 71% yield [22], respectively, Scheme 6.

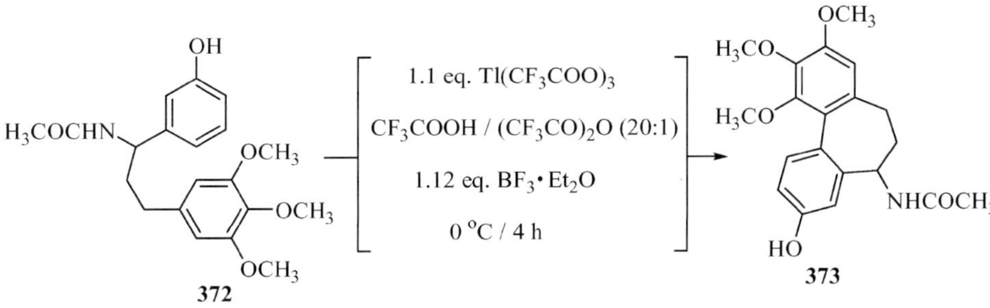

Scheme 6

Beside the biaryls, TTFA / TFA reagent has been applied in the synthesis of polyarenes by oxidative polymerization. Thus 3-alkylthiophenes were polymerized to the corresponding poly(3-alkyl)thiophenes [24] in fair yields.

In addition, several other one- and two-electron oxidants have been involved in the oxidative coupling reactions of mainly activated benzenes such as phenols, phenol-ethers, alkylated phenol-ethers, etc.

The following reagents are commonly used: CrO_3 in HOAc [25,26], $CH_3(n\text{-Bu})_3N$ MnO_4 [27,28], $Mn(OAc)_3$ [29,31], $Mn(acac)_3$ [32], MnO_2 [33], activated PbO_2 [34], $AgNO_3$ [35], Ag(II) ($AgNO_3$ in the presence of $K_2S_2O_8$) [36], $FeCl_3$ [37-47], $[Fe(DMF)_3Cl_2][FeCl_4]$ [48], $K_3[Fe(CN)_6]$ [49-54], $CuCl_2$ with various amines [55-61], $CuSO_4\text{-}Al_2O_3$ [62], $CuCl_2$ / $AlCl_3$ [63-67], $VOCl_3$ [68,69], VOF_3 [70-74], VOF_3 / $BF_3\cdot Et_2O$ [75], NH_4VO_3 / $HClO_4$ [76], $VO(acac)_2$ [77], $TiCl_4$ [78], $MoCl_5$ [79-81], $MoOCl_4$ [82], RuO_2 / CF_3COOH / $BF_3\cdot Et_2O$ [83-85], $Co(CF_3COO)_3$ [86], $(NH_4)_2[Ce(NO_3)_6]$ [87], $Pb(OAc)_4$ [88-93], $C_6H_5I(OAc)_2$ (PIDA) [94-97], $C_6H_5I(OOCCF_3)_2$ (PIFA) [96-99], $t\text{-}Bu_2O_2$ [100,101], $(CO_3t\text{-}Bu)_2$ [100], as well as electrochemical anodic oxidation [102-104]. Among less usual oxidants, nitronium salts, e.g. $NOBF_4$ [105], and nitrobenzene in the presence of aluminum(III) chloride (Scholl reaction system) [106], have also affected the oxidative coupling of arenes.

Chromium trioxide or $K_2Cr_2O_7$ in the presence of sulfuric acid, as widely employed classical oxidants, induce the oxidative phenolic coupling reactions. Since the phenols undergo oxidative coupling reactions at *ortho-* and *para-*positions to hydroxy group, the reactions with strong oxidants such as chromium(VI) reagents usually work well in the homo-couplings of substrates properly substituted in *ortho-* or *para-*position. In the case of phenols bearing no *ortho-* and *para-*substituents, the reaction seldom has practical synthetic utility. When more than one *ortho-* or *para-*positions to the phenolic group are not occupied, the *ortho-ortho*, *ortho-para*, and *para-para* coupling reactions occur, giving often unseparable mixtures. Among a number of by-products which arise form several possible side-reactions, the formation of quinones from either parent or biaryl derived phenols is noteworth [25]. In successful case, 2,4-di-*tert*-butyl-

5-methylphenol (**374**) was oxidatively coupled to biaryl **375** in 75% yield [25], Scheme 7.

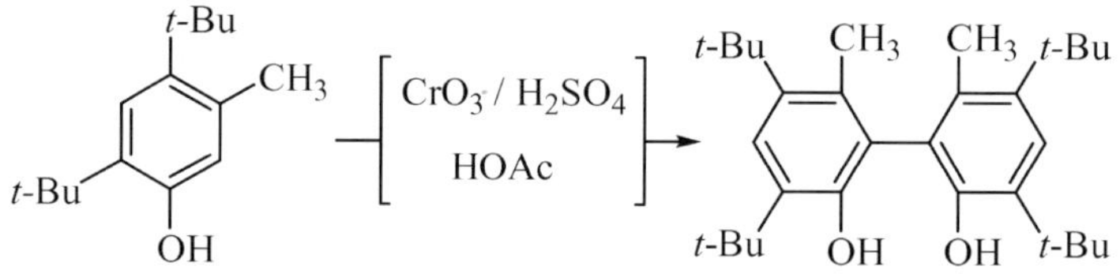

Scheme 7

Methyl(tri-*n*-butyl)ammonium permanganate, as well as various tetraalkylammonium permanganates, prepared separately or more conveniently *in situ* from equimolar quantities of $KMnO_4$ and quaternary ammonium salts, R_4NX (R = alkyl, X = HSO_4^-, Cl^-, Br^-), are soluble in organic solvents such as dichloromethane [27,28]. These reagents efficiently promote the oxidative couplings of phenols and phenol-ethers. Once again, only one *ortho*- or *para*-position to phenolic group must be unsubstituted to achieve high and synthetically useful yield of homo-coupling product. Thus 2-methoxy-4-methylphenol (**376**) was converted with $MeBu_3NMnO_4$ to biaryl **377** with a 81% yield [27], respectively, Scheme 8.

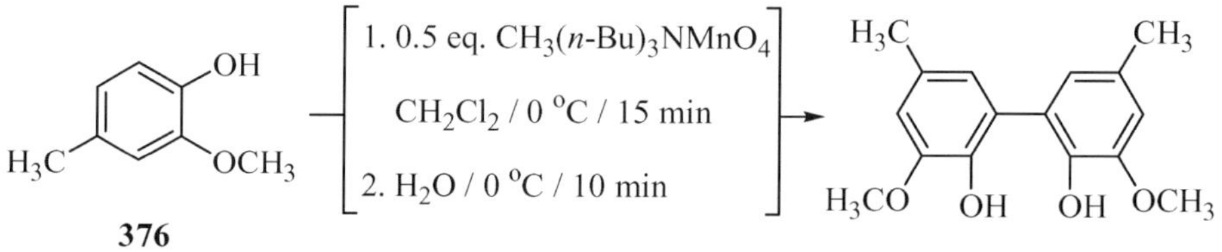

Scheme 8

These reagents are moderately strong oxidants, however, the presence of certain sensitive functional groups is not tolerated well. In this manner, aldehyde group, benzyl and allyl alcohols at room temperature or slightly below are usually oxidized by these reagents. Moreover, R_4NMnO_4 smoothly converts benzylethers and benzylamines to the corresponding benzoates and benzamides at room temperature or in refluxing dichloromethane within a few hours. Anyway, the R_4NMnO_4-based oxidative phenolic couplings apparently have great synthetic importance providing an effective access to quite complicated, mainly symmetrical biaryl structures.

Additionaly, inexpensive and easily available permanganate reagents allow to scale-up the potential oxidative coupling products.

Manganese(III) acetate [29-31] and -acetylacetonate [32] are widely used one-electron oxidants. The latter reagent, which, for instance, efficiently reacts with 2-naphthol (**378**) to give binaphthol (**4**) in 69% yield, possesses higher practical importance [32], Scheme 9.

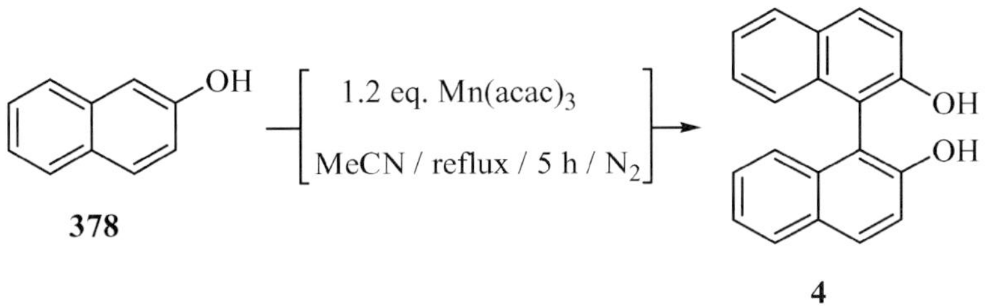

Scheme 9

Among other commonly used oxidants, MnO_2 and PbO_2 have been traditionally applied as heterogeneous reagents in the oxidative phenolic couplings affording a wide variety of alkaloid structures, almost always in very low yields [33,34].

Silver-based reagents have a little practical utility in the OCA reactions, still, an interesting application from the mechanistic point of view was described. Thus Effenberger and coworkers [35] performed the oxidative dimerization of 1,3,5-tripyrrolidylbenzene (**379**) with silver nitrate in acetonitrile where the dimeric complex **380** was obtained, which, upon treatment with strong base, gave biaryl **381** in very high yield, Scheme 10.

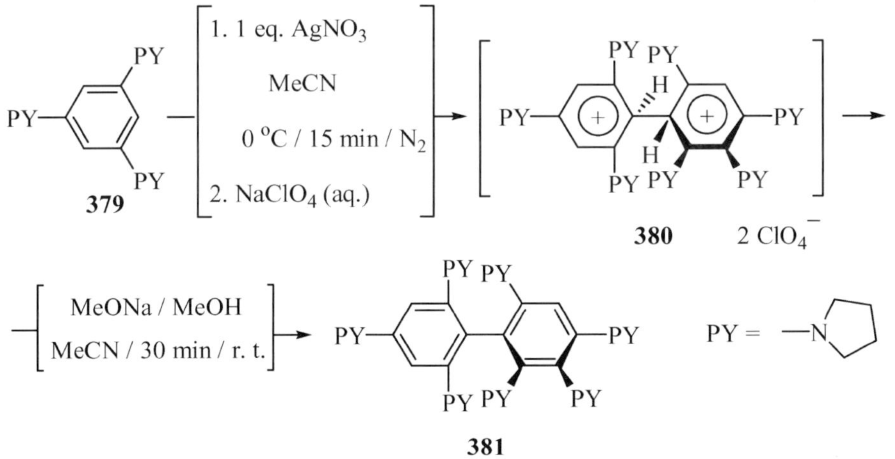

Scheme 10

Lewis acid, e.g. Ag^+, reacts with aromatic system to generate the aryl radical-cation which is capable of undergoing the further arylation to give arylcyclohexadienyl radical-cation. The latter is oxidized (-2 e^-) by suitable oxidant to form the corresponding biaryl. Some of above mentioned reagents are actually both Lewis acid and oxidants, serving as efficient reagents for overall transformation. Beside the silver(I) salts, silver(I) oxide has been traditionally used as one-electron heterogeneous oxidant for coupling reactions of phenols and phenol-ethers to give biaryls in low to moderate yields. Persulfate anion is capable to generate the Ag(II), also very powerful reagent which also has been employed in the OCA reactions, however, with limited success [36].

The most commonly employed reagents for oxidative coupling of electron-rich arenes are iron(III) salts acting as Lewis acids and one-electron oxidants in the same time [37-48]. Among them, iron(III) chloride has been successfully used wide-spread in the phenolic oxidative coupling reactions. Since iron(III) salts are one-electron oxidants, two moles are required to couple two aromatic rings. The hydrated iron(III) chloride reagents have to be distinguished from the anhydrous ones since the latter are stronger Lewis acids giving lower yields in the oxidative coupling reactions of more sensitive substrates. The reactions with anhydrous reagent were conducted either in dichloromethane solution [37], or under the solventless conditions with $FeCl_3$ previously adsorbed on acidic alumina [38]. Thus, 2-isopropoxyanisole (**382**) was readily cross-coupled with biaryl **383** to afford the triphenylene **384** in 72% yield [37], respectively, Scheme 11.

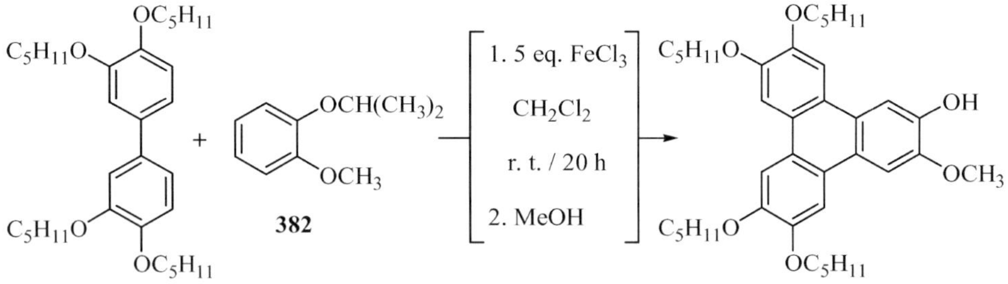

Scheme 11

The reaction is accompanied with the cleavage of *O*-isopropyl protecting group, which occured after addition of methanol. Anhydrous iron(III) chloride efficiently affects the room temperature oxidative coupling reactions of electron-rich aromatics other than phenols and phenol-ethers, e.g. thiophenes [39], as well as the cross-couplings, e.g. *N*-alkylcarbazoles with pyridine-*N*-oxides [40], etc.

Slightly less reactive and far more selective is a silica gel bounded $FeCl_3 \cdot 6H_2O$. The reagent is easily prepared by adding the oven-dried (150 °C) chromatographic silica gel to the previously prepared solution of $FeCl_3 \cdot 6H_2O$ in diethyl eter / methanol (9.5:0.5), followed by evaporation and drying (80 °C, high vacuum) [41]. Thus prepared moisture- and light-sensitive yellow powder was found to be excellent OCA reagent providing high yield of products derived from catechols, catechol-ethers or other properly disubstituted (*ortho-ortho* or *ortho-para*) mono-, di-, and tri-phenols [41,42]. For instance, 4-methyl-1,2-dimethoxybenzene (**385**) was converted to biaryl **386** in 95% yield [41], respectively, Scheme 12.

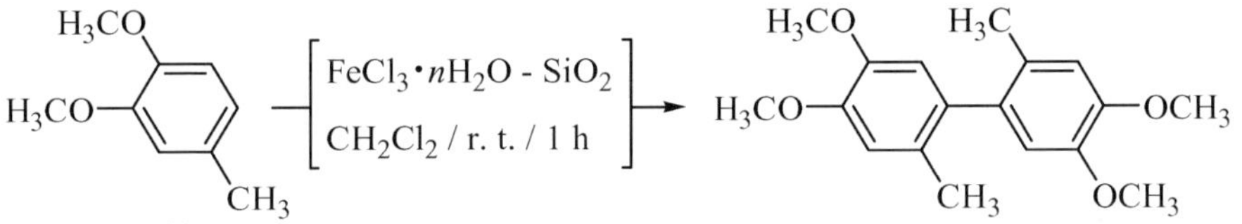

Scheme 12

Moreover, $FeCl_3 \cdot 6H_2O$-mediated oxidative coupling reactions have been performed under the solventless conditions by simple milling a mixture of the reagent and phenol to be coupled, and keeping the fine powder at slightly elevated temperatures. Thus 2,2'-binaphthol (**4**) was obtained by this method from 2-naphthol (**378**) in 95% yield [43], respectively, Scheme 13.

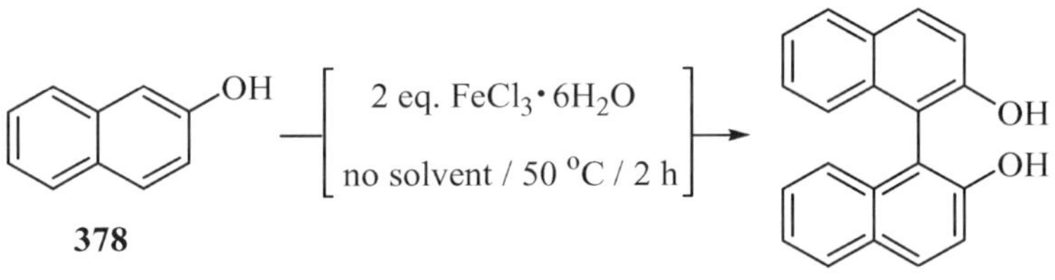

Scheme 13

In contrast, the same reaction in aqueous methanol proceeds much less efficiently giving the expected biaryl in 60% yield. However, further improvements have been achieved by performing the reaction in water as the only solvent (at 50 °C) when excellent yields of binaphthol (**4**) were obtained using $FeCl_3 \cdot 6H_2O$, $Fe(NO_3)_3 \cdot 9H_2O$, or $NH_4Fe(SO_4)_2 \cdot 12H_2O$. These methods offer a really simple and effective green-chemistry approach suitable for high-scale production of binaphthol (**4**) [44]. The solventless $FeCl_3 \cdot 6H_2O$-mediated oxidative coupling reactions can be additionally improved under microwave irradiation (250 W, 1-2 min) with almost retained excellent yields and within drastically reduced reaction times. In this fashion, binaphthol (**4**) was obtained from 2-naphthol in 93-96% yield within 1 min [45]. Ferric chloride hexahydrate was also employed in the oxidative cross-coupling reactions between two electronically different substrates, for example, 2-naphthol (**378**) and 2-naphthylamine (**387**) were reacted to afford binaphthyl **388** in 78% yield, with 20% of binaphthol (**4**) as a homo-coupling by-product [46], Scheme 14.

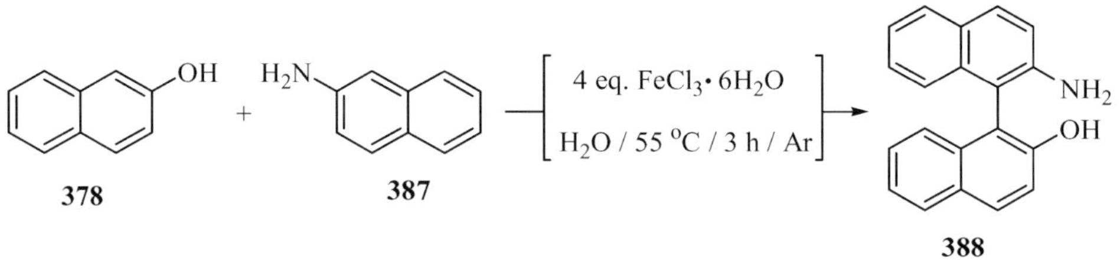

Scheme 14

Iron(III) chloride has readily affected the polymerization of benzene *via* radical-cation mechanism to poly-*p*-phenylene (**PPP**) in the presence of an equimolar amount of water [47]. An additional iron(III) complex which has been effective in the oxidative couplings of phenols and phenol-ethers is [Fe(DMF)$_3$Cl$_2$][FeCl$_4$]. This was prepared by adding DMF (1.5 eq.) to a solution of anhydrous FeCl$_3$ (1 eq.) in dry diethyl ether, accompanied with precipitation of the complex as a yellowish green powder. The complex is not hygroscopic and could be crystallized from dichloromethane or ethanol. The oxidative coupling reactions with this complex were conducted in refluxing (35 °C) two-phase water / diethyl ether mixtures to furnish good yields of coupling products derived from mono-, diphenols and alkoxyphenols [48]. Further Fe(III)-based reagent is potassium hexacyanoferrate(III), rather traditional oxidative phenolic coupling reagent [49-54]. The reactions with K$_3$[Fe(CN)$_6$] are usually carried out in two-phase system: organic solvent (toluene or chloroform) / water in the presence of two equivalents of the reagent and a suitable base, e.g. K$_2$CO$_3$ or NaHCO$_3$. The latter is providing the phenolic substrate into the aqueous phase as appropriate water-soluble phenolates since the reaction actually proceeds in the aqueous phase. In certain instances, the products may lose the aromatic character in one ring. Thus in the synthesis of naturally occuring alkaloid narwedine, compound **389** was oxidatively converted to arylcyclohexadienone **390** which, upon subsequent intramolecular Michael addition, gave bromo-narwedine **391** with a 40% yield [52], respectively, Scheme 15. When the substrate structure favours the *ipso*-attack of an intermediate aryl-radical, the *spiro*-structure is generated, rather than the biaryl containing a disfavourable membered ring. The yields are often low, but in the case of substrates bearing no oxidant-sensitive functionalities prone to uniform intermolecular homo-coupling or intramolecular cross-coupling reactions, fair yields can be achieved.

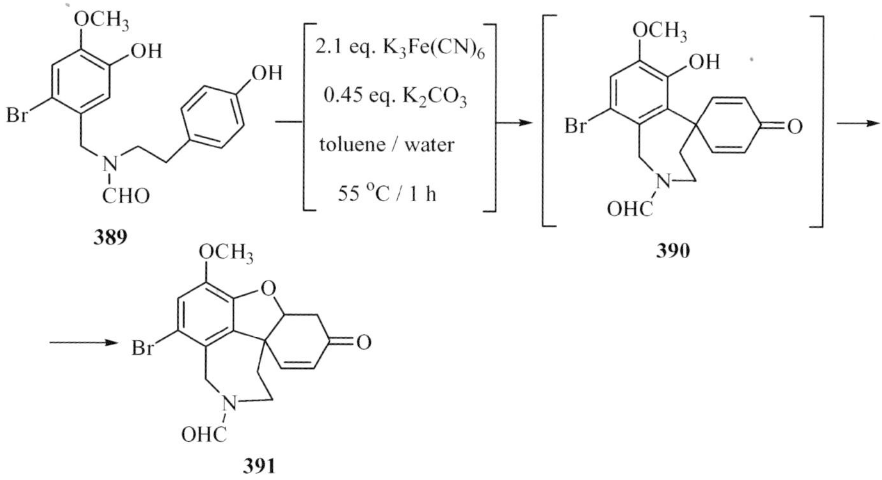

Scheme 15

The yields, if desired, can be significantly increased by employing buffered systems, e.g. less alkaline aqueous phase for more oxidant-sensitive substrates, vigorous high speed stirring which has a profound effect on the reaction yields, and the use of higher dilution factor [52].

Beside Fe(III)-containing reagents, copper(II) salts in connection with amines (four-fold excess to Cu^{2+}), due to their mild oxidizing properties, play an important role in the oxidative coupling reactions of electron-rich arenes, e.g. phenols [55-58], anilines [59], etc. Amines such as α-phenylethylamine [55], 1,2-ethylenediamine [56], tetraethylethylenediamine [56], tetramethylethylenediamine [60,61], isopropanolamine [57], benzylamine [57,59], α-methylphenylethylamine [57], t-butylamine [58], and other amines whose effectiveness is decreased in order: $R\text{-}NH_2 > R_2NH > R_3N$ form the copper(II) complexes which are efficient reagents for oxidative coupling of phenols. The reaction proceeds within the copper(II) coordinative sphere, as clearly shown through the yields in homo-coupling reaction of 2-naphthol as a function of molar ratio amine / Cu^{2+}. Thus the highest yields were obtained when the molar ratio was four, whereas at higher ratios amine vs. Cu^{2+}, the coordination of phenol to copper(II) is precluded, resulting with diminished yields [57]. The copper(II) salts are also capable to oxidize the phenols to biaryls in the presence of pyridine [56] or tetramethylethylenediamine [60,61] catalytically under an oxygen atmosphere. Moreover, the process has been carried out enantioselectively by using the copper(II) complexes of chiral amines, see Chapter 8. Beside homo-couplings, two electronically different substrates can be cross-coupled with good chemoselectivity. For example, mild reactive, and therefore highly selective Cu(t-BuNH$_2$)$_4$Cl$_2$, *in situ* prepared by mixing copper(II) chloride and t-butylamine (4 eq.) in methanol at room temperature or slightly above, readily affects a chemoselective cross-coupling reaction between two 2-naphthols having different redox-potentials. Thus 2-naphthol (**378**), bearing electron-donating group(s) in phenolic ring is oxidized more readily to generate an aryl-radical **392**, which further undergoes the arylation of electron-poor naphthol **393**. An intermediate radical **394**, formed in this manner, is additionaly oxidized to the dicyclohexadienone which gives more stable tautomeric unsymmetrical 2,2'-binaphthol (**395**) in 91% yield with only 4-5% of symmetrical homo-coupling by-products [39], Scheme 16.

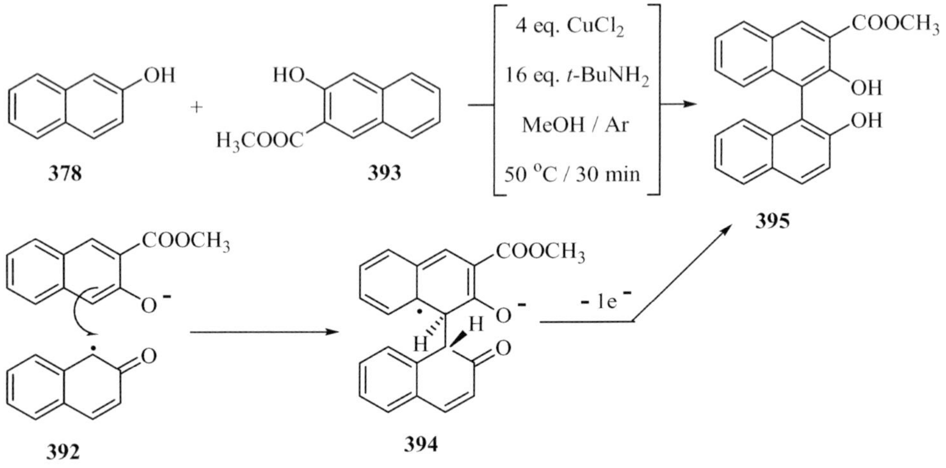

Scheme 16

Slightly more electron-rich substrate, 2-aminoanthracene (**396**), was also converted to its homo-coupled product **397** by applying Cu(BnNH$_2$)$_4$Cl$_2$ complex, however, accompanied with formation of the corresponding carbazole **398** in almost equimolar ratio [59], Scheme 17.

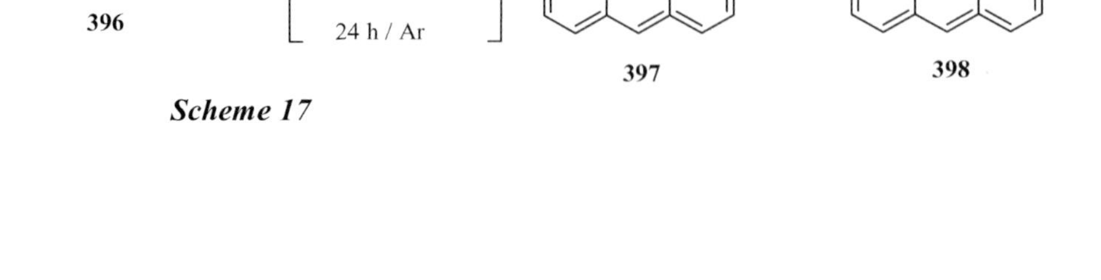

Scheme 17

Beside copper(II chloride, other simple copper salts such as copper(II) acetate, nitrate and sulfate have been employed in the oxidative coupling reactions [55]. The complex Cu(OH)Cl·TMEDA, which is readily prepared by mixing an equimolar amount of copper(I) chloride and tetramethylethylenediamine in 95% aqueous methanol under an oxygen atmosphere, proved to be an excellent catalyst (1-8 mol%) for the oxidative phenolic coupling reactions with oxygen or air as an ultimate oxidant [60,61]. Further selective heterogeneous copper(II)-based oxidant is $CuSO_4$ adsorbed on neutral alumina. The latter catalysed the homo-coupling of 2-naphthol (**378**) at 20 mol% catalyst loading in refluxing chlorobenzene at 140 °C for 8 h with air bubbling, to furnish the binaphthol (**4**) in 97% yield, respectively [62]. The oxidative coupling of arenes mediated by copper(II) chloride have been performed also in the presence of strong Lewis acids such as aluminum chloride [63-67]. Thus Kovacic's group [63] developed well known method for polymerization of benzene to poly-*p*-phenylene (**PPP**) with this system at equimolar ratio in the presence of a small amount of water, e.g. 10 mol% to $AlCl_3$, Scheme 18.

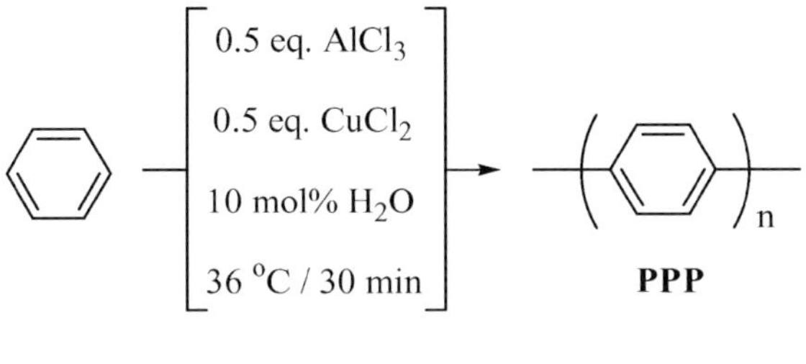

Scheme 18

Aluminum chloride reacts with an arene to give the arene radical-cation, which further follows the generally accepted oxidative coupling reaction mechanism, by an influence of copper(II) chloride as an oxidant to finally give a well defined polymeric product [65].

Among vanadium-based reagents, salts and complexes in the oxidation state +5 have been found as effective oxidants for the coupling reactions of electron-rich arenes. Holton's group [68] has introduced vanadium oxytrichloride as the first V(V) reagent, slightly less versatile than vanadium oxytrifluoride [69-75]. Quite recently, ammonium metavanadate in perchloric acid [76], and VO(acac)$_2$ [77] in the presence of oxygen were developed as more convenient and even catalytic vanadium systems. The reactions with vanadium reagents VOX_3, X = F, Cl are usually performed in ethereal solvents such as diethyl ether or tetrahydrofuran [68,69], in trifluoroacetic acid (TFA) [73,74], with or without TFA-anhydride added [69,71], in a mixture of TFA and dichloromethane [72], or more conveniently in dichloromethane with BF_3·Et_2O [75]. The vanadium oxytrihalides are capable of coupling polyphenols, polyphenol-ethers, alkylphenols, alkylphenol-ethers and polyalkoxyphenols to the respective biphenyls in moderate to good yields. Like Cu(II) and Fe(III)-reagents, vanadium(V) is one-electron oxidant, thus at least two molar equivalents of the latter are required per single aryl-aryl bond formed. Once again, the substrate must be appropriately disubstituted (*ortho-ortho*, *ortho-para*) to rich high yields of biphenyls. The free *ortho*-position to the phenolic hydroxy group is usually chlorinated with $VOCl_3$. Otherwise, at substrates with free, but sterically hindered positions, the *para-para* oxidative phenolic coupling reaction takes place to afford biphenyls in fair yields. In this manner, methoxycresols **376** and **399** were reacted with $VOCl_3$ to give *ortho*-chloro derivative **400** or *para-para* homo-coupling product **401**, when an *ortho*-position is too hindered for chlorination [69], Scheme 19.

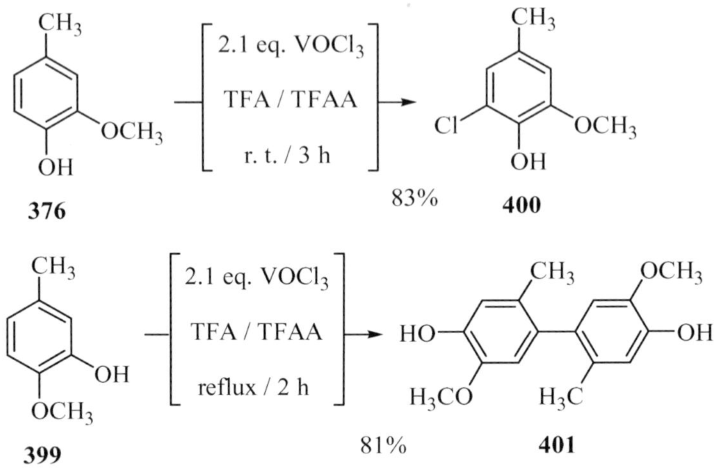

Scheme 19

Intramolecular oxidative coupling reactions affording the phenanthrene structure were effected with great success by the most common method applying vanadium oxytrifluoride in the presence of $BF_3 \cdot Et_2O$. For example, compound **402** was converted to phenanthrene **403** in 95% yield under very mild reaction conditions, respectively [75], Scheme 20.

In addition, convenient vanadium(V) reagent was prepared by mixing an aqueous solution of ammonium metavanadate with perchloric acid (more than 10 eq. to V) [76]. This reagent readily coupled various naphthol-ethers to the corresponding binaphthyls in good yields, yet, these substrates were more efficiently homo-coupled using iron(III) chloride-based reagents.

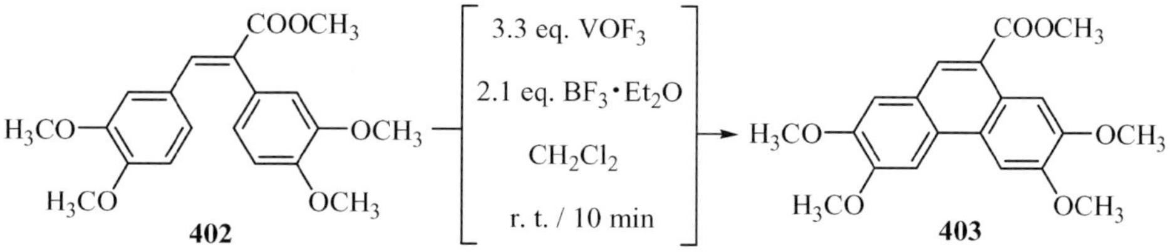

Scheme 20

Apart from phenols and phenol-ethers, vanadium reagents were effective in the oxidative couplings of electron-rich heterocycles. Thus poly-3-hexylthiophene (**404**), with average M_r 2.1 x 10^4 and head-to-tail content of about 65%, was prepared by oxidative polymerization of 3-hexylthiophene (**405**) with oxygen catalysed with vanadyl(IV)-acetylacetonate and iron(III) chloride in 60% yield [77], Scheme 21.

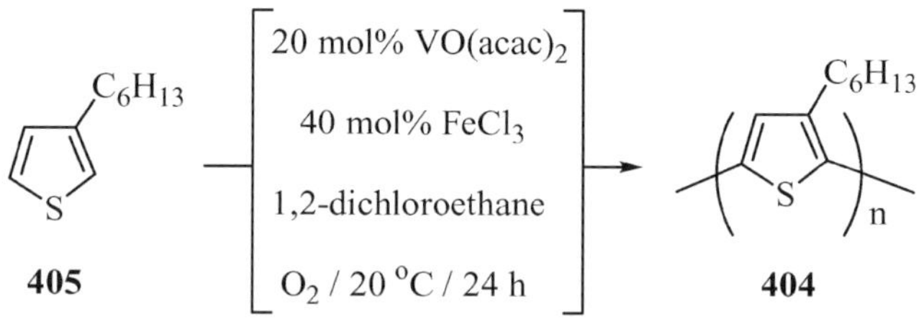

Scheme 21

The VO(acac)$_2$ improves the catalytic activity of iron(III) chloride causing higher yield and molecular weight of polymer. Neither VO(acac)$_2$, or FeCl$_3$, alone did not effect the reaction at such range of yield and degree of polymerization.

Whereas titanium based reagents in the oxidation states +2, +3 and +4 have found wide-spread and respective applications in organic chemistry, examples of titanium(IV) reagents, acting as oxidants, are quite rare. Nevertheless, electron-rich dialkylanilines or naphthidines can be oxidatively coupled with titanium(IV) chloride in the presence of triethylamine to give the *para-para* homo-coupling products, *N,N,N'N'*-tetraalkylbenzidines and naphthidines in good to excellent yields. Thus *N,N*-diethylaniline (**406**) gave *N,N,N',N'*-tetraethylbenzidine (**407**) in 92% yield [78], respectively, Scheme 22. However, the reactions with *N,N*-dimethylanilines furnish the *N*-monomethyl derivatives due to oxidative demethylation [78].

Molybdenum pentachloride smoothly reacts with 1,2-dialkoxybenzenes or 4-substituted-1,2-dialkoxybenzenes to give hexaalkoxytriphenylenes or tetraalkoxy-disubstituted biphenyls in good to excellent yields tolerating several labile groups including ester, acetal, triisopropylsilyl, etc [79,80].

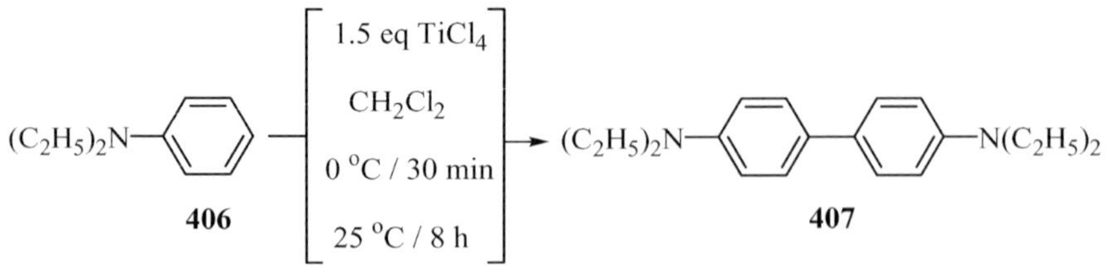

Scheme 22

For instance, compound **408** was dimerized to biphenyl **409** in almost quantitative yield [80], respectively, Scheme 23.

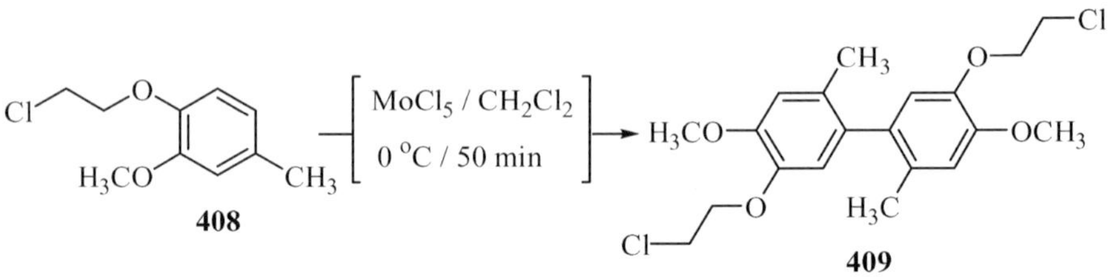

Scheme 23

Benzyl-, *p*-nitrobenzyl, *p*-methoxybenzyl and allyl-protecting groups were cleavaged, whereas acetate- and 2-methoxyacetate-masked phenols, *meta*- or *para*-disubstituted dialkoxybenzenes fail to undergo the reaction with $MoCl_5$. Since the oxidation state of outcoming molybdenum species is still unknown, one [80] (presumably Mo^{IV}) to two [79] (Mo^{III}) equivalents of $MoCl_5$ were employed per single aryl-aryl bond formed. Molybdenum pentachloride also effects the polymerization of benzene to poly-*p*-phenylene (**PPP**) under mild reaction conditions acting as Lewis acid and oxidant [81]. Among molybdenum(VI) reagents, $MoOCl_4$ in the presence of trifluoroacetic acid in chloroform solution, readily affects the oxidative phenolic coupling reactions in good yields [82].

Ruthenium(IV) tetrakis(trifluoroacetate), $Ru(TFA)_4$, *in situ* prepared by addition of $RuO_2 \cdot H_2O$ to a dichloromethane solution of trifluoroacetic acid (4 eq.), its anhydride (0.55-2 eq., dehydrating agent) and $BF_3 \cdot Et_2O$ (1-2 eq.), is also an effective oxidant [83-85]. This reagent smoothly oxidizes polyphenols and polyphenol-ethers to the respective biphenyls in high yields under very mild reaction conditions. Thus, compound **410** was converted to neoisostegan (**411**) in almost quantitative yield [85] Scheme 24. In compare to $VOCl_3$, VOF_3 or $Tl(TFA)_3$-mediated oxidative couplings, $Ru(TFA)_4$ is apparently more selective oxidant providing clean conversions and higher yields. However, the extreme expense of ruthenium chemicals seriously limits a wide applicability of this, otherwise very powerful synthetic method.

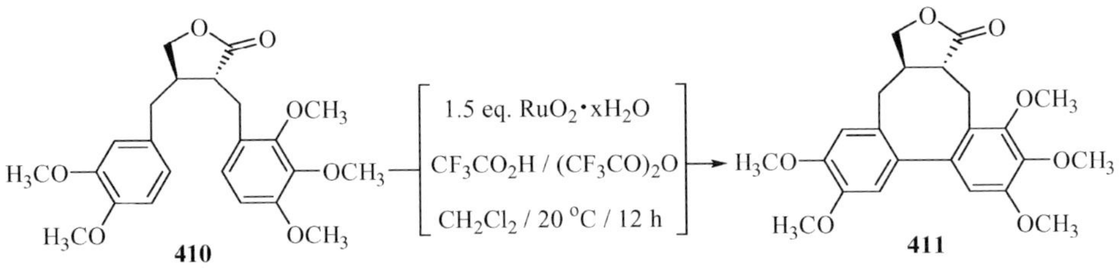

Scheme 24

Another useful one-electron oxidant is cobalt(III) trifluoroacetate which, in some instances, induces the oxidative coupling of arenes, but, with aromatic substitution, trifluoroacetoxylation, as the main side-reaction [86].

Among lanthanides, cerium(IV) salts and complexes are important reagents in both analysis and preparative organic chemistry. Cerium(IV) salts are reduced to Ce(III) acting as typical one-electron oxidants. Cerium(IV) sulfate, as the most common salt, is not suitable for synthetic purposes, because of its rather sparing solubility in water and other polar organic solvents. However, ammonium cerium(IV) nitrate (CAN) was found to be a versatile oxidizing agent in organic synthesis. Thus high yields of binaphthols were obtained in oxidative coupling of variously substituted 2-naphthols under very mild reaction conditions (r. t.) and within short reaction times [87]. The CAN-mediated reactions were performed in methanol or acetonitrile by employing an equimolar amount of the reagent per single aryl-aryl bond formed.

Popular two-electron oxidant lead(IV) acetate (LTA) [88-92], in the presence of $BF_3 \cdot Et_2O$ in dichloromethane, smoothly couples phenols and phenol-ethers to the respective biaryls in good to excellent yields [93]. Once again, at certain electron-poor substrates, the aromatic substitution - acetoxylation, became the major process.

The use of hipervalent iodine(III) reagents, iodosobenzene diacetate (PIDA) [94-98], or iodosobenzene bis(trifluoroacetate) (PIFA) [96-99] as stable crystalline substances has remarkably improved the oxidative couplings in many cases. This reagents are usually prepared by oxidation of iodobenzene with respective peracids or with sodium perborate in an excess of desired carboxylic acid, but are also commercially available. Other carboxylates are readily accessible by the ligand exchange reaction of acetates with the appropriate carboxylic acid [96,97]. For example, Bringmann's group [93] has described an elegant synthesis where binaphthol **412** was oxidized to naturally occuring anti HIV-1 agent Michellamine A (**413**) with PIFA in 82% yield, respectively, Scheme 25. Since PIFA and related reagents are two-electron oxidants, an equimolar amount of the latter is required per mole of biaryl formed. In comparison, LTA also gave high yield in the same reaction, Scheme 25.

PIFA and related reagents, as well as lead(IV) acetate, are compatible with unprotected phenolic hydroxy group.

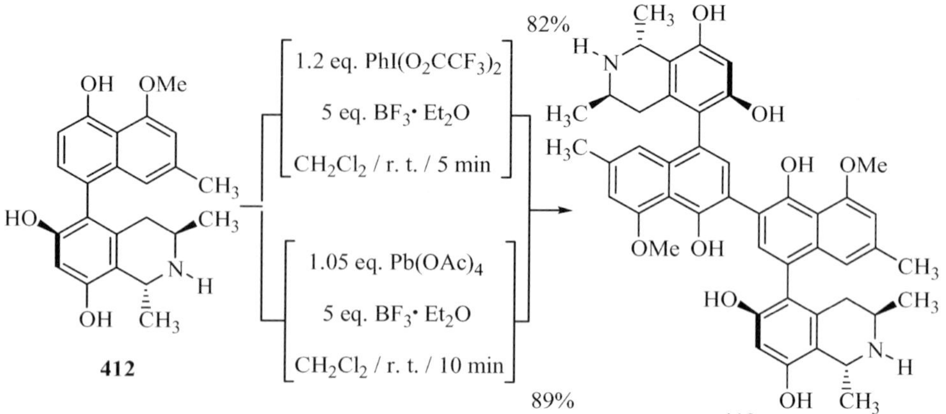

Scheme 25

Oxidative coupling reactions of electron-rich arenes have been accomplished with various peroxides [100,101]. Among them, di-*tert*-butyl peroxide is the most commonly used reagent. Either *ortho,ortho*- or *ortho,para*-disubstituted polyphenols and polyphenol-ethers can be homo-coupled in high yields by refluxing with an equimolar amount of the reagent in an inert solvent such as chlorobenzene. The peroxide-mediated coupling reaction apparently proceeds *via* the free-radical mechanism: generation of aryl radicals at unsubstituted, either *ortho*- or *para*-position to the phenolic hydroxy group, followed by free-radical arylation, and further oxidation of resulting arylcyclohexadienyl-radical. Namely, in the cotton-seed pigment gossypol (**2**) synthesis from naphthol **414**, the key-intermediate **415** was obtained in 92% yield [101], respectively, Scheme 26.

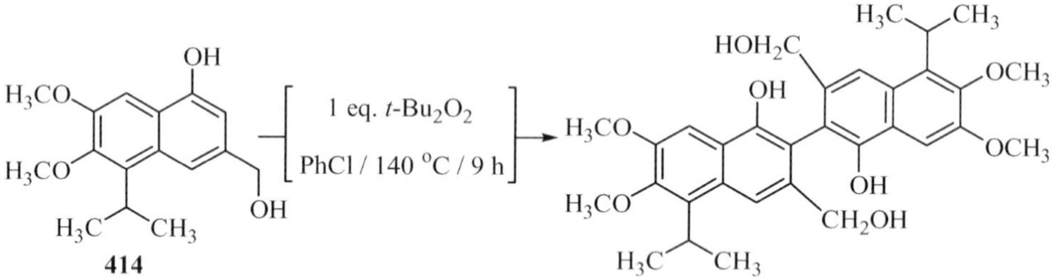

Scheme 26

The most important side-reaction which occurs during oxidative phenolic coupling reactions with *t*-Bu$_2$O$_2$ is the formation of diarylether. The latter is generated by reaction of phenoxy-radical (tautomeric form of α-hydroxyaryl radical) with the second molecule of arene. Oxidative coupling of electron-rich arenes involving di-*tert*-alkyl peroxides is important, environmental friendly, alternative synthetic method that avoids the use of any toxic metallic reagent. The oxidative coupling of arenes can also be effected electrochemically [102-104]. For example, laudanosine derivative **416** was converted by anodic oxidation at an anode potential of +0.95 V (100 mA; 2 Fmol^{-1}) *via* unstable intermediate **417** to phenanthrene **418** in 35% overall yield [103], Scheme 27.

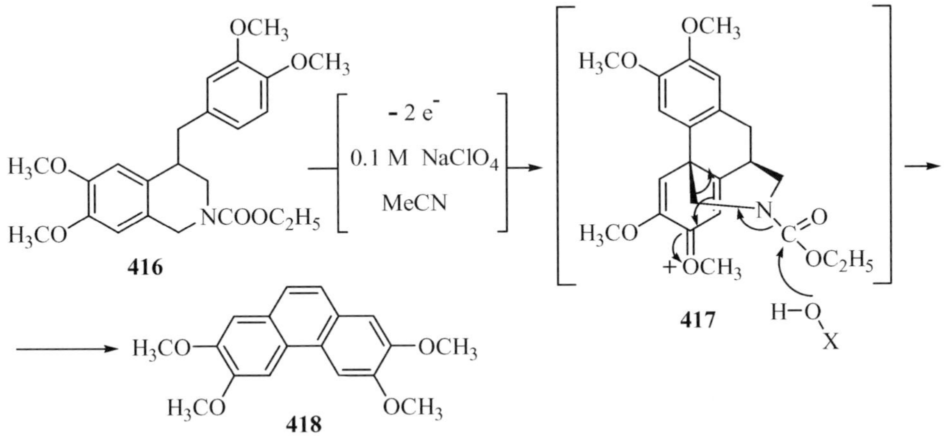

Scheme 27

Nitronium salts are capable to induce the OCA reaction *via* generation of aryl radical-cations, even with relatively electron-poor arenes, such as benzene [105]. Namely, NOBF$_4$ in catalytic amount (1-5 mol%) efficiently catalysed the oxidative coupling reactions of polyalkoxybenzenes in dichloromethane (containing 20% trifluoroacetic acid) under exposure to air affording the corresponding biaryls in almost quantitative yields [105]. Finally, oxidative couplings of arenes can be accomplished under the classical Scholl reaction conditions [106]. For example, 1-ethoxynaphthalene (**419**) was coupled to binaphthyl **420** by reaction with anhydrous aluminum chloride and nitrobenzene (as oxidant) in 65% yield, Scheme 28.

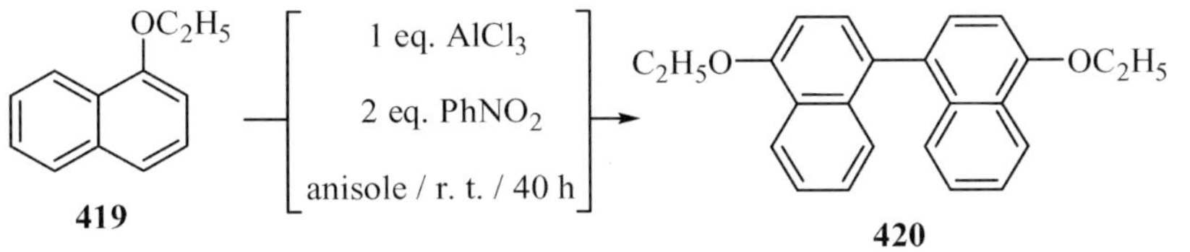

Scheme 28

The Scholl synthesis of biaryls proceeds mainly *via* the radical-cation mechanism, while in certain instances some evidences indicate, at least partially, cationic reaction pathway [106].

6.4. Selected synthetic procedures

> *6.4.1. Oxidative coupling reactions with thallium(III) trifluoroacetate: Preparation of 2,2'-dibromo-4,4',5,5'-tetramethoxybiphenyl (421) [12]*

To a solution of thallium(III) trifluoroacetate (5.50 g, 10 mmol) in trifluoroacetic acid (25 ml), 4-bromo-1,2-dimethoxybenzene (**422**, 4.34 g, 20 mmol) was added and the resulting mixture was stirred until a thallium(III) test was negative (the presence of unreacted Tl^{3+} causes the appearance of brown-black precipitate of Tl_2O_3 in the reaction of an aliquot with dilute aq. NaOH). Then, the reaction mixture was poured into water (150 ml) and extracted with chloroform (2x100 ml). The extracts were dried (MgSO$_4$), and passed through a short column of alumina and 3.50 g (81%) of pure 2,2'-dibromo-4,4',5,5'-tetramethoxybiphenyl (**421**), m.p. 159-160 °C, was obtained after evaporation and crystallization from toluene / petroleum ether (b.p. 100-120 °C).

> *6.4.2. Oxidative coupling reactions with methyltri-n-butylammonium permanganate: Preparation of 2,2'-dihydroxy-3,3'-dimethoxy-5,5'-dimethylbiphenyl (377) [27]*

2-Methoxy-4-methylphenol (**376**, 414 mg, 3.0 mmol) was dissolved in dichloromethane (10 ml) and a dichloromethane-solution of freshly prepared methyltri-*n*-butylammonium permanganate [28] (480 mg, 1.50 mmol) was added at

0 °C. Water (10 ml) was added after 10 min and the mixture was stirred for further 10 min. The organic phase was separated, dried (MgSO$_4$) and solvent was removed *in vacuo* to obtain the crude product (334 mg; 81%), which was further purified by preparative chromatography over silica gel with *n*-hexane / ethylacetate (3:1) as an eluent to obtain 223 mg (54%) of pure 2,2'-dihydroxy-3,3'-dimethoxy-5,5'-dimethyl biphenyl (**377**).

6.4.3. Oxidative cross-coupling of biphenyls with catechol-ethers using FeCl$_3$: Preparation of 2-hydroxy-3-methoxy-6,7,10,11-tetrapentyloxy triphenylene (384) [37]

To a stirred solution of 3,3',4,4'-tetra-*n*-pentyloxybiphenyl (**383**, 3.32 g, 6 mmol) and 1-isopropoxy-2-methoxybenzene (**382**, 1.66 g, 10 mmol, 1.67 eq.) in dichloromethane (50 ml), anhydrous iron(III) chloride (5.00 g, 30 mmol) was added slowly and the reaction mixture was stirred at room temperature overnight. Then, the mixture was poured into methanol (200 ml), and cooled to 0 °C. The solid was filtered off, washed with cold methanol, purified by column chromatography over silica gel using dichloromethane / petroleum ether (3:1), to give 2.67 g (72%) of pure **384** as a white solid, m.p. 118.5-119.5 °C.

6.4.4. Oxidative coupling reactions of 2-naphthols with FeCl$_3$: Preparation of 1,1'-binaphthyl-2,2'-diol (4) [43]

A mixture of 2-naphthol (**378**, 1.00 g, 7 mmol) and $FeCl_3 \cdot 6H_2O$ (3.80 g, 14 mmol, 2 eq.) was finely powdered by agate mortar and pestle. The mixture was then put in a test tube and kept at 50 °C for 2 h. Decomposition of the reaction mixture with dilute aqueous hydrochloric acid (10%, 50 ml), followed by extraction with dichloromethane (3x50 ml), drying of combined extracts (Na_2SO_4), filtration and evaporation gave the crude product. After additional purification by preparative chromatography on silica gel column with acetone / petroleum ether, 1.0 g (95%) of pure 1,1'-binaphthyl-2,2'-diol (**4**) was isolated, m.p. 217-218 °C.

6.4.5. *Oxidative cross-coupling of two electronically-different 2-naphthols with CuCl₂ in the presence of tert-butylamine: Preparation of 3-methoxy carbonyl-1,1'-binaphthyl-2,2'-diol (395) [58]*

Through a well stirred solution of 2-naphthol (**378**, 144 mg, 1 mmol), 3-methoxy carbonyl-2-naphthol (**393**, 202 mg, 1 mmol) and copper(II) chloride (538 mg, 4 mmol) in degassed methanol (20 ml) at room temperature, argon was bubbled for several minutes. Then, with a vigorous stirring, *tert*-butylamine (16 ml of 1 M solution in methanol, 16 mmol, 4 eq. to Cu^{2+}) was added dropwise, and the heterogeneous reaction mixture was heated at 50 °C for 30 min. The reaction mixture was cooled to approx. 10 °C, and decomposed with 6 M HCl. Methanol was evaporated and the product extracted with chloroform (3x20 ml). The pure product **395**, 313 mg (91%), was isolated by preparative chromatography on silica gel column (*n*-hexane / EtOAc).

6.4.6. *Oxidative coupling of phenols with VOF₃: Preparation of methyl 2,3,6,7-tetramethoxyphenanthrene-9-carboxylate (403) [75]*

To a stirred solution of methyl (*E*)-3,4-dimethoxy-α-(3',4'-dimethoxyphenyl methylene)benzeneacetate (**402**, 360 mg, 1 mmol) in dichloromethane / trifluoroacetic acid (25 ml; 2:3, v/v) at 0 °C, a solution of vanadium oxytrifluoride (350 mg, 2.8 mmol) in ethylacetate / trifluoroacetic acid (15 ml, 1:2, v/v) was added over 1 min. The mixture was stirred for 15 min, pored into citric acid (40 ml, 10% in water), the pH adjusted to 7.5 (with conc. aq. ammonia), and the organic phase separated. The aqueous phase was extracted with dichloromethane (2x15 ml) and combined organic phases washed with water (50 ml), dried (CaCl₂) and concentrated to dryness. The resulting brown solid (360 mg) was purified over silica gel with dichloromethane as an eluent to afford 264 mg (74%) of pure product **403** as fine crystals, m.p. 201-204 °C.

6.4.7. Oxidative coupling of catechol-ethers with MoCl₅: Preparation of 5,5'-bis(2-chloroethoxy)-4,4'-dimethoxy-2,2'-dimethylbiphenyl (409) [80]

To a solution of 1-(2-chloroethoxy)-2-methoxy-4-methylbenzene (**408**, 2.01g, 10 mmol) in dichloromethane (30 ml) cooled to 0 °C, molybdenum pentachloride (2.73 g, 10 mmol) was added. The reaction mixture was stirred at 0 °C for 50 min, then quenched with saturated aqueous NaHCO₃ solution (50 ml). Subsequent washings of the combined organic phases with water (100 ml), as well as brine (100 ml), and concentration provided the crude product that was purified by column chromatography on silica gel using cyclohexane / ethyl acetate as an eluent followed by recrystallization from ethyl acetate to give 1.98 g (99%) of pure product **409**, m.p. 132-134 °C.

6.4.8. Oxidative coupling of 2-naphthols with (NH₄)₂[Ce(NO₃)₆]: Preparation of 1,1'-binaphthyl-2,2'-diol (4) [87]

2-Naphthol (**378**, 0.36 g, 2.5 mmol) was dissolved in methanol (20 ml) at room temperature. Nitrogen was bubbled through the solution, into which ammonium cerium(IV) nitrate (1.39 g, 2.53 mmol) in methanol (15 ml) was added with stirring until the mixture turned reddish. The mixture was evaporated *in vacuo* and the residue was poured into water (100 ml) at 0 °C and filtered. The crude product was purified by silica gel chromatography using a petroleum ether / diethyl ether (3:1) as an eluent to afford 329 mg (92%) of pure 1,1'-binaphthyl-2,2'-diol (**4**) as a white crystalline solid, m.p. 215 °C.

6.4.9. Oxidative coupling of phenols with PhI(CF₃COO)₂ in the presence of BF₃·Et₂O: Preparation of Michellamine A (413) [93]

To a cooled (0 °C) suspension of compound **412** (31.6 µmol) in anhydrous dichloromethane (5 ml), BF₃·Et₂O (0.06 ml, 488 µmol, 15 eq.) was added. After stirring for 1 min, a solution of PhI(CF₃COO)₂ (17.0 mg, 39.5 µmol, 1.25 eq.) in anhydrous dichloromethane (2 ml) was added over a period of 5 min. After stirring for another 5 min, water (5 ml) was added, the phases were separated and the organic phase was extracted with water / methanol (8:2) several times. The combined aqueous layers were concentrated under reduced pressure and lyophilized. The residue was dissolved in dichloromethane / methanol (8:2), and the inorganic insoluble materials were filtered off through a short Celite column using the same eluent. The solvent was removed under reduced pressure, the residue was purified by HPLC giving 9.9 mg (82%) of pure Michellamine A (**413**), m.p. >220 °C (dec.); $[\alpha]_D^{20}$ -10.6 (c = 0.13, MeOH).

Note: The same procedure was carried out by using lead(IV) tetraacetate [88-92] giving the Michellamine A in slightly higher yield (89%) [93].

6.4.10. Oxidative coupling of phenols with t-Bu₂O₂: Preparation of 2,2'-binaphthol 415 [101]

6.4.10. *Oxidative coupling of phenols with t-Bu$_2$O$_2$: Preparation of 2,2'-binaphthol 415 [101]*

414 **415**

A solution of compound **414** (13.81 g, 0.05 mol) and di-*tert*-butyl peroxide (7.30 g, 0.05 mol, 1 eq.) in chlorobenzene (600 ml) was heated under reflux for 9 h. Evaporation of the solvent and crystallization from benzene gave 12.7 g (92%) of pure product **415**, m.p. 161-163 °C.

6.5. Conclusion

The *o*xidative *c*oupling of *a*renes (OCA) to biaryls or polyaryls is an important access to symmetrical, as well as certain classes of unsymmetrical structures. Concerning the OCA reaction mechanism, the structures of starting compounds have to be activated and substituted at proper position(s) to prevent the side-reactions and to provide high-yield of desired biaryls. In this Chapter, the overview of OCA reactions methodology is presented. A number of transitional metal, one-electron: Ag^{I}, Cu^{II}, Mn^{III}, Fe^{III}, Co^{III}, Ti^{IV}, Ce^{IV}, V^{V}, and two-electron: Tl^{III}, Pb^{IV} oxidants are effective reagents for these reactions. Among non-metallic reagents, peroxides, nitronium salts, as well as hypervalent iodonium compounds are especially interesting reagents. The latter play an important role in the high-yielding and very selective oxidative couplings of electron-rich arenes.

6.6. References

1. P. Kovacic and M. B. Jones, Chem. Rev. 87 (1987) 357.
2. R. VanHelden and G. Verberg, Recl. Trav. Chim. Pays-Bas 84 (1965) 1263.
3. M. O. Unger and R. A. Fouty, J. Org. Chem. 34 (1969) 18.
4. H. Iataaki and H. Yoshimoto, J. Org. Chem. 38 (1973) 76.

5.   F. R. S. Clark, R. O. C. Norman, C. B. Thomas and J. S. Willson, J. Chem. Soc., Perkin Trans. 1 (1974) 1289.

6.   T. Itahara, M. Hashimoto and H. Yumisashi, Synthesis (1984) 255.

7.   T. Yokota, S. Sakaguchi and Y. Ishii, Adv. Synth. Catal. 344 (2002) 849.

8.   M. Okamoto and T. Yamaji, Chem. Lett. (2001) 212.

9.   S. Mukhopadhyay, G. Rothenberg, D. Gitis and Y. Sasson, J. Org. Chem. 65 (2000) 3107.

10. I. H. Elson and J. K. Kochi, J. Am. Chem. Soc. 95 (1973) 5060.

11. S. F. Al-Azzawi and R. M. G. Roberts, J. Chem. Soc., Perkin Trans. II (1982) 677.

12. A. McKillop, A. G. Turrell, D. W. Young and E. C. Taylor, J. Am. Chem. Soc. 102 (1980) 6504.

13. E. C. Taylor, J. G. Andrade, G. J. H. Rall and A. McKillop, J. Am. Chem. Soc. 102 (1980) 6513.

14. E. C. Taylor, J. G. Andrade, G. J. H. Rall, I. J. Turchi, K. Steliou, G. E. Jagdmann and A. McKillop, J. Am. Chem. Soc. 103 (1981) 6856.

15. R. C. Cambie, G. R. Clark, P. A. Craw, P. S. Rutledge and P. D. Woodgate, Aust. J. Chem. 37 (1984) 1775.

16. J. S. Buckleton, R. C. Cambie, G. R. Clark, P. A. Craw, C. E. F. Rickard, P. S. Rutledge and P. D. Woodgate, Aust. J. Chem. 41 (1988) 305.

17. R. C. Cambie, P. A. Craw, P. S. Ruthledge and P. D. Woodgate, Aust. J. Chem. 41 (1988) 897.

18. J. K. Burden, R. C. Cambie, P. A. Craw, P. S. Rutledge and P. D. Woodgate, Aust. J. Chem. 41 (1988) 919.

19. A. A. Adesomoju, W. A. Davis, R. Rajaraman, J. C. Pelletier and M. P. Cava, J. Org. Chem. 49 (1984) 3220.

20. P. Magnus, J. Schultz and T. Gallagher, J. Chem. Soc., Chem. Commun. (1984) 1179.

21. P. Magnus, J. Schultz and T. Gallagher, J. Am. Chem. Soc. 107 (1985) 4984.

22. J. S. Sawyer and T. L. Macdonald, Tetrahedron Lett. 29 (1988) 4839.

23. M. A. Schwartz, B. F. Rose and B. Vishnuvajjala, J. Am. Chem. Soc. 95 (1973) 612.

24. J. Tormo, F. J. Moreno, J. Ruiz, L. Fajarí and L. Juliá, J. Org. Chem. 62 (1997) 878.

25. F. R. Hewgill and J. M. Stewart, J. Chem. Soc., Perkin. Trans. 1 (1988) 1305.

26. K. Yamamura, S. Ono and I. Tabushi, Tetrahedron Lett. 29 (1988) 1797.

27. M. Albrecht and M. Schneider, Synthesis (2000) 1557.

28. F. A. Marques, F. Simonelli, A. R. M. Oliviera, G. L. Gohr and P. C. Leal, Tetrahedron Lett. 39 (1998) 943.

29. B. B. Snider, Chem. Rev. 96 (1996) 339.

30. S. Uemura, T. Ikeda, S. Tanaka and M. Okano, J. Chem. Soc., Perkin Trans. 1 (1979) 2574.

31. K. Nyberg and L.-G. Wistrand, Chem. Scripta 6 (1974) 234.

32. M. J. S. Dewar and T. Nakaya, J. Am. Chem. Soc. 90 (1968) 7134.

33. B. Franck, J. Lubs and G. Dunkelmann, Angew. Chem. 79 (1967) 989.

34. W. E. Doering and M. Finkelstein, J. Org. Chem. 23 (1958) 141.

35. F. Effenberger, K.-E. Mack, R. Niess, F. Reisinger, A. Steinbach, W.-D. Stohrer, J. J. Stezowski, I. Rommel and A. Maier, J. Org. Chem. 53 (1988) 4379.

36. R. G. R. Bacon and A. R. Izzat, J. Chem. Soc. C (1966) 791.

37. R. J. Bushby and Z. Lu, Synthesis (2001) 763.

38. G. Cooke, V. Sage and T. Richomme, Synth. Commun. 29 (1999) 1767.

39. G. Barbarella, M. Zambianchi, R. Di Toro, M. Colonna, D. Iarossi, F. Goldoni and A. Bongini, J. Org. Chem. 61 (1996) 8285.

40. J. Zheng, C. Zhan, J. Qin and R. Zhan, Chem. Lett. (2002) 1222.

41. T. C. Jempty, L. L. Miller and Y. Mazur, J. Org. Chem. 45 (1980) 749.

42. H.-Y. Li, T. Nehira, M. Hagiwara and N. Harada, J. Org. Chem. 62 (1997) 7222.

43. F. Toda, K. Tanaka and S. Iwata, J. Org. Chem. 54 (1989) 3007.

44. K. Ding, Y. Wang, L. Zhang, Y. Wu and T. Matsuura, Tetrahedron 52 (1996) 1005.

45. L. Xu, F. Li, C. Xia and W. Sun, Synth. Commun. 33 (2003) 2763.

46. K. Ding, Q. Xu, Y. Wang, J. Liu, Z. Yu, B. Du, Y. Wu, H. Koshima and T. Matsuura, Chem. Commun. (1997) 693.

47. P. Kovacic and F. W. Koch, J. Org. Chem. 28 (1963) 1864.

48. S. Tobinaga and E. Kotani, J. Am. Chem. Soc. 94 (1972) 309.

49. C. G. Haynes, A. H. Turner and W. A. Waters, J. Chem. Soc. (1956) 2823.

50. D. A. Young, E. Young, D. G. Roux, E. V. Brandt and D. Ferreira, J. Chem. Soc., Perkin Trans. 1 (1987) 2345.

51. D. A. Chaplin, N. Fraser and P. D. Tiffin, Tetrahedron Lett. 38 (1997) 7931.

52. B. Küenburg, L. Czollner, J. Fröhlich and U. Jordis, Org. Proc. Res. & Dev. 3 (1999) 425.

53. T. Kametani, K. Yamaki, H. Yagi and K. Fukomuto, Chem. Commun. (1969) 425.

54. T. Kametani, C. Seino, K. Yamaki, S. Shibuya, K. Fukumoto, K. Kigasawa, F. Satoh, M. Hiiragi and T. Hayasaka, J. Chem. Soc. C (1971) 1043.

55. B. Feringa and H. Wynberg, Tetrahedron Lett. (1977) 4447.

56. K. Kushioka, J. Org. Chem. 48 (1983) 4948.

57. J. Brussee, J. L. G. Groenendijk, J. M. Koppele and A. C. A. Jansen, Tetrahedron 41 (1985) 3313.

58. M. Hovorka, J. Günterová and J. Závada, Tetrahedron Lett. 31 (1990) 413.

59. Š. Vyskočil, M. Smrčina, M. Lorenc, I. Tišlerová, R. D. Brooks, J. J. Kulagowski, V. Langer, L. J. Farrugia and P. Kočovský, J. Org. Chem. 66 (2001) 1359.

60. M. Noji, M. Nakajima and K. Koga, Tetrahedron Lett. 35 (1994) 7983.

61. B. H. Lipshutz, B. James, S. Vance and I. Carrico, Tetrahedron Lett. 38 (1997) 753.

62. T. Sakamoto, H. Yonehara and C. Pac, J. Org. Chem. 62 (1997) 3194.

63. P. Kovacic and A. Kyriakis, Tetrahedron Lett. (1962) 467.

64. L.-S. Wen and P. Kovacic, Tetrahedron 34 (1978) 2723.

65. J. J. Rooney and R. C. Pink, Proc. Chem. Soc. (1961) 142.

66. P. Kovacic and A. Kyriakis, J. Am. Chem. Soc. 85 (1963) 454.

67. P. Kovacic and J. Oziomek, J. Org. Chem. 29 (1964) 100.

68. M. A. Schwartz, R. A. Holton and S. W. Scott, J. Am. Chem. Soc. 91 (1969) 2800.

69. J. Quick and R. Ramachandra, Tetrahedron 36 (1980) 1301.

70. R. E. Damon, R. H. Schlessinger and J. F. Blount, J. Org. Chem. 41 (1976) 3772.

71. S. F. Dyke and P. Warren, Tetrahedron 35 (1979) 2555.

72. A. J. Liepa and R. E. Summons, J. Chem. Soc., Chem. Commun. (1977) 826.

73. S. M. Kupchan, A. J. Liepa, V. Kameswaran and R. F. Bryan, J. Am. Chem. Soc. 95 (1973) 6861.

74. T. Hirao, Chem. Rev. 97 (1997) 2707.

75. B. Halton, A. I. Maidment, D. L. Officer and J. M. Warnes, Aust. J. Chem. 37 (1984) 2119.

76. B. Hazra, S. Acharya, R. Ghosh, A. Patra and A. Banerjee, Synth. Commun. 29 (1999) 1571.

77. S. Yu, T. Hayakawa and M. Ueda, Chem. Lett. (1999) 559.

78. M. Periasamy, K. N. Jayakumar and P. Bharathi, J. Org. Chem. 65 (2000) 3548.

79. S. R. Waldvogel, Synlett (2002) 622.

80. B. Kramer, R. Fröhlich, K. Bergander and S. R. Waldvogel, Synthesis (2003) 91.

81. P. Kovacic and R. M. Lange, J. Org. Chem. 28 (1963) 968.

82. S. M. Kupchan and A. J. Liepa, J. Am. Chem. Soc. 95 (1973) 4062.

83. Y. Landais, A. Lebrun, V. Lenain and J.-P. Robin, Tetrahedron Lett. 28 (1987) 5161.

84. Y. Landais, D. Rambault and J.-P. Robin, Tetrahedron Lett. 28 (1987) 543.

85. Y. Landais and J.-P. Robin, Tetrahedron Lett. 27 (1986) 1785.

86. J. K. Kochi, R. T. Tang and T. Bernath, J. Am. Chem. Soc. 95 (1973) 7114.

87. P. Jiang and S. Lu, Synth. Commun. 31 (2001) 131.

88. F. Wessely, J. Kotlan and W. Metlesics, Monatsch. Chem. 85 (1954) 69.

89. D. L. Allara, B. C. Gilbert and R. O. C. Norman, Chem. Commun. (1965) 319.

90. J. B. Aylward, J. Chem. Soc. B (1967) 1268.

91. R. O. C. Norman, C. B. Thomas and J. S. Willson, J. Chem. Soc. B (1971) 518.

92. G. W. K. Cavill and D. H. Solomon, J. Chem. Soc. (1955) 1404.

93. G. Bringmann, W. Saeb, J. Mies, K. Messer, M. Wohlfarth and R. Brun, Synthesis (2000) 1843.

94. D. G. Vanderlaan and M. A. Schwartz, J. Org. Chem. 50 (1985) 743.

95. Y. Kita, M. Arisawa, M. Gyoten, M. Nakajima, R. Hamada, H. Tohma and T. Takada, J. Org. Chem. 63 (1998) 6625.

96. P. J. Stang and V. V. Zhdankin, Chem. Rev. 96 (1996) 1123.

97. V. V. Zhdankin and P. J. Stang, Chem. Rev. 102 (2002) 2523.

98. D. Krikorian, V. Tarpanov, S. Parushev and P. Mechkarova, Synth. Commun. 30 (2000) 2833.

99. H. Tohma, H. Morioka, S. Takizawa, M. Arisawa and Y. Kita, Tetrahedron 57 (2001) 345.

100. D. R. Armstrong, R. J. Breckenridge, C. Cameron, D. C. Nonhebel, P. L. Pauson and P. G. Perkins, Tetrahedron Lett. 24 (1983) 1071.

101. V. I. Ognyanov, O. S. Petrov, E. P. Tiholov and N. M. Mollov, Helv. Chim. Acta 72 (1989) 353.

102. P. Bird, M. Powell and M. Sainsbury, J. Chem. Soc., Perkin Trans. 1 (1983) 2053.

103. A. J. Majeed, P. J. Patel and M. Sainsbury, J. Chem. Soc., Perkin Trans. 1 (1985) 1195.

104. T. Osa, A. Yildiz and T. Kuwana, J. Am. Chem. Soc. 91 (1969) 3994.

105. F. Radner, J. Org. Chem. 53 (1988) 704.

106. G. A. Clowes, J. Chem. Soc. C (1968) 2519.

7. MISCELLANEOUS METHODS FOR SYNTHESIS OF BIARYLS

7.1. The Motherwell synthesis of biaryls

Motherwell's group developed an original approach to the synthesis of biaryls based on the intramolecular free-radical *ipso*-substitution reaction. This includes the reaction of *ortho*-iodo (or bromo) phenols, *N*-methylanilines or *N*-methylbenzamides with various arylsulfonyl chlorides to give the respective sulfonates **XXX**, *N*-methyl sulfonamides **XXXI**, or *N*-acylsulfonamides **XXXII**. The latter are reacted with tri-*n*-butyltin hydride in the presence of azobisisobutyronitrile (AIBN) in refluxing benzene to produce spirocyclic intermediate **XXXIII**, which, upon extrusion of sulfur dioxide, afforded the *ortho*-hydroxy- **XXXIV**, methyamino- **XXXV**, or *N*-methyl carboxamido-substituted biphenyls **XXXVI** in moderate to good yields [1-4], Scheme 1.

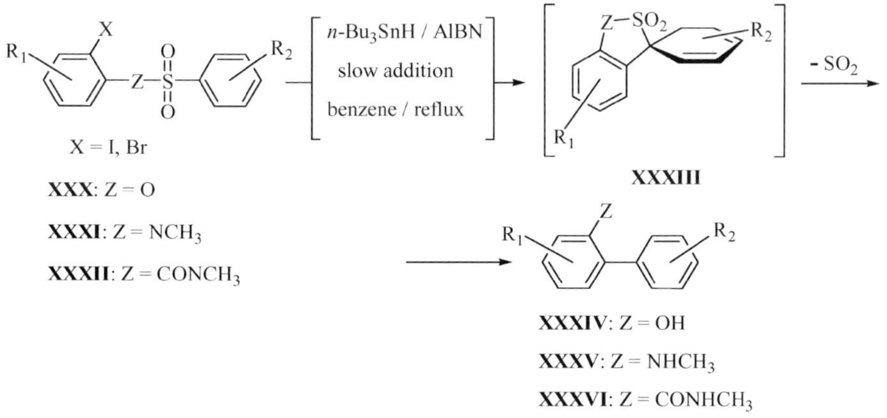

X = I, Br

XXX: Z = O

XXXI: Z = NCH$_3$

XXXII: Z = CONCH$_3$

XXXIII

XXXIV: Z = OH

XXXV: Z = NHCH$_3$

XXXVI: Z = CONHCH$_3$

Scheme 1

The reactions are performed by slow addition (10-15 h) of benzene solution of tri-*n*-butyltin hydride (1.3 eq.) and AIBN (0.7 eq.) to the boiling benzene solution of the corresponding Motherwell's precursor. Tri-*n*-butyltin hydride in the presence of radical initiator such as AIBN is well known reagent for the generation of aryl radicals from aryl iodides and bromides [5]. Side-reactions in this approach are the formation of direct free-radical addition and dehalogenation products. The formation of the latter side-product is minimized by slow addition of tin hydride. However, the formation of direct free-radical addition, the Pschorr-type product is a serious side-reaction. For example, compound **423** at 75% conversion gives 34% of biaryl **424** as *ipso*-product (Motherwell reaction pathway), and 39% of biaryl **425**, as the product of direct free-radical arylation, analogously to the Pschorr reaction pathway [1], Scheme 2.

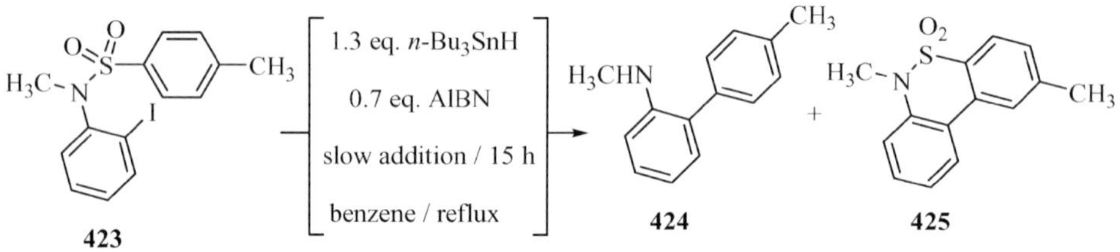

Scheme 2

However, at compounds substituted at one or both *ortho*-positions in the arylsulfonyl-moiety, the direct free-radical addition can be completely minimized or avoided, and thus providing an excellent access to sterically hindered *ortho*-di- and tri-substituted biaryls [2,3]. In this fashion, compound **426** was converted to the sterically encumbered biaryl **427** in 63% yield [2], respectively, Scheme 3.

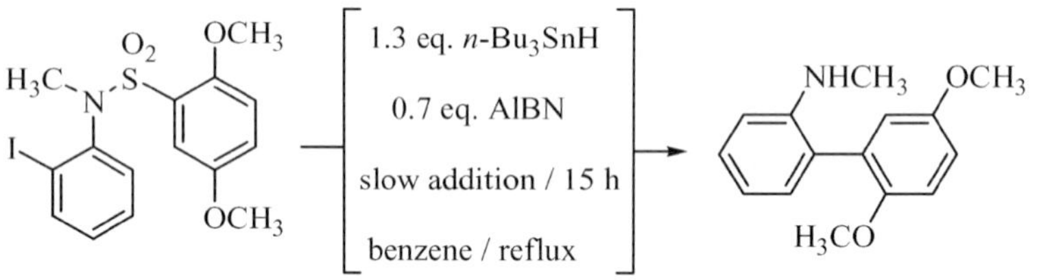

Scheme 3

Moreover, the powerful directing effect was observed with *ortho*-methoxycarbonyl group which favours the *ipso*-attack [3]. Further improvements have been introduced by using the diazonium salts as parent substrates, similarily to the Pschorr reaction [4].

Whereas substrates unsubstituted in the *ortho*-position to the diazonium group have given slightly decreased chemoselectivity of Motherwell versus Pschorr products, arylsulfonates, and *N*-metyl-*N*-arylsulfonamides substituted at adjacent position to the diazonium group gave an excellent selectivity and good to high yields of resulting 2-hydroxybiphenyls (**XXXIV**), and *N*-methyl-2-aminobiphenyls (**XXXV**) [4]. For instance, the treatment of diazonium salts **XXXVII** with titanium(III) chloride, as the most efficient reductant for diazonium salts, generates aryl radicals, which subsequently undergo highly selective *ipso*-substitution, Motherwell reaction pathway. The efficacy of this method is shown in the Table 1.

Table 1. Synthesis of 2-hydroxybiphenyls (**XXXVIII**) by the Motherwell method [4]

Aryl	Yield of XXXVIII (%)	Yield of XXXIX (%)
4-MeOC$_6$H$_4$	68	4
4-MeC$_6$H$_4$	69	8
2-FC$_6$H$_5$	68	2
mesityl	82	-
2-thienyl	60	-
1-naphthyl	74	-

The Motherwell method for synthesis of biaryls is apparently an important and perspective approach to the synthesis of *ortho*-substituted biaryl structures, still, further achievements with different models concerning substituent effects, as well as more efficient methodology are necessary.

Methodology similar to Pschorr and Motherwell reactions, was employed in the synthesis of phenanthrene-type structures, where an intramolecular free-radical arylation was accomplished by the reaction of bromo-*cis*-stilbene **428** with tri-*n*-butyl tin hydride, giving **429** in 85% yield [6], respectively, Scheme 4.

Related tin hydride-mediated free-radical cyclizations of iodo- or bromo-derivatives such as **428** have been widely used in the synthesis of naturally occuring products.

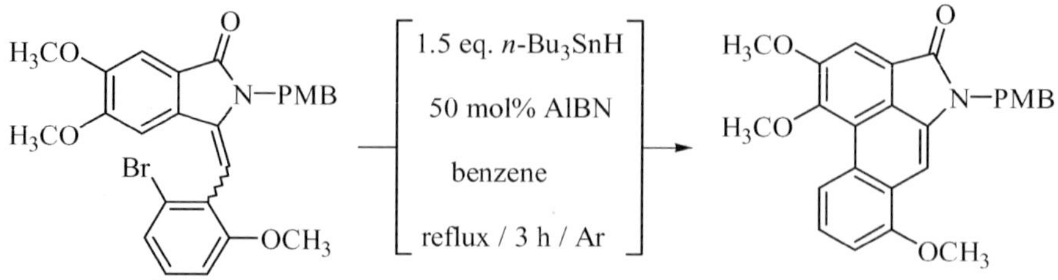

Scheme 4

7.2. Free-radical arylations of arenes with arylhydrazines and arylboronic acids

Demir's group has found that manganese(III) acetate acts as very selective oxidant for the generation of aryl radicals from arylhydrazines [7,8] and arylboronic acids [9]. When the oxidation is conducted in a liquid arene, the free-radical arylation takes place to give the biaryls in the Gomberg-Bachmann-Hey (GBH) fashion, but in high to excellent yields. Contrarily to the classical methods for oxidation of arylhydrazines in liquid arenes, mediated by silver(I) oxide, mercury(II) oxide, or manganese(IV) oxide, as well as novel approaches involving lead(IV) acetate, $(NH_4)_2[Ce(NO_3)_6]$, and $Co(acac)_3$ [7], the $Mn(OAc)_3$-promoted oxidation of arylhydrazines proceeds slowly to generate the free aryl radicals in very low concentrations. This provides the higher selectivity, cleaner reactions, and higher yields of biaryls. The methodology is very simple and includes the refluxing of arylhydrazine with manganese(III) acetate (3 eq.) in refluxing liquid arene such as benzene [7], thiophene [8], furan [9] for 0.5 to 10 h what resulted in high yields of unsymmetrical biaryls. Since the starting manganese(III) acetate was in dihydrate form, the azeotropic removal of water with benzene was used to obtain an anhydrous reagent, and the arylhydrazines were added as hydrochlorides. The yields of selected examples by Demir's method are presented in the Table 2. This method is valuable, slightly improved alternative to the phase transfer GBH reaction which offers a simple access to the unsymmetrical biaryls bearing no free-radical sensitive groups. For instance, the ortho-methyl group has survived the Demir's conditions and afforded the expected biaryl in good yield. On the other hand, ptGBH reaction of 2-tolyldiazonium tetrafluoroborate gave the corresponding indazole as the major product, by intramolecular free-radical attack to the adjacent methyl group.

Additonal improvement has been introduced by using the arylboronic acids as aryl radical precursors. Arylboronic acids are reacted with anhydrous manganese(III) acetate (3 eq.) in refluxing liquid arene, e.g. benzene, thiophene, or furan, to give unsymmetrical biaryls in moderate to excellent yields. Free-radical sensitive groups in arylboronic acid, e.g. formyl substituent, are well tolerated, Table 3.

Table 2. Manganese(III) acetate-mediated synthesis of unsymmetrical biaryls from arylhydrazines and liquid arenes [7,8]

R	Arene (Ar)	Yield (%)
H	benzene (Ph) [a]	75
2-Br	benzene (Ph) [a]	70
2-O$_2$N	benzene (Ph) [a]	68
2-CH$_3$	benzene (Ph) [a]	75
H	furan (2-furyl) [b]	60
2-Br	furan (2-furyl) [b]	53
2-Br	thiophene (2-thienyl) [b]	65
4-CH$_3$O	furan (2-furyl) [b]	30

[a] The reaction mixture was refluxed for 4 h.
[b] The reaction mixtures were refluxed for 5-10 h.

Table 3. Synthesis of biaryls by arylation of arenes with arylboronic acid under an influence of manganese(III) acetate [9]

R	Arene (Ar)	Yield (%)
H	benzene (Ph)	95
2-CH$_3$O	benzene (Ph)	82
2-CHO	benzene (Ph)	40
3-CHO	benzene (Ph)	60
4-CHO	benzene (Ph)	80
4-CH$_3$O	thiophene (2-thienyl)	66
2-CHO	thiophene (2-thienyl)	50
4-CH$_3$O	furan (2-furyl)	34
2-CHO	furan (2-furyl)	19

7.3. Homo-coupling reactions of miscellaneous arylmetallic reagents to biaryls

Variuos arylmetallic reagents as formal aryl-carbanion donors can be oxidatively coupled to give symmetrical biaryls. Some of these reactions have been described at a relevant place in the text. For example, the homo-coupling of aryllithiums with copper(II) chloride is well known modern alternative to the Ullmann reaction, see Chapter 2 [10,11]. Herein, analogous reactions with other arylmetallic reagents are described, Scheme 5.

$$Ar-M \xrightarrow{\text{oxidant}} Ar-Ar$$

M = Li, MgX, ZnX, SnR$_3$, B(OH)$_3$, etc.

Scheme 5

Aryllithiums, beside copper(II) halides, have been coupled with iron(III)-acetyl acetonate as an efficient one-electron oxidant. In this manner, compound **430** was lithiated in 2-position to form the aryllithium **431**, which was subsequently coupled with Fe(acac)$_3$ to afford bithiophenyl **432** in 57% overall yield [10], Scheme 6.

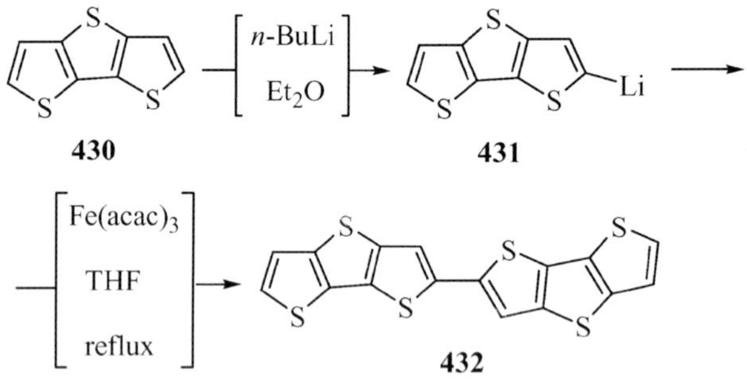

Scheme 6

Generally, all metallic one- or two-electron oxidants, as well as salts of metals which form thermally unstable diarylmetals, can affect the homo-coupling reaction of the given organometallic. However, relatively small number of methods have been developed for practical synthetic utility. The homo-coupling reaction of aryl Grignard reagents was more extensively studied. Namely, in the model reaction of phenylmagnesium iodide to biphenyl, the yields (%) obtained using the well established coupling reagents were: FeCl$_2$ (98), CoBr$_2$ (98), NiBr$_2$ (100), RuCl$_3$ (99), RhCl$_3$ (97.5), PdCl$_2$ (98) [12], TiCl$_4$ (82, with PhMgBr) [13], and TlBr (92, with

PhMgBr) [14]. The reactions were performed by adding an anhydrous metal halide to the aryl Grignard solution at 0 °C followed by stirring at room temperature. The yields of biaryls from *para*- and *meta*-substituted aryl Grignard reagents are generally good to excellent, but at *ortho*-substituted substrates yields are only moderate employing halides of metals with lower atomic weight. For example, mesitylmagnesium bromide (**125**) was coupled with cobalt(II) bromide to give bimesityl (**433**) in 20% yield [12], whereas 2-methoxyphenylmagnesium bromide (**124**) gave 2,2-dimethoxybiphenyl (**92**) in only 35% yield [13], Scheme 7.

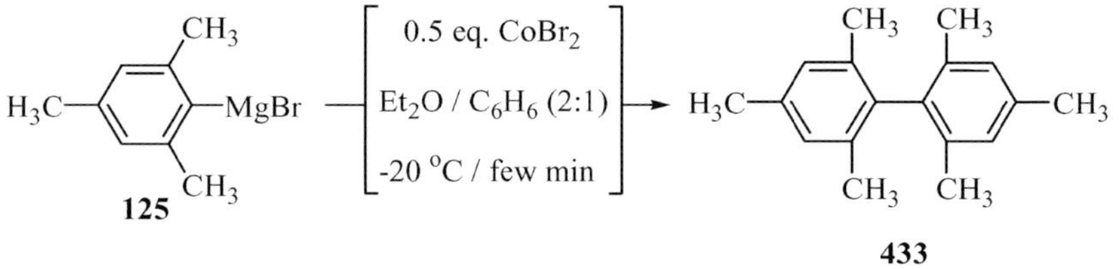

Scheme 7

The homo-coupling reaction promoted by titanium(IV) chloride was also applied at substrates bearing *N,N*-dialkylamido, ester, and nitrile functionalities when higher order magnesium reagent, *n*-Bu$_3$MgLi was used as metallating reagent. The efficacy of this reaction is shown in the Table 4.

Table 4. Synthesis of symmetrical biaryls from aryl iodides and bromides *via* the metallation with *n*-Bu$_3$MgLi followed by TiCl$_4$-mediated coupling [13]

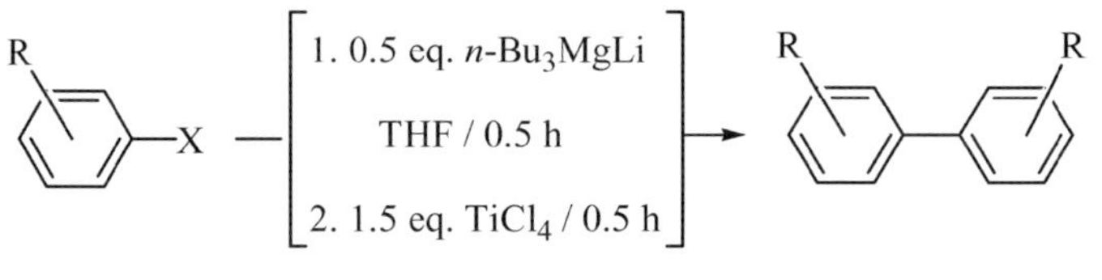

R	X	Yield (%)
4-CH$_3$O	Br [a]	80
3-CF$_3$	Br [a]	62
4-Br	Br [a]	54
3-CH$_3$O	I [b]	73
4-Et$_2$NCO	Br [a]	72
4-*n*-BuO$_2$C	I [b]	63
3-NC	Br [a]	43
4-NC	Br [a]	62

[a] The metallation was conducted at 0 °C, coupling at -78 °C and then at 0 °C.
[b] The metallation was carried out at -78 °C, coupling at -78 °C and then at 0 °C.

Another classical coupling reagent is thallium(I) bromide providing high-yielding approach to the symmetrical biaryls derived from *para*- and *meta*-substituted aryl Grignard reagents, while the reactions with *ortho*-occupied substrates failed [14]. The only products in these cases are dehalogenation parent arenes and diarylthallium bromides [14]. Examples of TlBr-promoted homo-couplings are listed in the Table 5. There were attempts to affect the cross-coupling reactions of arylthallium(III) dibromides with aryl bromides, but the selectivity of such reactions was barely higher than is the statistic ratio of three possible biaryls [14]. However, under the photochemical conditions, diarylthallium(III) trifluoroacetates are versatile reagents for arylation of liquid arenes to produce unsymmetrical biaryls in good to excellent yields [15].

Table 5. Synthesis of symmetrical biaryls by thallium(I) bromide-promoted homo-coupling reaction of aryl Grignard reagents [14]

R	Yield (%)
3-CH$_3$	97
3-*n*-Pr	65
4-CH$_3$O	99
3-Cl	74
4-Me$_2$N	76

Arylzincs are converted to symmetrical biaryls by copper(II) halides [16], as well as by *N*-chlorosuccinimide [17]. The method is more important than the homo-coupling of aryllithiums since, in a number of cases, higher yields were obtained, presumably because of more effective transmetallation of intermediate diarylzincs to copper(II). In this manner, tetra-*ortho*-bromobiphenyl (**434**) was lithiated with *n*-BuLi, and subsequently converted to the zinc species **435**, stable at low temperatures. The latter is reacted with copper(II) chloride to biphenylene **436** in 72% yield, accompanied with formation of intermolecular coupling product **437** in very small amount [16], Scheme 8. In contrast, the direct homo-coupling of bis-lithiated compound **434** with copper(II) chloride furnished the compound **436** in only 15% yield, with predominant formation of intermolecular coupling product **437** in 61% yield [16].

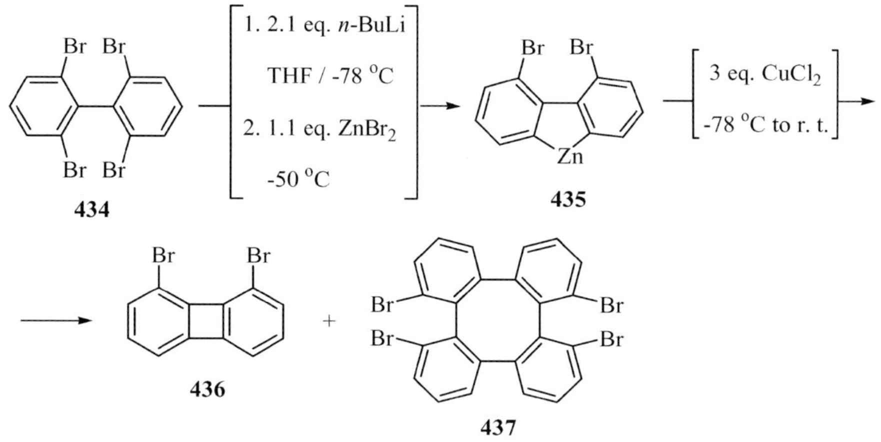

Scheme 8

Additonal simple method for performing the homo-coupling reaction of arylzincs under mild conditions is Takagi's $Pd(PPh_3)_4$ or $Pd(PPh_3)_2Cl_2$-catalysed route from arylzinc iodides in the presence of N-chlorosuccinimide (NCS) [17]. Alternatively, arylzincs can be prepared by metallation of relatively acidic arenes, e.g. thiophene at 2-position, followed by transmetallation with zinc(II) iodide [17], or by direct reaction of activated zinc dust with reactive aryl iodides and bromides, see Chapter 4. Selected examples of Takagi's results are given in the Table 6.

Table 6. Homo-coupling of arylzinc halides catalysed by palladium complexes in the presence of N-chlorosuccinimide (NCS) [17]

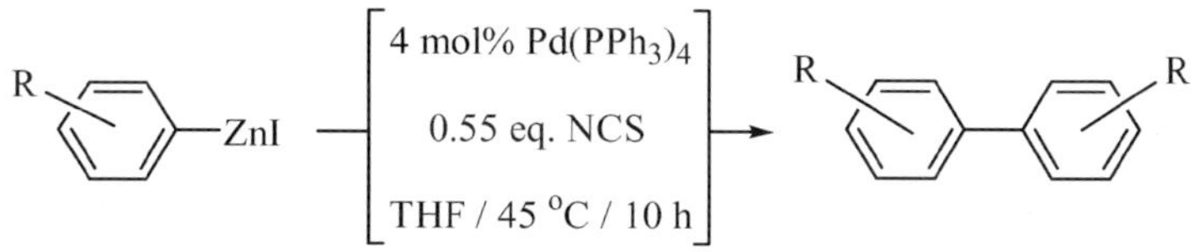

R	Yield (%)
2-CH$_3$O$_2$C	85
3-CH$_3$O$_2$C	93
4-C$_2$H$_5$O$_2$C	80
2-NC	88
2-Cl	93
2-CH$_3$CO$_2$, 5-CH$_3$	85 [a]

[a] Reaction time: 16 h

Among less conventional organometallics, arylzirconiums have been coupled in Ni(dppe)Cl$_2$-catalysed reaction to afford biaryls in moderate yields. For example, 2-bromophosphinine **438** was lithiated with *n*-BuLi, followed by transmetallation with ZrCp$_2$Cl$_2$ to form **439**. The latter smoothly underwent the homo-coupling reaction in the presence of Ni(dppe)Cl$_2$ to give diphosphinine **440** with a 55% yield [18], Scheme 9.

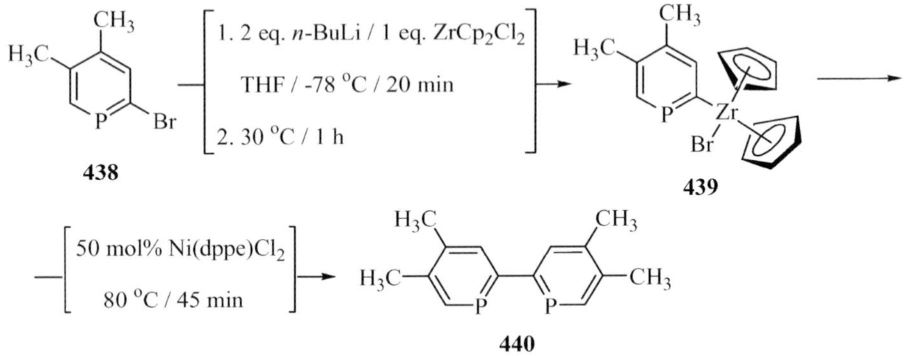

Scheme 9

One of the most important arylmetalloid reagents, arylboronic acids, have been converted to the symmetrical biaryls by the several methods [19-23]. The oldest approach includes *in situ* formation of arylboronic acid [19] *via* reaction of aryl iodides or bromides with magnesium (1 eq.) and diborane (1.5 eq. as BH$_3$), followed by treatment with methanolic solution of potassium hydroxide, and subsequently aqueous silver nitrate. In the first phase, presumably the corresponding arylborane is formed, which, upon the hydrolysis with KOH, gave arylboronic acid (or its cyclic anhydride) [20], followed by oxidative coupling with silver(I) oxide affording symmetrical biaryls in good yields, if the substituents were in *para*- or *meta*-positions. In contrast, *ortho*-substituted substrates furnish biaryls in low yields, for example, 2-bromotoluene, gave 2,2'-dimethylbiphenyl in only 17% yield [20]. The method, concerning its simplicity, is interesting one-pot three-step conversion, Table 7.

The modern methods for homo-coupling reaction of arylboronic acids, greatly clearified through the results from Moreno-Mañas's group, involve the palladium-catalysed processes [21]. This is the most important side-reaction in almost all Suzuki-Miyaura reactions, see Chapter 5. Various palladium complexes, such as Pd(PPh$_3$)$_4$ [21,22], Pd(dba)$_2$ [21], Pd(OAc)$_2$ [21], Pd(OAc)$_2$ / 2PPh$_3$ [23], Pd(PPh$_3$)$_2$Cl$_2$ [23], as well as 10% Pd-C [21] have affected the homo-coupling of arylboronic acids, under relatively mild reaction conditions, stirred in toluene solution at room temperature for 2-3 days [21].

Table 7. Synthesis of symmetrical biaryls *via in situ* generation of arylboronic acids from aryl iodides and bromides followed by the homo-coupling reaction with silver(I) nitrate [19,20]

R	X	Yield (%)
H	Br	59
4-CH$_3$	Br	70
3-CH$_3$O	Br	52
3-Cl	I	60
4-Me$_2$N	Br	71

However, practical synthetic method for performing the homo-coupling of arylboronic acids to the respective symmetrical biaryls is accomplished with copper(II) nitrate as an ultimate oxidant in good to high yields, Table 8.

Table 8. Palladium-catalysed homo-coupling reaction of arylboronic acids to symmetrical biaryls in the presence of copper(II) nitrate as an oxidant [22]

R	Yield (%)
H	80
2-F	61
4-Cl	72
2-CH$_3$	68
4-CH$_3$	71

Apart from arylboronic acids, tetraarylborates, under photochemical conditions [24] or in the presence of a stoichiometric amount of diphenylsilicon dichloride, under an oxygen atmosphere, give biaryls in good to high yields [25]. Namely, tetramethyl ammonium tetra(4-tolyl)borate (**441**) was coupled to 4,4'-dimethylbiphenyl (**442**) in 96% yield, respectively, Scheme 10.

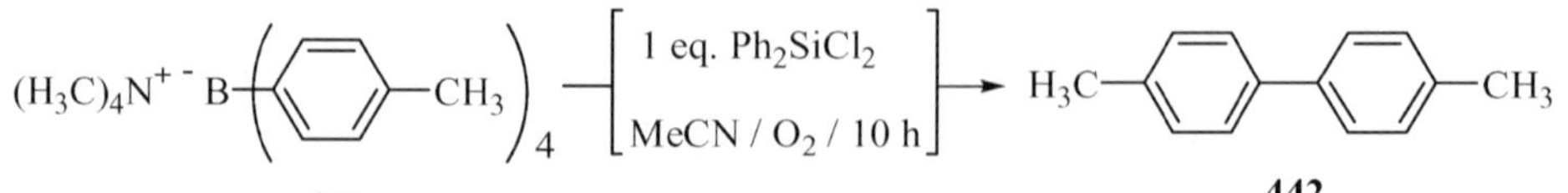

Scheme 10

Aryltrimethyl- or tri-*n*-butylstannanes as common Stille reagents are also successfully homo-coupled under palladium catalysed reactions in the presence of ethyl 2,3-dibromophenylpropionate (**443**) [26], or more conveniently by oxidation with copper(I) salts, e.g. CuCl [27], or copper(II) salts, e.g. Cu(NO$_3$)$_2$·3H$_2$O [28,29], which proceed smoothly at room temperature in tetrahydrofuran or DMF to afford symmetrical biaryls in excellent yields. The Cu(NO$_3$)$_2$-mediated homo-coupling works well also with diaryldimethyl(or *n*-butyl)stannanes [30]. Moreover, the reaction can be accomplished with a catalytic amount of copper(II) chloride or manganese(II) bromide (10 mol%) in the presence of iodine as stoichiometric oxidant [31]. For example, compound **443**, acting as an oxidant, converts the phenyltri-*n*-butylstannane (**184**) to biphenyl (**8**) in 86% yield [26], Scheme 11.

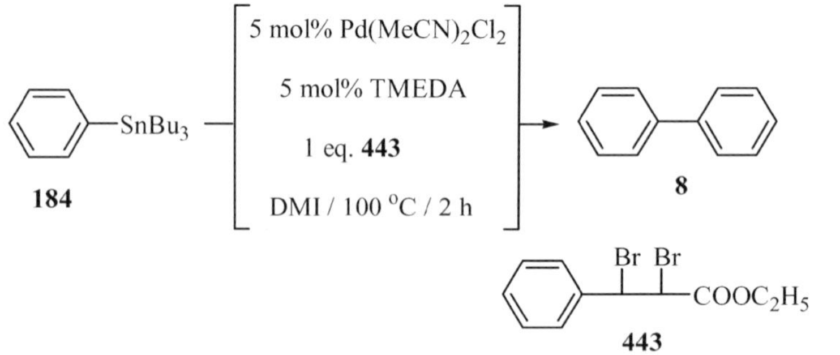

Scheme 11

More efficient copper(I) chloride-induced homo-coupling reaction of aryltrimethyl stannanes is accomplished by simple stirring in DMF at room temperature. Namely, bis-stannane **444** was reacted with five-fold molar excess of CuCl to give biaryl **445** in 91%, yield [27], respectively, Scheme 12.

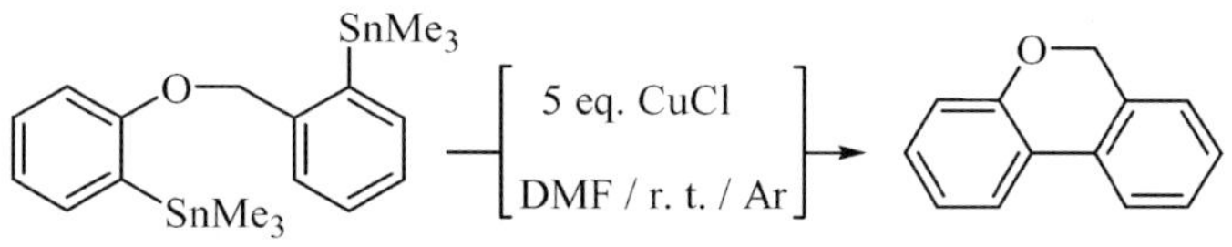

Scheme 12

The copper(II) nitrate-mediated homo-coupling reaction works efficiently also with strained substrates such as cyclophanes. For instance, compound **446** was cyclized to the cyclophane **447** in 41% yield [28], Scheme 13.

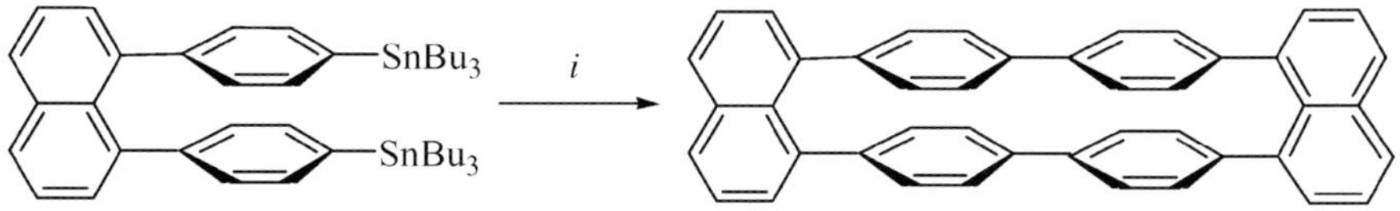

i: 2.2 eq. $Cu(NO_3)_2 \cdot 3H_2O$ / THF / r. t. / 1 h

Scheme 13

Diaryldialkylstannanes also smoothly react with $Cu(NO_3)_2$ to give the homo-coupled products in good to excellent yields, Table 9. Atempts to carry out the cross-coupling reactions of two different diarylstannanes resulted in the formation of all three possible biaryls in practically statistic ratio [30].

Table 9. Copper(II) nitrate induced homo-coupling reaction of diaryl(dimethyl)stannanes [30]

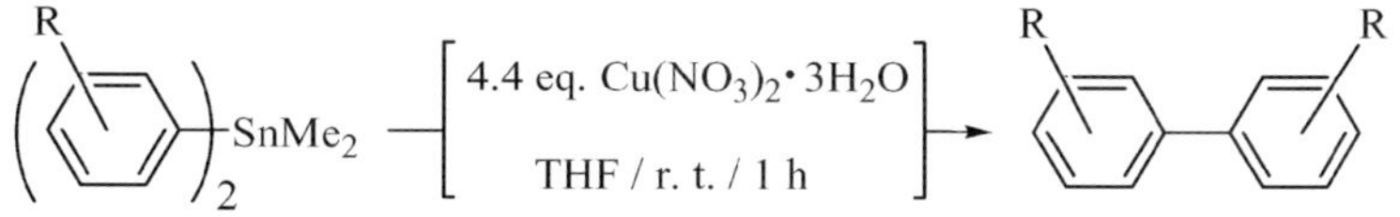

R	Yield (%)
H	97
4-CH$_3$	92
3-CH$_3$	78
2-CH$_3$	82
2-CH$_3$O	49
3-Cl	81

The homo-coupling reaction with only catalytic amount of copper(II) chloride or manganese(II) bromide has been affected in polar solvents such as DMF or NMP with iodine as an ultimate oxidant [31]. Namely, 2-thienyltri-*n*-butylstannane (**173**) was converted to 2,2'-bithienyl (**98**) with CuCl$_2$ in 72% yield, whereas the MnBr$_2$-catalysed reaction resulted with a 83% yield [31], Scheme 14.

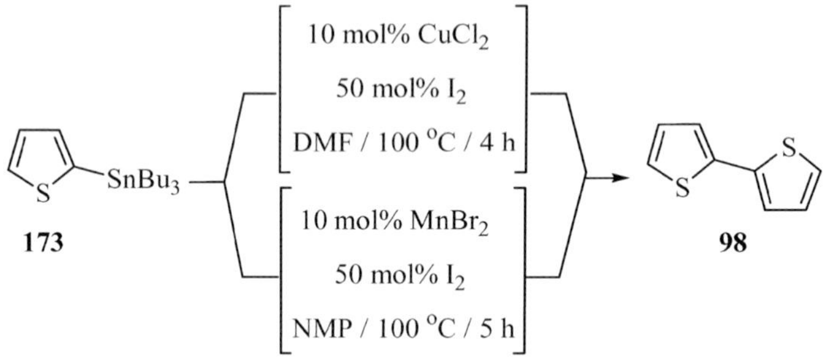

Scheme 14

Similarly to arylstannanes, aryldimethylsilicon chlorides [26,32,33], aryldiethylsilicon chlorides [33] and diarylsilicon difluorides [26] have been successfully coupled to the corresponding symmetrical biaryls in good to high yields using copper [32,33] or palladium [26] catalysts. Thus phenyldimethylsilicon chloride (**448**) was converted to biphenyl (**8**) in 73% yield [32], Scheme 15.

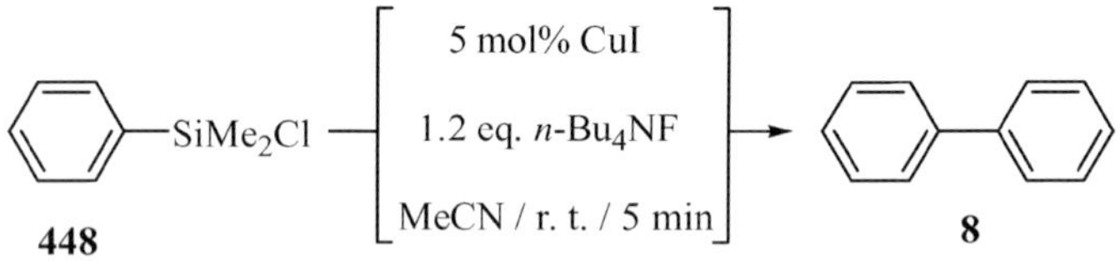

Scheme 15

The reaction is facilitated by the polar solvents, e.g. DMF [33], DMSO [33], or acetonitrile [32], and tetraalkylammonium fluorides as organic solvent-soluble sources of fluoride anion, which convert the aryldimethylsilicon chlorides to the respective fluorides. The latter smoothly undergo the transmetallation reaction from silicon to copper, what finally resulted in the formation of biaryls in high yields. Beside copper(I) iodide, other Cu(I) salts such as CuCl, CuBr and CuOTf have been employed, however, CuI works the most efficiently with ArSiMe$_2$X [32], whereas copper(I) chloride and triflate are suitable in the reactions with ArSiEtF$_2$, and ArSiF$_3$

[33]. Both above described strategies involving arylstannane and arylsilicon reagents proceed thanks to the readiness of the transmetallation reaction from tin or silicon to copper.

Further important organometallics which offer ready access to the homo-coupling products are arylmercuric halides. This can be easily prepared by several ways, but the most convenienty from the corresponding aryldiazonium salts *via* the Nesmeyanov reaction, see Chapter 4. Arylmercuric chlorides derived from electron-rich arenes such as thiophene, which easily undergoes the direct mercuration reaction at 2-position, can be readily coupled with an equimolar amount of palladium(II) chloride in the presence of triethylamine as the base in high yield [35], Scheme 16.

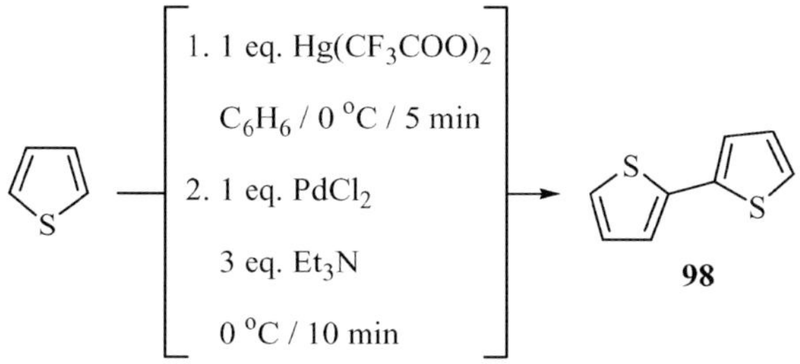

Scheme 16

Practical synthetic method for the catalytic homo-coupling reaction of arylmercury halides includes the heating with copper powder (4 eq.) in refluxing pyridine facilitated by palladium(II) chloride (10 mol%) [34]. A number of biaryls were prepared by this method, as listed in the Table 10. Although the presence of free-amino and acetamido groups is well tolerated, the reactions with unprotected carboxylic acids, phenols, and significant steric encumbrances completely failed. It can not be excluded that the use of more effective palladium catalyst(s) would not improve the failures with sterically hindered arylmercury halides. Nevertheless, reactions with substrates bearing a single *ortho*-group to form respective hindered di-*ortho*-substituted biaryls proceed in good yields.

Additional noteworth organometalloid reagents which readily undergo the homo-coupling reaction are diaryltellurium dichlorides. These reagents, derived from electron-rich arenes, e.g. anisole, *N,N*-dialkylanilines, or alkylthiophenols, can be prepared by direct electrophilic aromatic substitution with tellurium(IV) chloride [36,37], or from aryl Grignard reagents and TeCl$_4$. For example, anisole reacts with TeCl$_4$, to form 4-methoxyphenyltellurium(IV) chloride (**449**), which subsequently undergoes further substitution reaction with anisole to give di(4-methoxyphenyl) tellurium(IV) chloride (**450**). The latter, upon treatment with Raney-nickel, copper, or palladium, gives 4,4'-dimethoxybiphenyl (**77**) in good to high yields. The reaction proceeds with formation of di(4-methoxyphenyl)tellurium (**451**) which is decomposed to biaryl **77** and tellurium, analogously to the reductive elimination of other diarylmetallic species [36,37], Scheme 17.

Table 10. Palladium-catalysed homo-coupling reaction of arylmercury halides in the presence of copper powder [34]

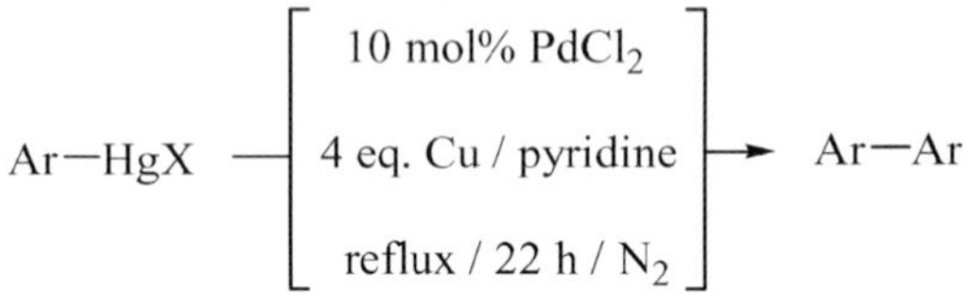

$$Ar-HgX \quad \longrightarrow \quad \begin{bmatrix} 10\ mol\%\ PdCl_2 \\ 4\ eq.\ Cu\ /\ pyridine \\ reflux\ /\ 22\ h\ /\ N_2 \end{bmatrix} \quad \longrightarrow \quad Ar-Ar$$

Ar	X	Yield (%)
Ph	OAc	86
2-CH$_3$OC$_6$H$_4$	OAc	84
4-CH$_3$OC$_6$H$_4$	OAc	90
4-H$_2$NC$_6$H$_4$	OAc	76
4-AcNHC$_6$H$_4$	OAc	69
1-naphthyl	Cl	47
mesityl	OAc	0

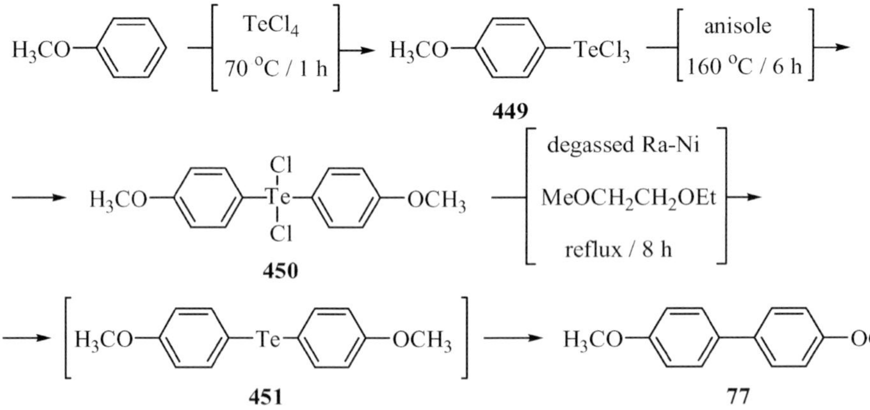

Scheme 17

The electrophilic aromatic substitution reaction of less reactive arenes such as benzene with tellurium(IV) chloride can be accelerated by adding the Lewis acid, e.g. aluminum chloride. Namely, refluxing the mixture of TeCl$_4$ and AlCl$_3$ (3 eq.) in benzene afforded diphenyltellurium dichloride in 59% yield, respectively [37]. Diaryltelluriums can be also prepared by reduction of diaryltellurium dichlorides with zinc [36], or from triarylbismuth reagents by heating with tellurium in refluxing mesitylene [38]. Unfortunately, an extremely bad odour of organotelluriums makes this route rather unatractive.

7.4. The Meyers synthesis of biaryls and Related reactions

The nucleophilic aromatic substitution of *ortho*-methoxy group in aryloxazolines with aryllithiums or aryl Grignard reagents proceeds smoothly with formation of 2-arylphenyloxazolines, which, upon acid hydrolysis, give *ortho*-arylbenzoic acids in good to excellent yields. This route to *ortho*-arylbenzoic acids is well known as Meyers synthesis of biaryls, although the strategy has been used in the synthesis of a wide variety of *ortho*-substituted benzoic acids. The Meyers route includes transformation of (un)substituted 2-methoxybenzoic acid **XXXX** to its acid chloride **XXXXI**, followed by reaction with 2-amino-2-methyl-1-propanol (**452**) to give amide **XXXXII**. The latter is cyclized to oxazoline **XXXXIII** by treatment with thionyl chloride. Oxazoline **XXXXIII** reacts with aryllithium or aryl Grignard reagent to produce the *ortho*-arylated oxazoline **XXXXIV**. This is further hydrolyzed with acid to the respective biphenic acid **XXXXV** [39,40], Scheme 18.

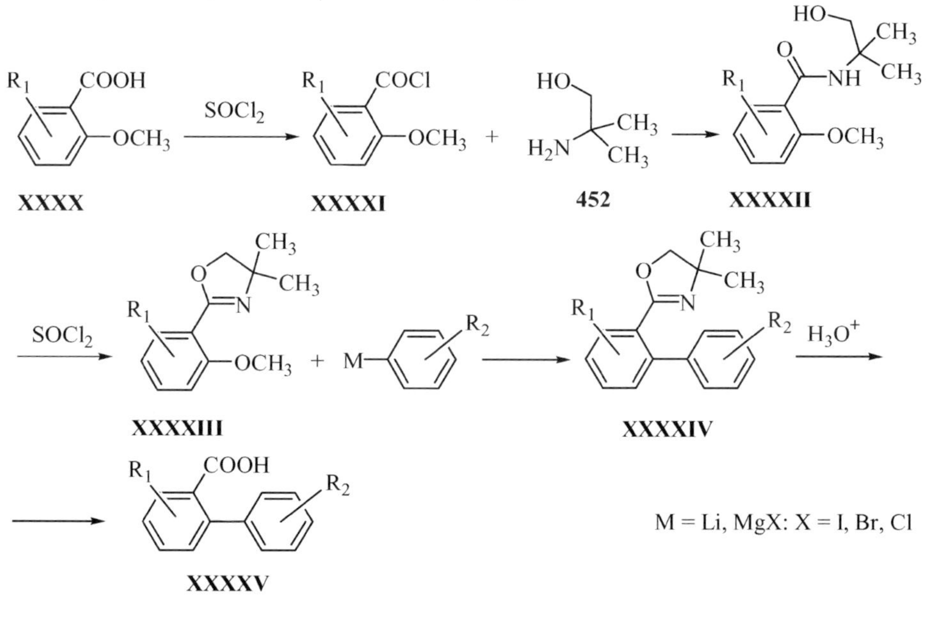

Scheme 18

2-Amino-2-methyl-1-propanol (**452**) is employed rather than simple ethanolamine since the former forms an oxazoline without acidic hydrogens in the α-position to the double C=N bond. In contrast, simple ethanolamine would give the respective oxazoline with relatively acidic properties, causing several side-reactions, e.g. protonation of organometallic reagent. Selected examples of the synthesis of biphenic acids by Meyers route are shown in the Table 11. Thus obtained oxazolines are generally hydrolyzed by boiling in 4-5 M aqueous hydrochloric acid for 16-24 h in high yields, usually higher than 75%. In difficult cases, an alternative method involves the reaction of oxazoline with five-fold excess of methyl iodide (r. t. / 24 h). The latter quaternizes the oxazoline, which then readily undergoes the base-catalysed saponification with a mixture of 20% aqueous sodium hydroxide and methanol (1:1) - often for 15-24 h.

Table 11. Efficacy of the Meyers synthesis of biaryls [39,40]

R_1	M	R_2	Temp. (°C)	Yield (%)
CH_3O	Li	H	-45	100
CH_3O	Li	NMe_2	-45	66
CH_3O	Li	2-CH_3O	-45	13
CH_3O	MgBr	H	25	95
H	MgBr	H	25	95

The Meyers route, which has been used with great success in a number of impressive natural product synthesis [41-45], plays an important role in the synthesis of relatively sterically hindered biaryls bearing the carboxy group in the *ortho*-position. To illustrate the general applicability of the Meyers synthesis of biphenic acids, herein, an interesting example is presented, where 2,3,5-trimethoxybenzoic acid (**453**) was transformed into the oxazoline **454** with 2-amino-2-methyl-1-propanol (**452**) and thionyl chloride in 94% yield [43]. The oxazoline **454** was reacted with the Grignard reagent **455**, prepared from 2,6-dimethoxy-4-methylphenyl bromide, affording the respective tetra-*ortho*-substituted biaryl **456** in 92% yield. The reaction of oxazoline **456** with methyl iodide, followed by saponification, furnished the highly encumbered, tetra-*ortho*-substituted biaryl, isolated as methyl ester **457**, in 95% yield [43], respectively, Scheme 19.

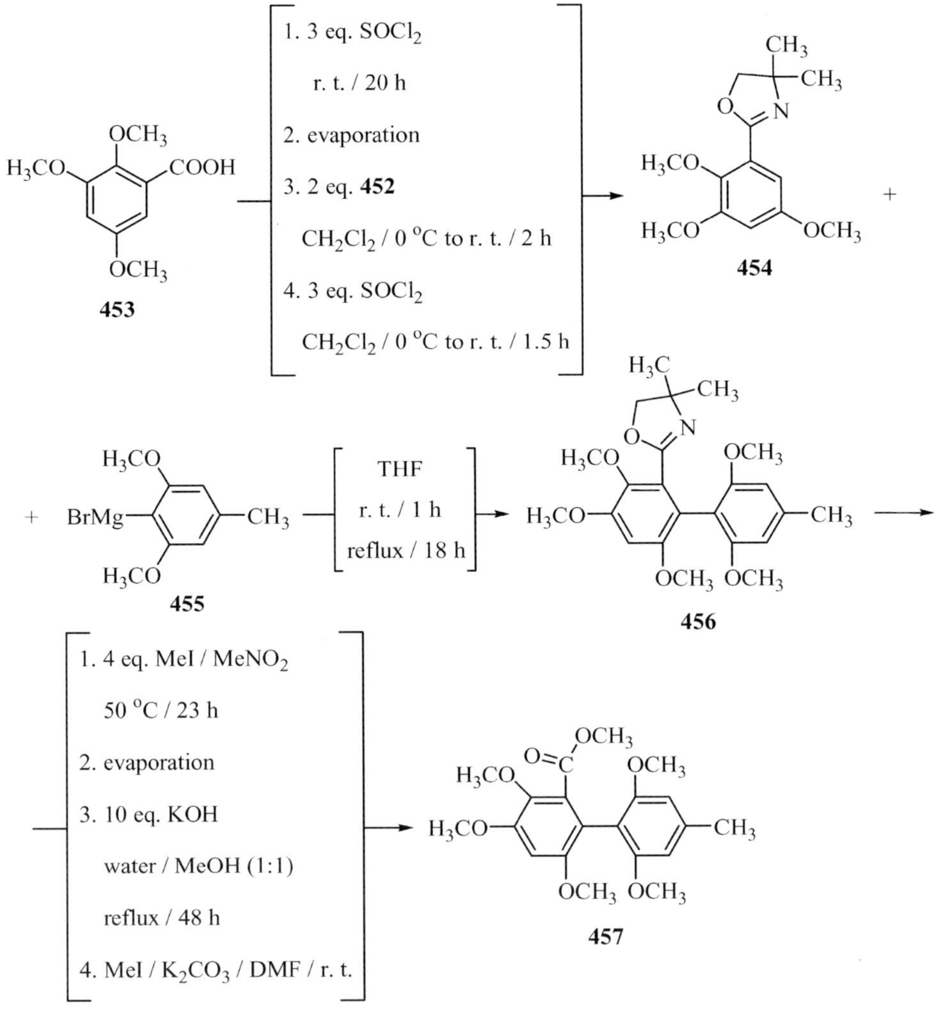

Scheme 19

Analogously to the Meyers approach, other authors have introduced related methods based on electron-withdrawing substituents which activate the leaving group in the *ortho*-position, e.g. methoxy, chloro, etc., to nucleophilic substitution. Miyano and coworkers [46] found that sterically hindered 2-methoxybenzoic acid esters of 2,6-di-*tert*-butyl-4-methylphenol (**458**), well known and inexpensive antioxidant, react with aryl Grignard reagents by substitution of *ortho*-methoxy group to give the *ortho*-arylbenzoic acid esters. In this manner, 2-methoxybenzoic acid (**459**) was esterified with **458** to form an ester **460** in 92% yield, which readily reacted with 4-tolyl magnesium bromide (136) affording the biaryl 461 in 91% yield. The cleavage of encumbered ester 461 was accomplished by sodium methoxide in the presence of polar cosolvents such as HMPA, DMF, or NMP to produce the respective biaryl 462, an important pharmaceutical intermediate, in almost quantitative yield, Scheme 20.

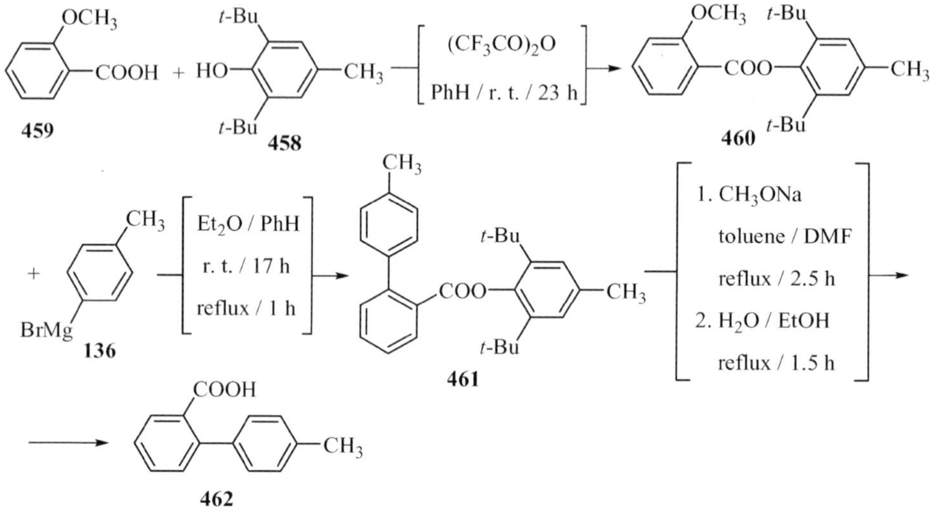

Scheme 20

An aldehyde function can be transformed to the *N*-cyclohexylimino (or *N-tert*-butyl) group, which retains electron-withdrawing properties, and becomes much less reactive to nucleophilic attack at electrophilic C=N carbon because of significant steric hindrances. Such imines activate the *ortho*-halogen atom to the nucleophilic displacement acting as alternatives to Meyers oxazoline or Miyano's sterically hindered esters. Thus Moreau's group [47] reported the successful synthesis of biaryl 112, based on the reaction of imine 463 with 4-tolylmagnesium bromide (136) by subsequent treatment with hydroxylamine to furnish the respective oxime 464 in 71% yield. The latter is converted by dehydration in boiling formic acid to the biaryl 112 in 91% yield, Scheme 21. The crucial step, nucleophilic substitution, as stated, is catalysed by anhydrous manganese(II) chloride. Unfortunately, this interesting reaction and the role of $MnCl_2$ have not been investigated in detail.

Somewhat different approach to the synthesis of *ortho*-substituted biaryls was reported by Julia and coworkers [48]. They found that *tert*-butylsulfonyl group can be replaced by aryl-carbanions originated from Grignard reagents in $Ni(acac)_2$-catalysed process with formation of biaryls in moderate to high yields. Aryl-*tert*-butylsulfones are readily obtained by quenching the aryllithiums with di-*tert*-butyl disulfide, followed by oxidation, e.g. $NaBO_3$ in HOAc. In the simplest case, phenyl-*tert*-butylsulfone (**465**) reacts with 4-tolylmagnesium bromide (**136**, 2 eq.) to produce 4-methylbiphenyl (**139**) in 80% yield [48], Scheme 22.

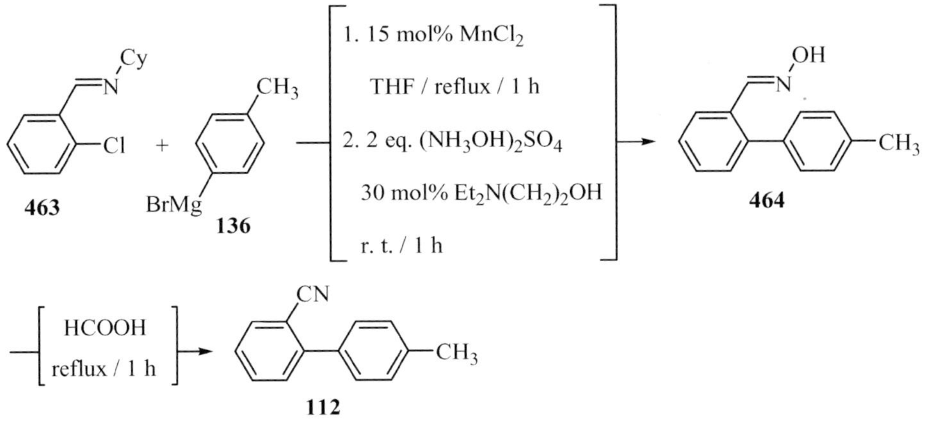

Scheme 21

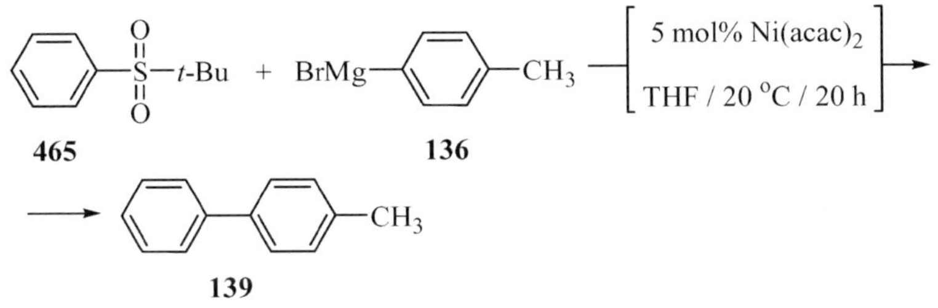

Scheme 22

Since *tert*-butylsulfone group acts as strong director of *ortho*-lithiation, the Julia's method can be useful in the synthesis of various *ortho*-substituted biaryls. Finally, arenes such as pyridine do not require the presence of electron-withdrawing substituent for the reaction with nucleophiles. Pyridines react with aryllithiums at 2-position, which is electrophilic, *via* nucleophilic addition, to give the respective 2-arylpyridines (Chichibabin reaction manner). In this fashion, the reaction of lithiated ferrocene with pyridine results in formation of α-pyridylferrocene in 24% yield [49].

7.5. Diaryliodonium salts and other hypervalent iodine compounds in the synthesis of biaryls

Aryliodonium salts, $Ar_2I^+ X^-$ [50], and some other hypervalent aryliodonium compounds take a part in further important alternative cross-coupling reactions with arylboronic acids [51], aryltrifluoroborates [52], triarylbismuth(V) compounds [53], and diaryltellurium dichlorides [54] to afford biaryls in good to excellent yields. For example, hypervalent iodine compound **466**, readily produced from 2-iodobenzoic acid, reacts with arylboronic acids in palladium-catalysed process in Suzuki-Miyaura-fashion to give the 2-arylbenzoic acids in high yields [51], Table 12.

Table 12. Synthesis of 2-arylbenzoic acids from 1-hydroxy-1,2-benziodoxol-3(1*H*)-one (**466**) and arylboronic acids [51]

R	Yield (%)
H	83
2-CH$_3$	72
4-CH$_3$	80
2-CH$_3$O	67
4-CH$_3$O	87
4-Cl	79

Under the same reaction conditions, the arylboronate esters also undergo the reaction with similar success [51]. Among unsymmetrical diaryliodonium salts, only the electron-rich aryl group is transfered to palladium giving the biaryl product [52], Table 13. Under the essentialy same conditions, Koser's reagent analogues, hydroxy(tosyloxy)iodoarenes, smoothly affect the reaction with potassium aryltrifluoroborates in high yields [52]. In the same fashion, the palladium-catalysed reaction of diaryliodonium salts with triarylbismuth(V) chlorides, fluorides or acetates readily proceeds under mild reaction conditions in solvents such as acetonitrile, DMF or DME to form biaryls in high yields. Thus iodonium salt **467** was reacted with triphenylbismuth(V) dichloride (**468**) to give 4-methoxybiphenyl (**78**) in 85% yield [53], respectively , Scheme 23.

Diaryltellurium dichlorides, Ar_2TeCl_2, react under the similar reaction conditions, however, the presence of sodium methoxide (3 eq.) is necessary for *in situ* generation of diaryltellurium dimethoxides, $Ar_2Te(OMe)_2$ from Ar_2TeCl_2, and thus the reaction solvent must be the mixture of acetonitrile and methanol (1:1) to insure the enough solubility of the base. For example, the reaction of iodonium salt **469** with di(4-methoxyphenyl)tellurium dichloride (**450**) afforded the biaryl **470** in 86% yield [54], Scheme 24.

Table 13. Synthesis of biaryls from diaryliodonium salts and potassium aryltrifluoroborates [52]

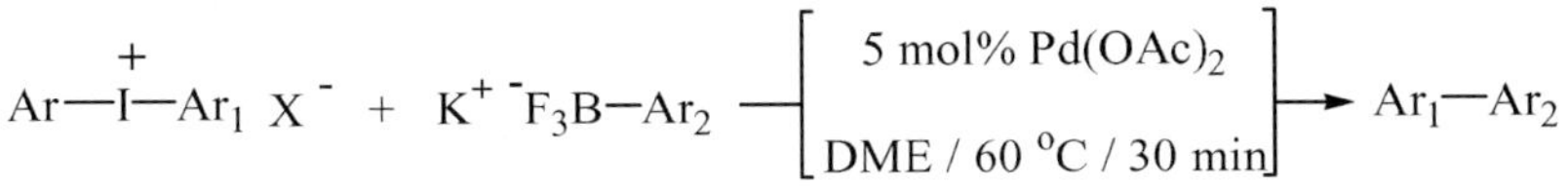

$$Ar-\overset{+}{I}-Ar_1 \ X^- \ + \ K^+ \ ^-F_3B-Ar_2 \quad \left[\begin{array}{c} 5 \ mol\% \ Pd(OAc)_2 \\ \hline DME \ / \ 60 \ ^\circ C \ / \ 30 \ min \end{array} \right] \longrightarrow \ Ar_1-Ar_2$$

Ar	Ar_1	X	Ar_2	Yield (%)
Ph	Ph	BF_4	Ph	99
$4\text{-}CH_3C_6H_4$	$4\text{-}CH_3C_6H_4$	BF_4	$2\text{-}CH_3OC_6H_4$	84
2-thienyl	Ph	OTs	$2\text{-}CH_3OC_6H_4$	83
Ph	Ph	BF_4	$2\text{-}CH_3OC_6H_4$	87
$4\text{-}CH_3OC_6H_4$	Ph	BF_4	$4\text{-}CH_3C_6H_4$	97

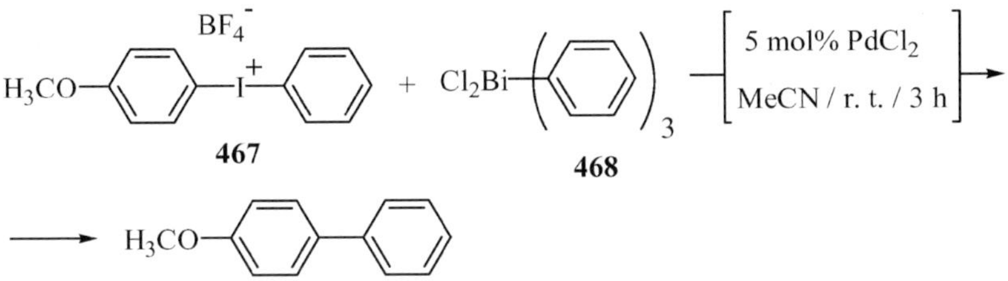

Scheme 23

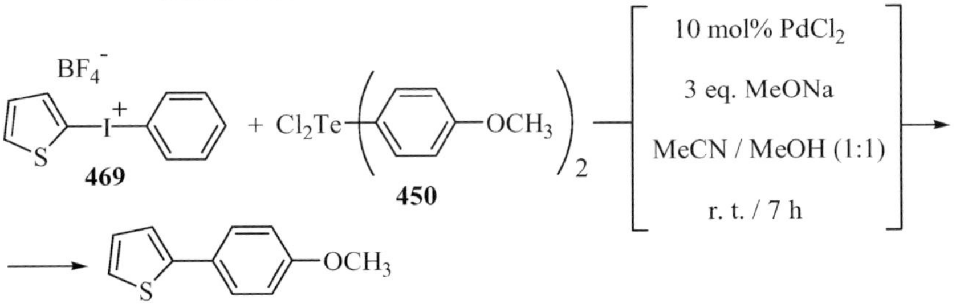

Scheme 24

7.6. Palladium-catalysed arylation of arenes with aryl halides and sulfonates

Arenes react with aryl halides or triflates to form a new aryl-aryl bond in the process well known as a dehydrohalogenation coupling reaction, nicely reviewed by Echavarren's group [55]. This reaction has been used mainly to achieve an intramolecular cyclization [55-61], but also an intermolecular arylation can be accomplished [62,63]. Thus cyclizations of aryl halides, such as **Ia**, containing an aromatic ring in a suitable position to form penta- or hexa-membered ring, tricyclic compounds type **IIa**, as well as intermolecular arylation of electron-rich arenes **XXXXVI**, e.g. thiophenes or furans, bearing an electron-withdrawing group, afforded the respective biaryl **IIb** in good to excellent yields, Scheme 25.

Intramolecular cyclization:

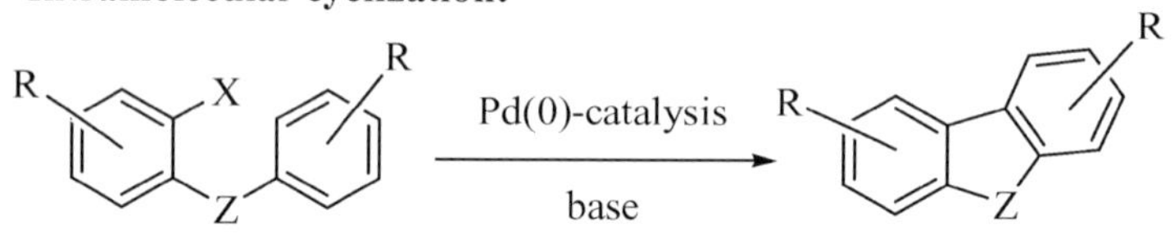

Ia   IIa

Intermolecular arylation:

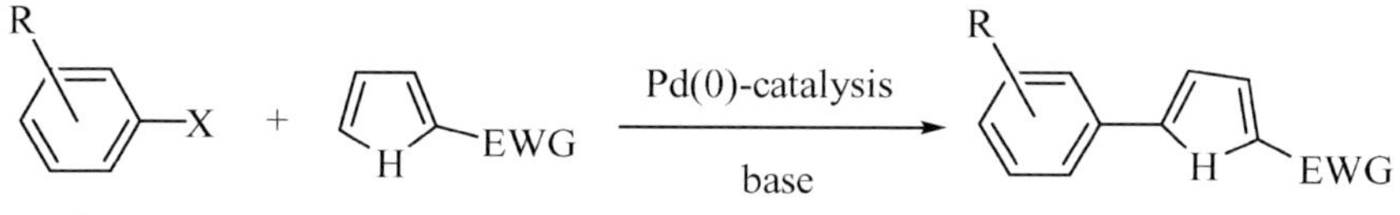

I   XXXXVI   IIb

X = I, Br, Cl, OTf

H = O, S

Z = CH_2, -$(CH_2)_2$-, O, S, NH, -CONH-, -COO-, -SO_2NH-, etc.

EWG = electron-withdrawing group: CHO, CN, COOR, etc.

Scheme 25

This transformations have been carried out by using palladium complexes (usually 5-10 mol%) as the catalysts: Pd(OAc)$_2$ [55-57], Pd(PPh$_3$)$_2$Cl$_2$ [55], Pd(MeCN)$_2$Cl$_2$ [55], Pd(PCy$_3$)$_2$Cl$_2$ [55], PdCl$_2$ [58], Pd(PPh$_3$)$_4$ [59], Pd(OAc)$_2$ / PPh$_3$ [60], Pd(OAc)$_2$ / P(2-tolyl)$_3$ [60], and Pd$_2$(dba)$_3$ / P(t-Bu)$_3$ [61]. The presence of a stoichiometric amount of the base is required in order to neutralize the hydrogen halides or triflic acid liberated during the course of reaction. For this purpose, M_2CO_3 (M = Na, K, Cs, Ag) [55,56,58,60], $NaHCO_3$ [55], MOAc (M = Na, K) [55,60], KOPh [55], DBU [57], KOt-Bu [59], EtNi-Pr$_2$ [60], and K_3PO_4 [61] were commonly employed bases (1.2-8 eq.) in generally polar solvent such as acetonitrile [55], DMF [55], DMAc [55-57], or dioxane [61]. The reactions are effected at relatively high, usually refluxing temperatures of above mentioned polar solvents, within the range 80-190 °C. The reaction mechanism of this reaction, as generally accepted, includes the oxidative addition of an aryl halide (or triflate) to the catalytically active palladium(0) complex, with subsequent intramolecular electrophilic aromatic substitution of arylpalladium(II) halide to the adjacent aromatic ring to give the diarylpalladium(II) complex. The latter, upon the reductive elimination, gives the corresponding biaryl accompanied with regeneration of catalytically active palladium(0) species. For example, compound **471** reacts with palladium(0) complex to give the oxidative addition product **472**, which undergoes the intramolecular electrophilic aromatic substitution to form the diarylpalladium(II) **473**. This, by reductive elimination of Pd(0) complex, furnishes the tricyclic compound **474** in 80% yield [56], Scheme 26.

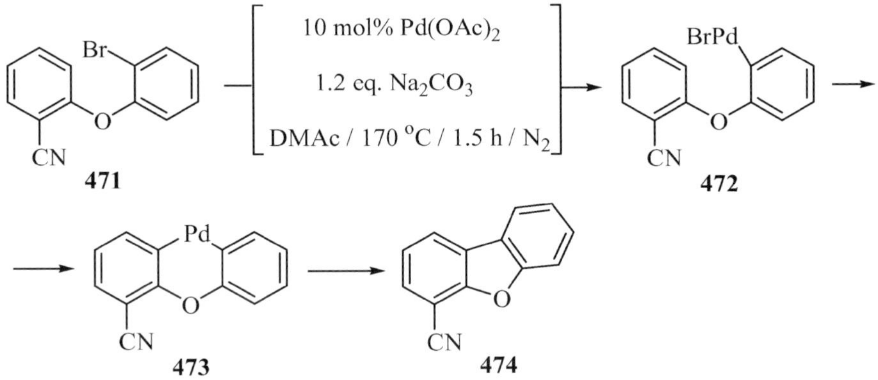

Scheme 26

However, in certain instances, the C-H activation of the adjacent aromatic ring by the influence of arylpalladium(II) halide must also be taken in account [55]. The palladium(0) species are generated *in situ* by reduction of palladium(II) salts, $Pd(OAc)_2$ or $PdCl_2$ with organic materials, e.g. reaction solvent, at relatively high temperatures in the presence of phosphine ligands. In the case of ligandless methods, n-Bu$_4$NBr or n-Bu$_4$NHSO$_4$ were added to stabilize the nano-sized palladium clusters, analogously to the ligandless Suzuki-Miyaura methods, see Chapter 5. Concerning the reaction mechanism, the reaction proceeds easily with electron-rich aromatics. Thus the cyclization reaction of phenol **475** is accomplished at 95 °C to give the respective *ortho*-arylation product **476** (87%), as well as *para*-arylation product **477** (2-5%) [59], Scheme 27.

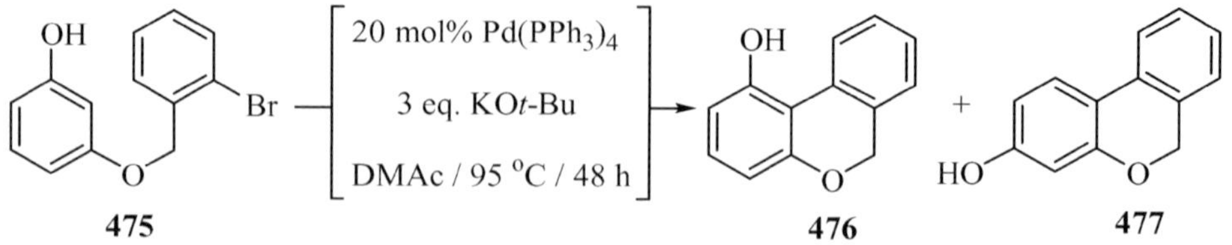

Scheme 27

Among the intermolecular arylations, specially useful are those with thiophenes and furans. The latter electron-rich heterocycles bearing an electron-withdrawing group readily undergo the arylation with aryl iodides and bromides to give the respective biaryls in good to excellent yields. In this fashion, 4-nitro-iodobenzene (**225**) reacts with furfural (**478**) to produce the biaryl **479** in 88% yield, respectively, Scheme 28.

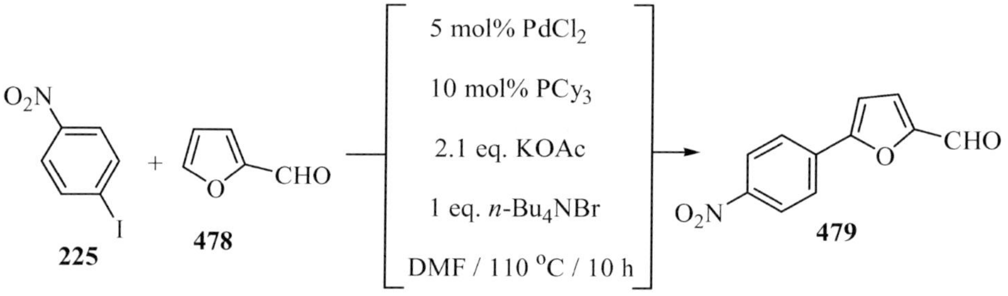

Scheme 28

The least reactive aryl chlorides smoothly effected the intramolecular cyclization in the presence of the palladium complex of electron-rich tri-*tert*-butylphosphine, *in situ* formed by mixing the Pd$_2$(dba)$_3$ (2.5 mol%) and P*t*-Bu$_3$ (10 mol%), under relatively mild reaction conditions (refluxing dioxane / 10 eq. K$_3$PO$_4$) [61]. Similarly to palladium complexes, the ruthenium complexes, Ru(PPh$_3$)$_3$Cl$_2$, [Ru(η^6-C$_6$H$_6$]$_2$Cl$_2$, or [Ru(COD)Cl$_2$]$_2$ in the presence of triphenylphosphine (4 eq. to Ru) at 2.5 mol%-catalyst loading, act as efficient catalysts for the arylation of 2-phenylpyridine (**150**) with bromobenzene resulting in formation of mono-*ortho*-arylated biaryl **480** (71%) as the major product, and di-*ortho*-arylated compound **481** (11%) [62], Scheme 29.

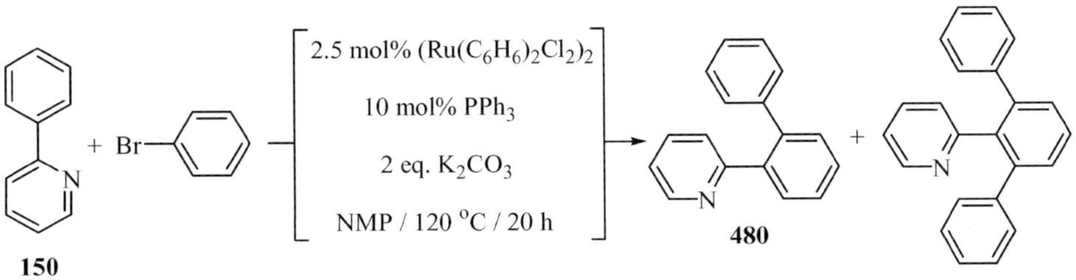

Scheme 29

The palladium-catalysed arylation of arenes with aryl halides and triflates is of a great importance in the synthesis of a wide variety of tri- and polycyclic aromatic systems, and is one of the simplest ways to arylated thiophenes, furans and certain other heterocycles [55-63].

7.7. Synthesis of biaryls involving aryllead(IV) tricarboxylates

Aryllead(IV) tricarboxylates, $ArPb(OOCR)_3$ [64], undergo the arylation of electron-rich arenes under very mild reaction conditions, usually at room temperature within a few hours, to produce the respective biaryls in moderate to high yields. The reaction involves the electrophilic aromatic substitution, and with mono-substituted benzenes, all three isomeric biaryls are obtained, with *ortho*-isomer as the major product [65-69], Scheme 30.

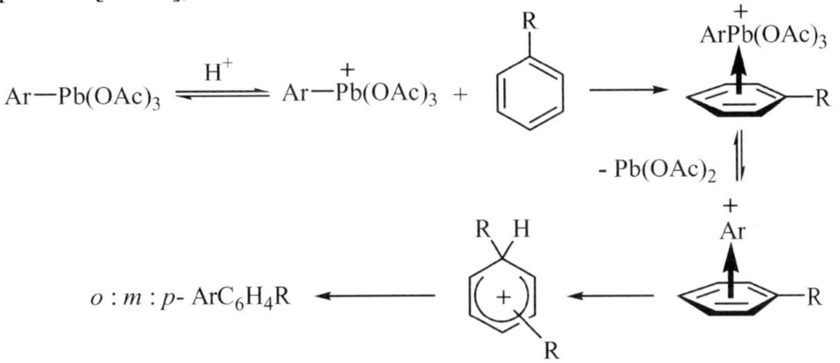

Scheme 30

Concerning the reaction mechanism, the factors having the profound influence on the reactions are the nucleophilicity of arene and electrophilicity of aryllead(IV) reagent. Electron-donating groups increase an electron-density of the arene, and thus enhance its reactivity.

The electrophilicity of lead in the aryllead(IV) tricarboxylates can be increased by replacing an acetate group with more electron-withdrawing substituents, e.g. trifluoroacetate. This can be accomplished by performing the reaction with ArPb(OAc)$_3$ in more acidic acetic acid derivatives: chloroacetic, dichloroacetic, trichloroacetic or trifluoroacetic acid. In this manner, phenyllead(IV) triacetate (**482**) reacts with anisole to give the isomeric methoxybiphenyls **91**, **483**, **78** in 32% yield [65], Scheme 31.

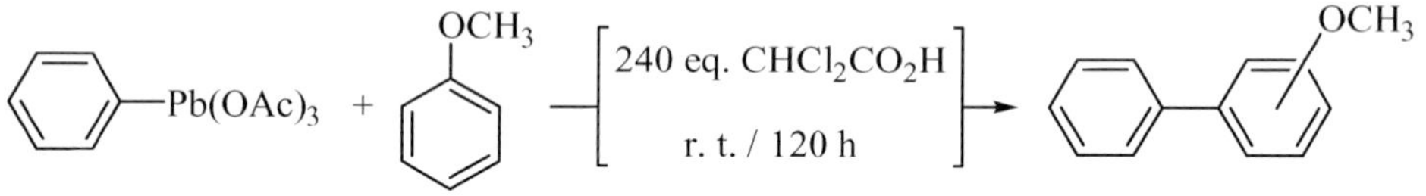

482

91: 2-OCH$_3$, 47.2%

483: 3-OCH$_3$, 0.4%

78: 4-OCH$_3$, 52.4%

Scheme 31

The aryllead(IV) tricarboxylates can be prepared, among several possible routes, the most conveniently by the reaction of aryltri-*n*-butylstannanes [66], or arylboronic acids [67] with lead(IV) acetate in the presence of mercury(II) trifluoroacetate as the catalyst, or, in the case of electron-rich arenes such as polyalkoxybenzenes, by direct plumbylation with Pb(OAc)$_4$ or its derivatives, Scheme 32.

From the synthetic point of view, aryllead(IV) compounds are efficient reagents for arylation of electron-rich aromatics such as polyalkylbenzenes, polyalkoxybenzenes, thiophenes, furans, phenols and anilines. Namely, the reaction of 4-methoxylead(IV) triacetate (**484**) with mesitylene (**485**) in the presence of trifluoroacetic acid resulted in formation of the biaryl **486** with a 84% yield [65], respectively, Scheme 33.

Beside haloacetic acids, this reaction can be strongly accelerated by aluminum chloride, trifluoroacetate, Al(OCOCF$_2$CF$_3$)$_3$, or Al(OCOCF$_2$CF$_2$CF$_3$)$_3$ (8 eq. to ArPb(OAc)$_3$). These fluorinated carboxylic salts were readily prepared by careful addition of the given acid to resublimed aluminum chloride [65]. These catalysts allow to perform the arylation with less reactive arenes such as toluene and benzene. The main side-reaction which occurs during the aryllead(IV) tricarboxylates involving reactions in trifluoroacetic acid is the protodeplumbylation. This reaction produces the parent arene by protonation of aryllead(IV) compound, and may become the major process [65,66].

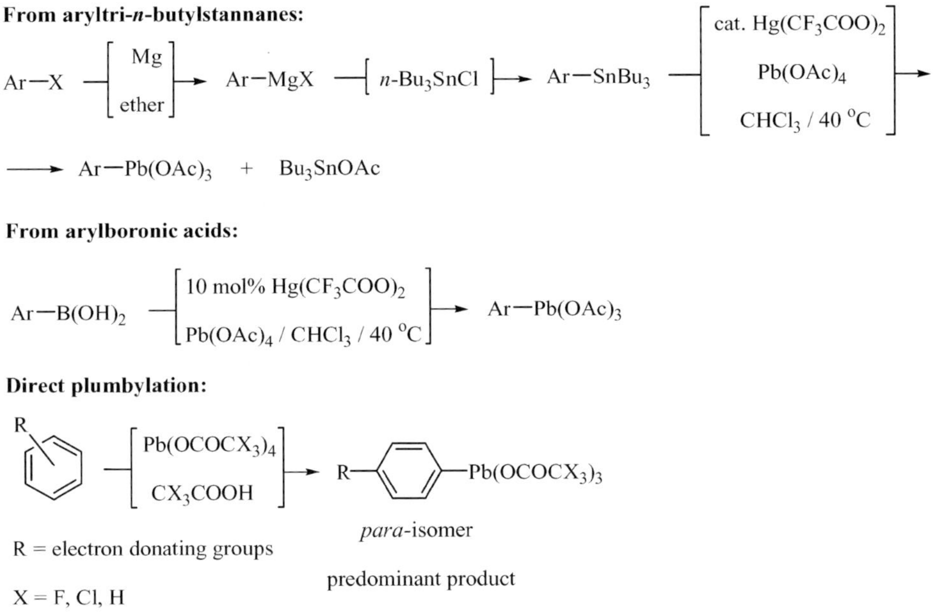
From aryltri-n-butylstannanes:
Ar—X
Mg
ether
Ar—MgX
n-Bu3SnCl
Ar—SnBu3
cat. Hg(CF3COO)2
Pb(OAc)4
CHCl3 / 40 °C
Ar—Pb(OAc)3 + Bu3SnOAc
From arylboronic acids:
Ar—B(OH)2
10 mol% Hg(CF3COO)2
Pb(OAc)4 / CHCl3 / 40 °C
Ar—Pb(OAc)3
Direct plumbylation:
R
Pb(OCOCX3)4
CX3COOH
R
Pb(OCOCX3)3
para-isomer
predominant product
R = electron donating groups
X = F, Cl, H

Scheme 32

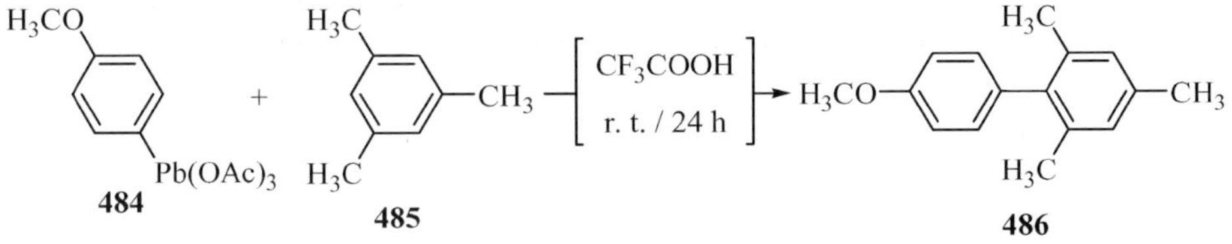
H3CO
Pb(OAc)3
484
H3C
CH3
H3C
485
CF3COOH
r. t. / 24 h
H3CO
H3C
CH3
H3C
486

Scheme 33

Phenols are reacted with aryllead(IV) tricarboxylates in the presence of pyridine to give, predominantly the *ortho*-, but also considerable amount of *para*-arylated products [66]. In this fashion, 4-methoxyphenyllead(IV) triacetate (**484**), upon reaction with 2,6-dimethylphenol (**487**), gave *ortho*-arylated product **488** in 75-90% yield where no *para*-arylation occured. Fully substituted phenol **458** bearing very bulky *ortho*-substituents furnished the *para*-arylated compound **489**, as sole product in 30% yield, in both cases with the loss of aromaticity of phenolic ring [66], Scheme 34.

Anilines can be arylated with aryllead(IV) tricarboxylates at nitrogen in the presence of copper(II) acetate under very mild conditions. However, the *C*-arylation has been effected upon previous treatment of unprotected aniline with *tert*-butylmagnesium chloride, followed by reaction with aryllead(IV) tricarboxylates in the presence of the strong base, e.g. DABCO.

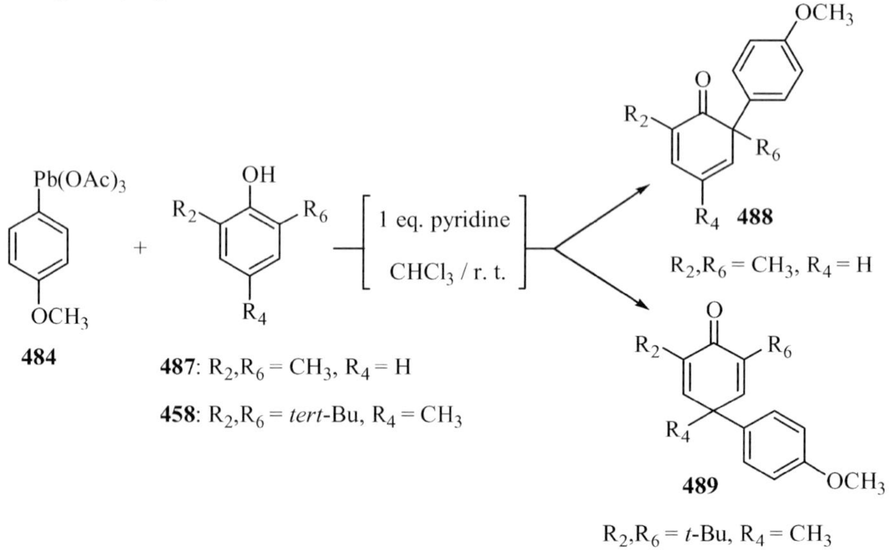

Scheme 34

Thus 2,4,6-trimethylaniline (**490**) was reacted with phenyllead(IV) triacetate (**482**) in the presence of a catalytic amount of copper(II) acetate affording the *N*-phenyl derivative **491** in 94% yield [66]. Contrarily, the *C*-arylation of, for example, 3,5-dimethylaniline (**492**) was accomplished by treatment with *t*-BuMgCl to give the magnesium amide **493**, which subsequently reacts with aryllead compound **494**, *via* lead-amide **495**, to produce the respective 2-aminobiphenyl **496** in 88% yield [68], Scheme 35.

Apart from electrophilic aromatic substitution of activated arenes with aryllead(IV) tricarboxylates, the latter undergo the palladium-catalysed cross-coupling reactions with diaryliodonium salts furnishing the biaryls in good yields. Namely, the reaction of phenyllead(IV) triacetate (**482**) and diaryliodonium salt **467** was effected with $Pd_2(dba)_3$ in the presence of sodium methoxide as the base under very mild reaction conditions, to give 4-methoxybiphenyl (**78**) with a 75% yield [69], Scheme 36.

Analogously to the cross-coupling reactions with other organometallics, the unsymmetrical diaryliodonium salt donates only the electron-rich aryl-group (e.g. 4-methoxyphenyl), while the electron-poor aryl group (e.g. phenyl) forms the corresponding aryl iodide (e.g. iodobenzene). Sodium methoxide promotes the reaction by converting the starting $ArPb(OAc)_3$ to the $ArPb(OMe)_2OAc$, which rapidly undergoes the transmetallation reaction of aryl group from lead to palladium [69].

N-arylation:

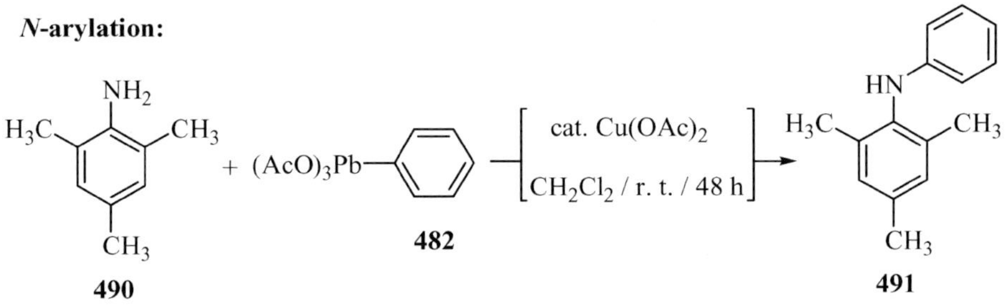

482   **491**

490

C-arylation:

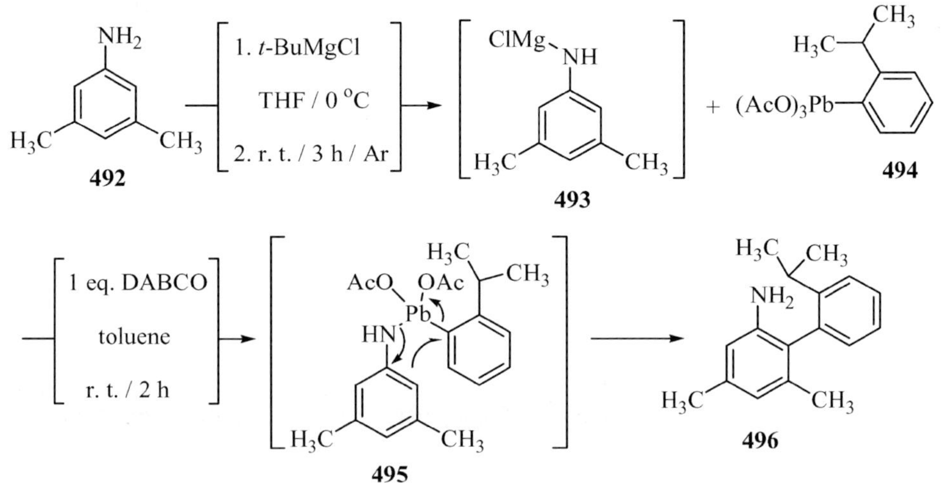

492   **493**   **494**

495   **496**

Scheme 35

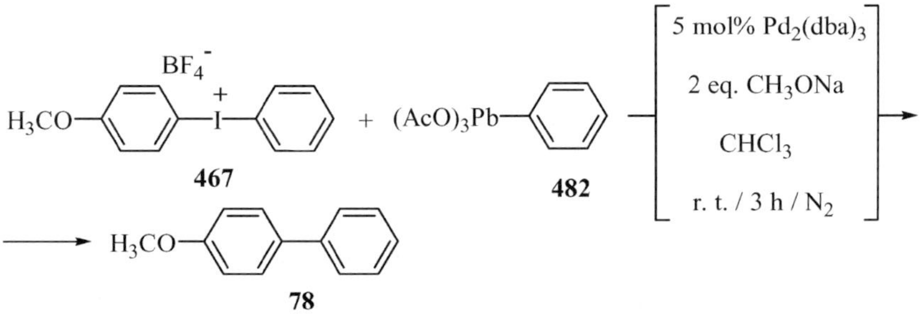

467   **482**

78

Scheme 36

Whereas the aryllead(IV) tricarboxylates in the above mentioned arylation reactions of electron-rich arenes act as aryl-cation donors, in the palladium-catalysed cross-coupling reaction with diaryliodonium salts, they play the role as nucleophilic counterparts.

7.8. Synthesis of biaryls involving arylbismuth and arylantimony reagents

Additional organometallics useful in the synthesis of biaryls are arylbismuth reagents: triarylbismuth(III), Ar_3Bi, diarylbismuth(III) chlorides, acetates or trifluoroacetates, Ar_2BiX (X = Cl, OAc, $OCOCF_3$), triarylbismuth(V) dichlorides, Ar_3BiCl_2, and triarylbismuth(V) carbonates, Ar_3BiCO_3 [70-73]. Whereas the former, bismuth(III) reagents are excellent precursors for symmetrical biaryls [71,72], the latter bismuth(V) reagents undergo the arylation reaction of phenols to give the α-arylated phenols [71], Scheme 37.

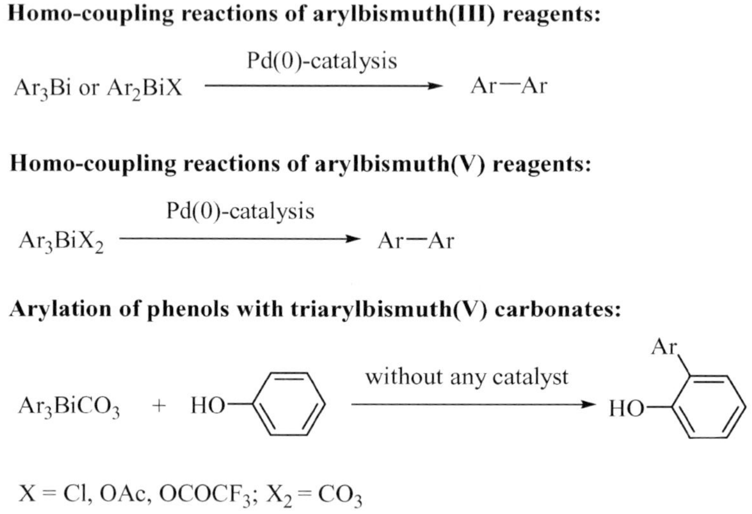

Homo-coupling reactions of arylbismuth(III) reagents:

$$Ar_3Bi \text{ or } Ar_2BiX \xrightarrow{\text{Pd(0)-catalysis}} Ar-Ar$$

Homo-coupling reactions of arylbismuth(V) reagents:

$$Ar_3BiX_2 \xrightarrow{\text{Pd(0)-catalysis}} Ar-Ar$$

Arylation of phenols with triarylbismuth(V) carbonates:

$$Ar_3BiCO_3 \quad + \quad HO-\langle\text{aryl}\rangle \xrightarrow{\text{without any catalyst}} HO-\langle\text{aryl-Ar}\rangle$$

X = Cl, OAc, $OCOCF_3$; $X_2 = CO_3$

Scheme 37

The triarylbismuth(III) reagents, Ar_3Bi, are prepared by adding the corresponding aryl Grignard reagents to the THF solution of bismuth(III) chloride [70]. The diarylbismuth(III) chlorides are obtained by the reaction of triarylbismuth(III) reagents with $BiCl_3$ in dry ether at 0 °C [71]. Triarylbismuth(V) dichlorides are readily formed by careful adding the dichloromethane solution of sulfuryl chloride to the cooled (-78 °C) solution of triarylbismuth(III) reagents in the same solvent [71]. Their carbonate derivatives are prepared by simple treatment the acetone solution of triarylbismuth(V) dichlorides with aqueous potassium carbonate [70].

The triarylbismuth(III) reagents, Ar₃Bi, smoothly undergo the homo-coupling reaction in the presence of a stoichiometric amount of palladium(II) acetate (1 eq. to Ar₃Bi) and triethylamine (2 eq.), in tetrahydrofuran or far more efficiently in HMPA, to afford the symmetrical biaryls in almost quantitative yields [72]. For example, tri(4-methoxy phenyl)bismuth (**497**) is converted to its homo-coupling product, 4,4'-dimethoxy biphenyl (**77**) in 98% yield [72], respectively, Scheme 38.

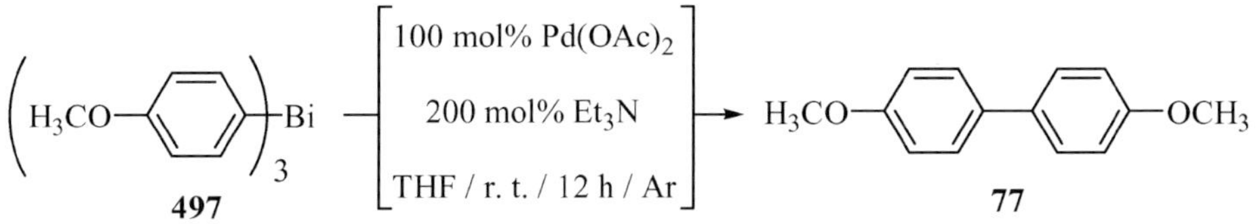

Scheme 38

Employing exactly the same procedure, diarylbismuth(III) chlorides, Ar₂BiCl, and triarylbismuth(V) reagents, Ar₃BiX₂ (X= Cl, OAc, OCOCF₃; X₂= CO₃) readily proceed the reaction, but with a catalytic amount of palladium(II) acetate (5 mol%) and triethylamine (10 mol%) [72]. However, the performing in an air or oxygen atmosphere allows to effect the reaction with catalytic palladium loading, e.g. 5 mol%, also in excellent yields. Since the oxygen solubility is much higher in methanol than in THF, this catalytic method was conducted in the former solvent. As the methanol is protic solvent, the protodebismuthation side-product was isolated, however, in amounts not higher than 10% [73]. Thus tri(4-chlorophenyl)bismuth (**498**) gave 4,4'-dichlorobiphenyl (**269**) in 92% yield, accompanied with formation of chlorobenzene (4%) as reduction by-product [73], Scheme 39.

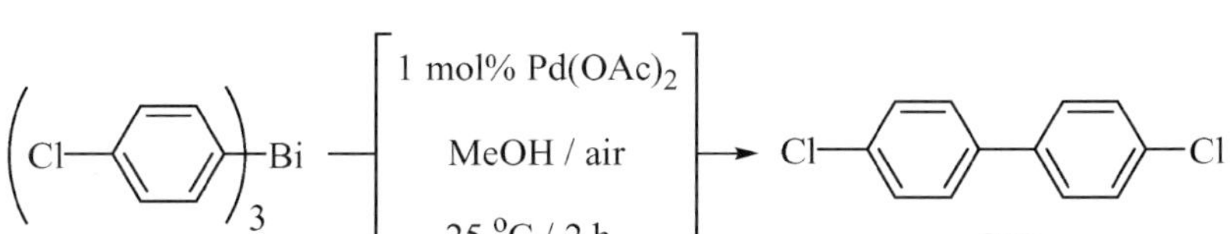

Scheme 39

Unfortunately, the method works well only with *para*- and *meta*-substituted triarylbismuth compounds, while the *para*-substituted ones failed to give the reaction or resulted in very low yields [73]. Arylation of phenols and other electron-rich aromatics, e.g. indole, with triarylbismuth(V) dichlorides, carbonates and related compounds is readily affected under neutral conditions, by simple stirring in dichloromethane solution at room temperature under a nitrogen atmosphere [71], or in the presence of the bases such as sodium hydride or *tert*-butyltetramethyl guanidine [66]. Phenols bearing electron-withdrawing substituents are predominantly arylated at oxygen to give the diarylethers, whereas those with electron-donating groups are mostly *ortho-C*-arylated [66]. In this manner, triphenylbismuth(V) carbonate (**499**, 1.5 eq.), upon the reaction with 2-naphthol (**378**, 1 eq.), resulted in formation of 1-phenyl-2-naphthol (**500**) in 76% yield under very mild reaction conditions [71], Scheme 40.

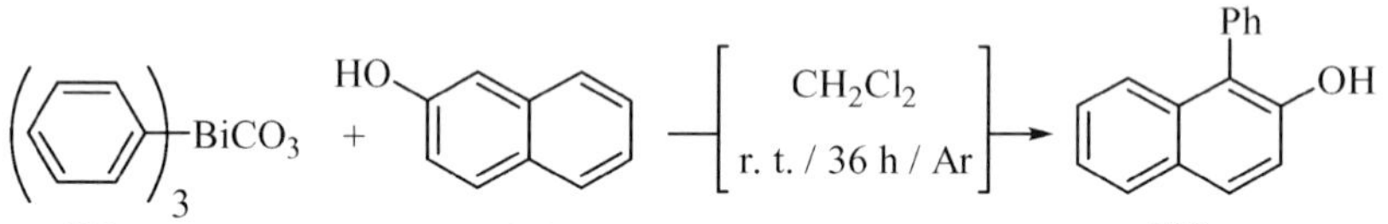

Scheme 40

Analogously to triarylbismuth(III) compounds, triarylantimony(III) reagents, Ar$_3$Sb, smoothly undergo the homo-coupling reaction to symmetrical biaryls in excellent yields, but using stoichiometric Pd(OAc)$_2$-protocol, see Scheme 37. However, triarylantimony(V) dichlorides, Ar$_3$SbCl$_2$, under the same reaction conditions, gave the respective biaryls with unaltered high yields, at only catalytic palladium loading, 5 mol% [72]. The synthesis of biaryls involving various arylbismuth and arylantimony reagents, because of ready transmetallation reaction from bismuth or antimony to palladium, is very promising area, where new, interesting and powerful alternative methods may arise.

7.9. Synthesis of biaryls *via* benzidine rearrangement

Further unique approach to the synthesis of biaryls is the acid-catalysed rearrangement of *N,N'*-diarylhydrazines to *para,para*-diaminobiaryls [74-88]. In the simplest case *N,N'*-diphenylhydrazine (hydrazobenzene, **501**) is rearranged to 4,4'-diaminobiphenyl (**502**) [74]. Since the latter is well known under the trivial name benzidine, the reaction is called *benzidine rearrangement* (in older literature, the hydrazo-rearrangement). This reaction is an intramolecular process [75,76] proceeding by protonation of the starting hydrazine to give **501a** with the new aryl-aryl bond formation by coupling at *para,para*-positions, *via* the concerted transitional state such as **503** [88], and subsequent tautomerisation of the imine **502a** to give the benzidine (**502**), Scheme 41.

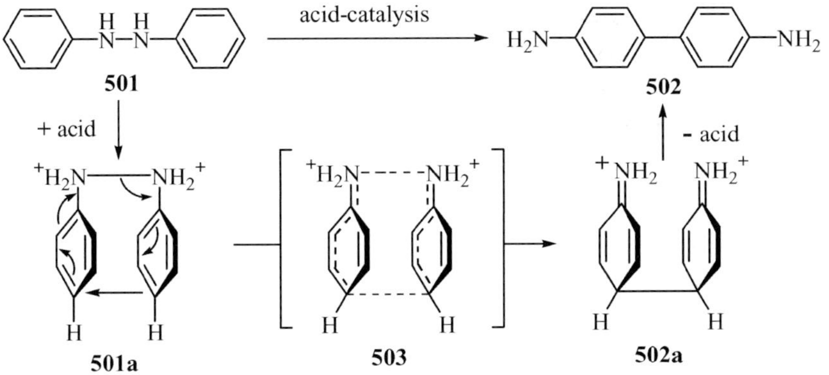

Scheme 41

The reaction is usually conducted in aqueous ethanol or dioxane solution of the starting *N,N'*-diarylhydrazine in the presence of hydrochloric, sulfuric, or perchloric acid, or mixture of hydrochloric acid and zinc chloride, under relatively mild conditions, room or slightly elevated temperatures. In this fashion, passing the hydrogen chloride through the ethanolic solution of 2,2'-dimethoxyhydrazobenzene (**504**) at room temperature, 3,3'-dimethoxybenzidine (**505**) was obtained in 73% yield, accompanied with the formation of small amounts of 2,2'-dimethoxyazobenzene (**506**, 12%) and *o*-anisidine (**259**) as side-products [81], Scheme 42.

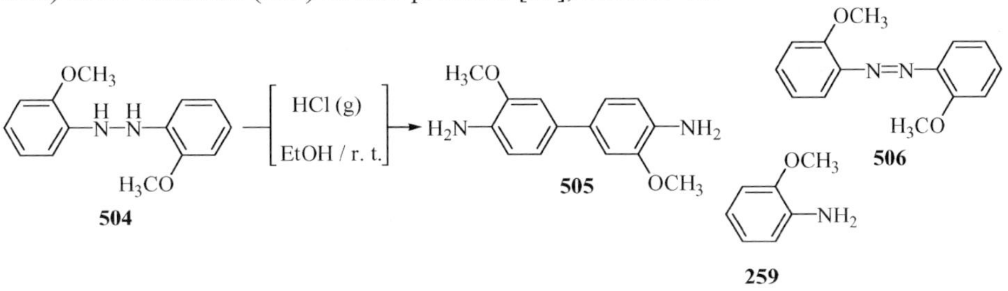

Scheme 42

2,2'-Hydrazonaphthalene (**507**) gives 2,2'-diamino-1,1'-binaphthyl (**508**) in 95% yield with some amount of carbazole **509** (5%) [82,83], Scheme 43.

The benzidine rearrangement is also effective approach to the synthesis of unsymmetrical biaryls, if the starting material is unsymmetrically substituted *N,N'*-diarylhydrazine, ArNHNHAr'. In this manner, the reaction has been successfully used in the phenyl-heteroaryl series [85-87]. Namely, compound **510** underwent the benzidine rearrangement with four-molar excess of phthalic anhydride to the respective biaryl **511**, whose amino groups are directly *in situ* protected as phthaloyl groups, in 75% overall yield, respectively, Scheme 44.

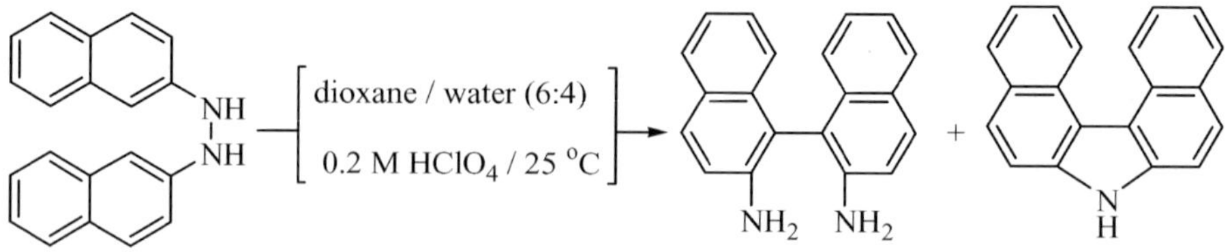

507   508   509

Scheme 43

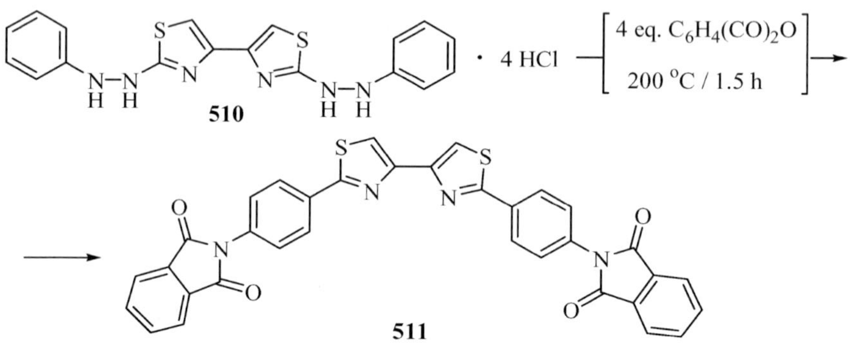

Scheme 44

More interestingly, using the benzidine rearrangement, cyclic *N,N'*-diphenylhydrazine **512** was converted to the respective strained *N,N'*-decamethylenebenzidine (**513**) in 4% yield [86], Scheme 45.

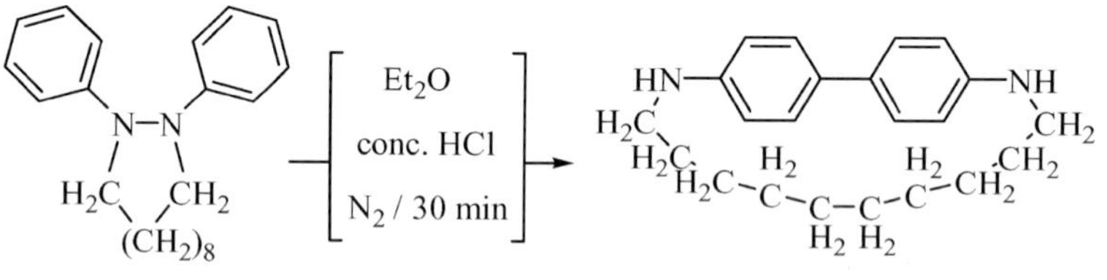

512   513

Scheme 45

Analogously to the benzidine rearrangement, hydrazo-biaryls do undergo the reaction furnishing the tetraaryls in good yields. For example, bis[4-(2-furyl)phenyl]diazane (**514**) by [9,9]-sigmatropic shift gave tetraaryl **515** in 75% yield [89], Scheme 46.

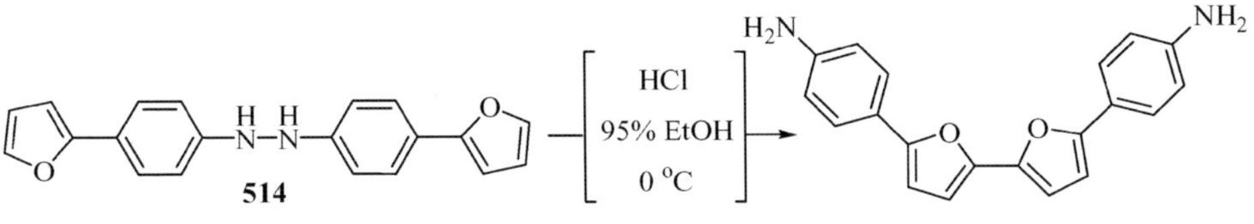

Scheme 46

Beside *para,para*-direction, the rearrangement always produces a small amount of *ortho,para*-product, probably *via* non-concerted, forbidden suprafacial [3,5]-pathway [88]. For example, hydrazobenzene (**501**), beside benzidine (**502**), produces very small, but isolable, amount of *ortho,para*-product, diphenyline (**516**).

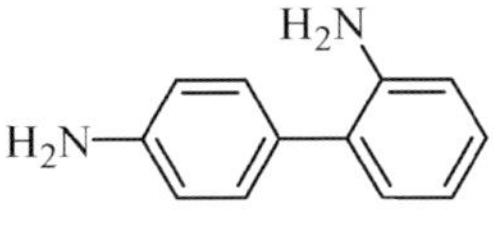

516

When the *para*-position is substituted, the formation of diarylamine becomes the major process, known as the *semidine rearrangement*. Further side-reaction is the oxidation of starting *N,N'*-diarylhydrazine to azo-compound by dissolved oxygen [77], often catalysed by traces of metals, e.g. copper(II) salts [78]. Symmetrical *N,N'*-diaryl hydrazines, as starting materials for the benzidine rearrangement, are readily available materials by reduction of the corresponding nitroaromatics with zinc dust in the presence of sodium hydroxide (for example, refluxing methanol / water, 4:1) [81]. The unsymmetrical *N,N'*-diarylhydrazines are obtained by the condensation of appropriate arylamines with arylnitroso compounds.

In conclusion, the 4,4'-diaminobiaryls are efficiently obtained by the benzidine rearrangement of *N,N'*-diarylhydrazines, which are readily available from *ortho-* and *meta*-substituted anilines as starting materials.

7.10. Photochemical synthesis of biaryls

The aryl-aryl bond formation can be realized in the photochemical manner by irradiation of aryl iodides or bromides such as **Ia** which, by photo-dehydrohalogenation, give biaryls **IIa** [90-92], Scheme 47. However, the photochemical synthesis of biaryls have also been performed as an intermolecular process [93]. The reactions are usually performed by irradiation of the given substrate in benzene or cyclohexane solution in the quartz-glass vessel, under the UV-light of high-pressure mercury vapour lamp (power range: 100-500 W). Triethylamine is often added to the reaction mixture, serving as the traping agent for hydrogen bromide or iodide, liberated during the course of reaction.

Intramolecular cyclization:

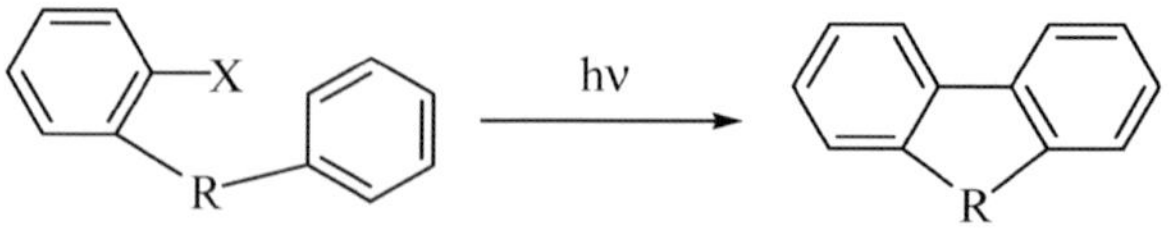

Ia   IIa

Intermolecular arylation:

$$Ar—X \quad + \quad H—Ar' \quad \xrightarrow{\text{hv}} \quad Ar—Ar'$$

X = I, Br

R = -CH$_2$-, -(CH$_2$)$_2$-, -CH$_2$NHCH$_2$-, -CH$_2$NRCH$_2$-, etc.

Scheme 47

The yields vary significantly, what depends on a number of factors: the structure of substrate, the power of UV-lamp, possible side-reactions. Nevertheless, the reaction is very useful for the synthesis of highly-functionalyzed tricyclic systems [90-92]. Thus iododibenzylamine **517** was converted to dibenzazepine **518** in 57% yield at 87% conversion using the medium-pressure Hg-lamp [90], Scheme 48.

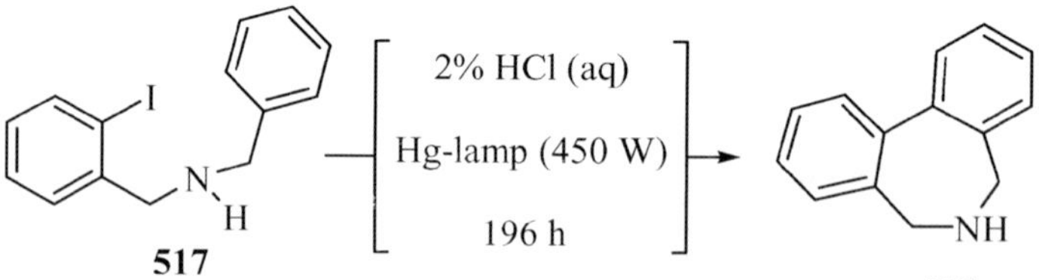

Scheme 48

Among intermolecular reactions, an interesting example is arylation of thiophene with diiodopyrrole **519** where the respective bithiophene **520** was obtained in 78% yield [93], respectively, Scheme 49.

Photochemical synthesis of biaryls have been intensively employed in the synthesis of natural products due to generally mild conditions, relatively high efficacy and simple methodology.

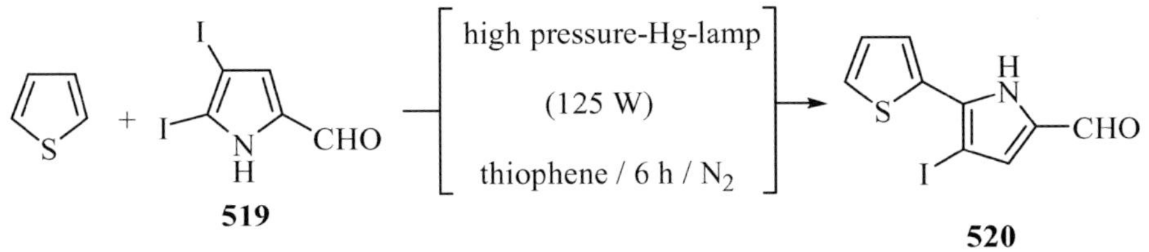

Scheme 49

7.11. Other methods for synthesis of biaryls

The title "other methods" does not mean that these methods are unimportant, but rather includes a few specific approaches to the synthesis of biaryls. First of these involves the benzynes as reactive species, capable to react with a number of arenes, including themselves, to give the respective biaryls in various yields [94-97]. As the benzyne precursors, benzenediazonium carboxylate (**59a**) [94], diphenyliodonium-2-carboxylate (**521**) [96], benzothiadiazole dioxide (**522**), and 1-aminobenzotriazole (**523**) have been used. More conveniently, the benzynes have been *in situ* generated upon the treatment of anthranilic acid (**58**) with alkyl nitrites (*via* **59a**), Scheme 50.

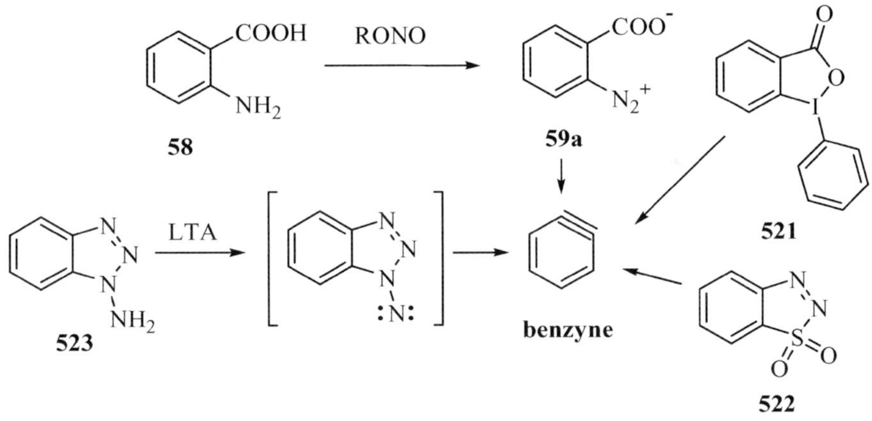

Scheme 50

The benzyne involving reactions have been employed in the synthesis of various biphenylenes. Namely, 3-amino-2-naphthoic acid (**524**) was converted to the corresponding diazonium salt **525** with isopentyl nitrite in the presence of a catalytic amount of trifluoroacetic acid in 76% yield. The latter was decomposed in boiling 1,2,3-trichloropropane to give dibenzo[*b,h*]biphenylene (**526**) in 2.7% isolated yield [95], Scheme 51.

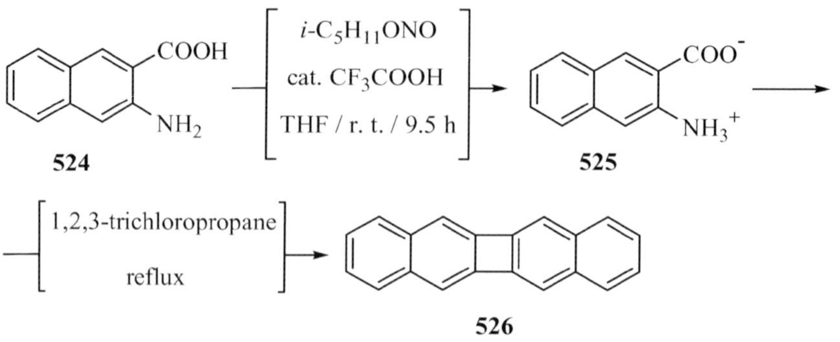

Scheme 51

Biehl's group [97] has developed an efficient syntesis of benzo[4,5]thieno [2,3-*b*]pyridines and appropriate naphthyl-analogues by reaction of *in situ* generated benzyne with Barton esters. For example, when the acetone solution of anthranilic acid (**58**) was added dropwise to the refluxing mixture of the Barton ester **527** and isopentyl nitrite in dichloromethane, after an additional 3-h refluxing period, the respective benzo[4,5]thieno[2,3-*b*]pyridine (**528**) was obtained in 52% yield, Scheme 52.

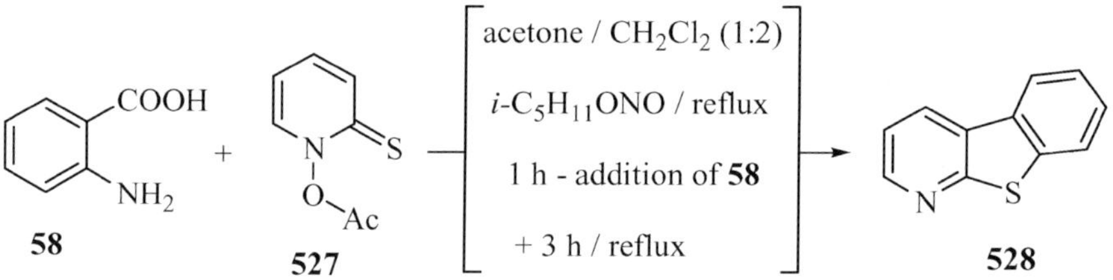

Scheme 52

Using this method, several substituted analogues were prepared in moderate yields [97]. The benzynes can be stabilized by forming the complexes with zirconocene, as shown in the Buchwald's group work [98]. Thus 2-bromotoluene (**346**) reacts with the strong base, e.g. *n*-BuLi, to generate the aryllithium, which is further trapped with Cp$_2$Zr(CH$_3$)Cl to give the arylzirconium **529**. The latter undergoes the elimination of methane to generate the zirconocene-benzyne complex **530**. Complexes like **530**, smoothly react with aryl bromides, e.g. 4-chloro-bromobenzene (**178**), in the palladium(0)-catalysed reaction to produce the *ortho*-iodobiaryls, e.g. **531**, derived form the cross-coupling at less hindered carbon of three-membered zirconocene complex **530**, in moderate to high yields. The carbon-zirconium bond in the resulting complex **531** is cleavaged by addition of iodine, accompanied with formation of synthetically versatile *ortho*-iodobiaryls, in this particular case, the biaryl **532** with a 64% overall yield [98], Scheme 53.

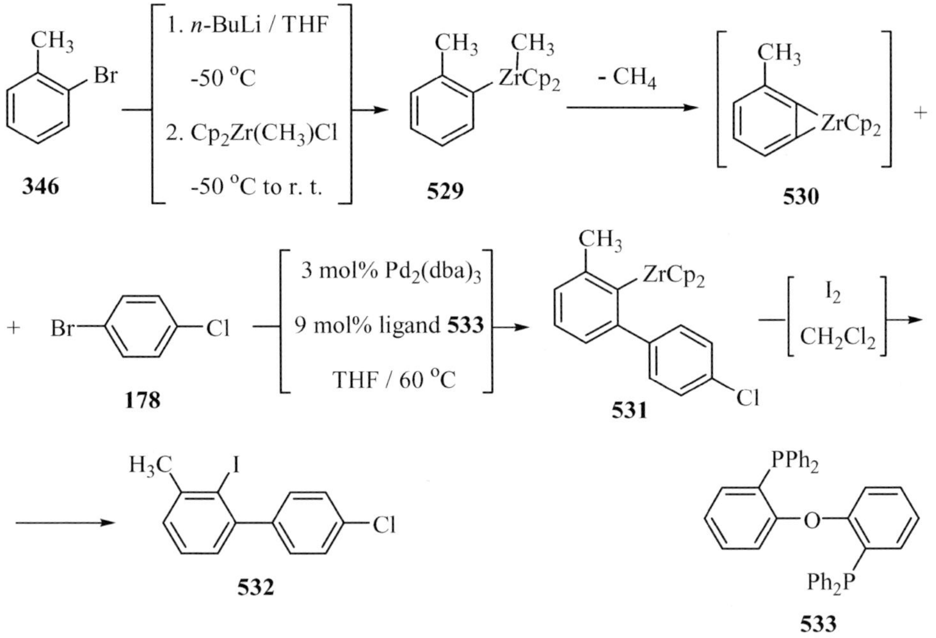

Scheme 53

Among further specific methods, Newkome's group [99] has found that ethyl pyridin-3-carboxylate (**534**) underwent the dehydrogenative coupling at unhindered α-position to give the diethyl 2,2'-bipyridine-5,5'-dicarboxylate (**535**) in 35% yield, upon refluxing with 10% Pd-C under reduced pressure, Scheme 54.

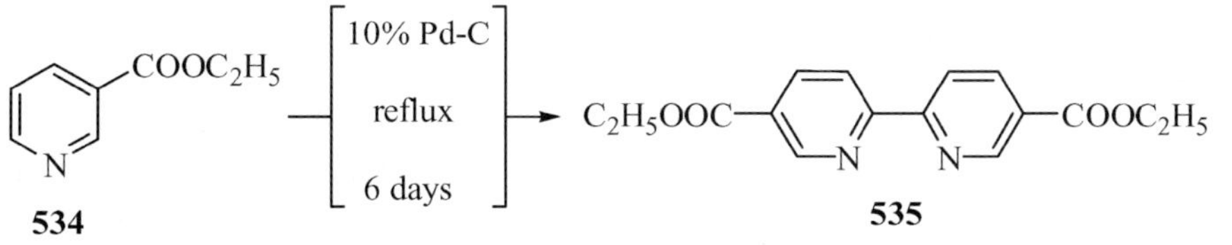

Scheme 54

Following interesting and valuable method for synthesis of polyaromatics, including well known cancerogenic structures, is the trichlorosilane-mediated palladium-catalysed homo-coupling reaction of 1,4-epoxy-1,4-dihydroarenes to biaryls [100]. For example, the reaction of 7-oxabenzonorbornadiene (**536**) with trichlorosilane in the presence of zinc powder and palladium catalyst, under very mild reaction conditions, afforded 2,2'-binaphthyl (**166**) with a 95% isolated yield [100], respectively, Scheme 55.

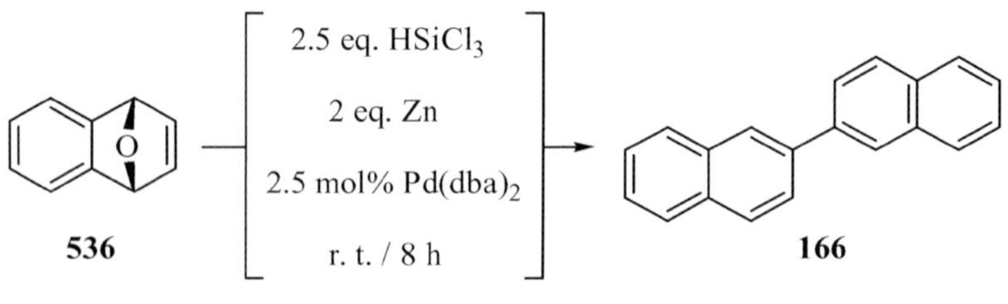

Scheme 55

The method is very effective in the synthesis of several naphthalene, phenanthrene, and other polyphenyl-dimers [100].

7.12. Selected synthetic procedures

7.12.1. The Motherwell synthesis of biaryls: Preparation of 2N-methyl-4',6-dimethyl-1-aminobiphenyl (537) [4]

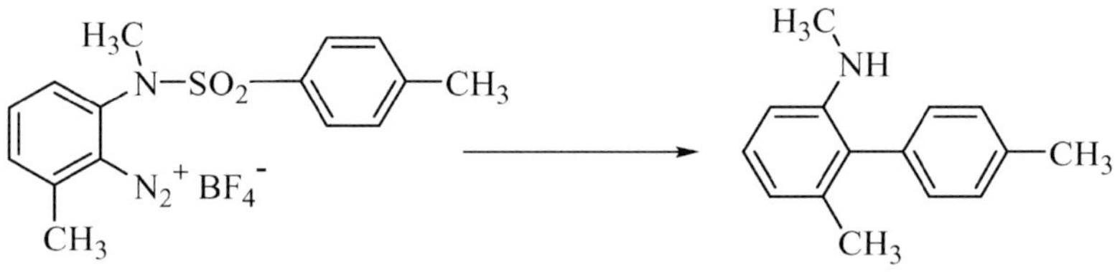

A solution of titanium(III) chloride (4 ml, 4 mmol) in 2 M aqueous hydrochloric acid was added dropwise to a solution of **538** [4] (390 mg, 1 mmol) in acetone (4 ml) at 0 °C. After the addition, the reaction mixture was stirred for 1 h at 0 °C, water (25 ml) and sodium hydrogencarbonate were added, and the mixture was extracted with dichloromethane (3x20 ml). The organic layers were washed with brine, dried over MgSO$_4$, and evaporated. The colourless oil was purified by column chromatography on silica gel eluated with petrolether and diethyl ether (9:1) giving 226 mg (50%) of **537** as a colourless oil that crystallized on standing, m.p. 73-74 °C (diethyl ether).

7.12.2. Free-radical arylation of arenes with arylhydrazines: Preparation of 2-(4-bromophenyl)furan (539) [8]

Br—⟨benzene⟩—NHNH$_3^+$ Cl⁻   +   ⟨furan⟩   ⟶   Br—⟨benzene⟩—⟨furan⟩

540   **539**

To a mixture of manganese(III) acetate dihydrate (563 mg, 2.1 mmol) in furan (10 ml) was added 4-bromophenylhydrazinium chloride (**540**, 156 mg, 0.7 mmol) and the resulting mixture refluxed until the TLC showed the dissapearance of **540** (about 5 h). After the completion of the reaction, the mixture was filtered through a pad of silica gel using *n*-hexane or petroleum ether as an eluent. Concentration under reduced pressure furnished 101 mg (65%) of pure 2-(4-bromophenyl)furan (**539**) as white crystals, m.p. 80-82 °C.

Note: The same procedure works well starting from arylboronic acids to afford the biaryls in slightly better yields [9].

7.12.3. Homo-coupling reaction of aryl Grignard reagents with FeCl$_2$: Preparation of biphenyl (8) [12]

⟨benzene⟩—MgI   ⟶   ⟨benzene⟩—⟨benzene⟩

221b   **8**

To a solution of phenylmagnesium iodide (**221b**, 0.03 mol in 50 ml of THF), cooled to 0 °C, anhydrous iron(II) chloride (1.27 g, 0.01 mol) was added. The solution turned black, and a black solid deposited. The mixture was stirred for one hour, and then hydrolyzed by filtration through a sintered glass plate into iced solution of ammonium chloride (5%, 200 ml). The mixture was extracted with dichloromethane (3x50 ml), combined organic layers dried (Na$_2$SO$_4$), filtered, and evaporated to give 1.51 g (98%) of pure biphenyl (**8**), m.p. 71-72 °C.

7.12.4. Titanium tetrachloride-mediated homo-coupling of aryl Grignard reagents: Preparation of 4,4'-dicyanobiphenyl (541) [13]

NC—⟨benzene⟩—Br   ⟶   NC—⟨benzene⟩—⟨benzene⟩—CN

272   **541**

To a solution of butylmagnesium bromide (1.0 ml, 1.0 M THF solution, 1.0 mmol) in THF (5 ml) was added butyllithium (1.3 ml, 1.6 M hexane solution, 2.0 mmol) at 0 °C and the mixture was stirred for 10 min. The solution was cooled to -40 °C, and a solution of 4-bromobenzonitrile (**272**, 0.36 g, 2.0 mmol) in THF (3 ml) was added dropwise. The mixture was stirred for 0.5 h. In another flask, THF (10 ml) was placed and cooled to -40 °C. Titanium tetrachloride (0.33 ml, 3.0 mmol) was added and the mixture was stirred for 10 min. To this suspension, the THF solution of arylmagnesate prepared above was added dropwise. The reaction mixture was stirred for 0.5 h at -40 °C and then poured into saturated aqueous ammonium chloride solution (20 ml) and extracted with ethyl acetate (3x10 ml). The organic layers were dried over Na_2SO_4 and concentrated *in vacuo* to give a white solid. Washing this solid with *n*-hexane gave 0.13 g (62%) 4,4'-dicyanobiphenyl (**541**); IR (nujol) 2222, 1605, 817 cm^{-1}.

7.12.5. *The Meyers synthesis of biaryls: Preparation of biphenyl-2-carboxylic acid (543) [40]*

A mixture of 2-methoxybenzoic acid (**459**, 50.00 g, 0.33 mol) and thionyl chloride (117.3 g, 0.98 mol) was stirred at 25 °C for 24 h. The excess of thionyl chloride was removed *in vacuo*, and the residue was distilled (b.p. 68 °C, 0.05 mmHg), yielding 51.4 g of acid chloride as a colourless oil. A solution of the acid chloride in dichloromethane (75 ml) was added dropwise to 2-amino-2-methyl-1-propanol (**452**, 53.7 g, 0.6 mol) in dichloromethane (125 ml) at 0 °C. After stirring for 2.5 h at 25 °C, the solution was filtered and filtrate evaporated to give 68.3 g of the crystalline amide. The latter (25 g) was treated dropwise with 40.2 g of thionyl chloride and magnetically stirred. The solution was then poured into dry diethyl ether (150 ml), and the oxazoline hydrochloride precipitated and removed by filtration. The salt was neutralized with 20% sodium hydroxide, and the alkaline solution was extracted with diethyl ether, dried (MgSO₄), and concentrated to give an oil (21 g, 83%), which crystallized on standing, m.p. 66-68 °C. An analytical sample was purified by recrystallization from *n*-hexane, m.p. 68-69.5 °C; IR (KBr) 1635 cm⁻¹. To the solution of thus obtained oxazoline **544** (1.19 g, 5.8 mmol) in dry THF (10 ml), the solution of phenylmagnesium bromide (**221a**, 6.00 mmol) in THF was added at 25 °C under nitrogen. The reaction mixture was stirred for 20 h at 25 °C, and then quenched in saturated aqueous ammonium chloride solution, extracted with diethyl ether (3x30 ml), dried (K₂CO₃), and concentrated. The product was purified by column chromatography on silica gel using ethyl acetate / *n*-hexane as an eluent. From combined product fractions, 1.38 g (95 %) of biphenyl oxazoline **545** was obtained. The latter (1.26 g, 5 mmol) was subjected to acid-catalysed hydrolysis by refluxing the solution in 4 M HCl (100 ml) for 24 h. After cooling, the heterogeneous mixture was extracted with diethyl ether (3x30 ml). The ethereal extracts were washed with water and saturated brine, dried (MgSO₄), and concentrated to give products of acceptable purity. Further purification was achieved by recrystallization from ethanol / water, to give 743 mg (75%) of pure biphenyl-2-carboxylic acid (**543**), m.p. 113-114 °C.

7.12.6. Homo-coupling of arylboronic acids: Preparation of 2,2'-binaphthyl (166) [22]

To a solution of naphthalene-2-boronic acid (**542**, 172 mg, 1 mmol), sodium acetate (410 mg, 5 mmol), and copper(II) nitrate (490 mg, 3 mmol) in 95% ethanol (5 ml) was added Pd(PPh₃)₄ (116 mg, 0.1 mmol, 10 mol%) and the resulting mixture was stirred for 8 h at ambient temperature. Then, the solvent was removed *in vacuo*, and the residue was chromatographed on silica gel with *n*-hexane. From the combined product fractions, 103 mg (81%) of pure 2,2'-binaphthyl (**166**) as white crystals was obtained, m.p. 158-160 °C.

Note: Copper(II) nitrate is also efficient reagent for the homo-coupling reaction of aryltrialkylstannanes at molar ratio ArSnR₃ : Cu(NO₃)₂ = 1 : 5 to give high yields of symmetrical biaryls in dry DMF or THF [27,28].

7.12.7. Palladium-catalysed arylation of arenes with aryl halides: Preparation of 2-cijanodibenzofuran (474) [56]

471 → 474

2-Bromo-2'-cyanodiphenylether (**471**, 2.74 g, 10 mmol), anhydrous sodium carbonate (1.3 g), and palladium(II) acetate (0.22 g, 1 mmol, 10 mol%) in *N,N*-dimethyl acetamide (20 ml) were heated at 170 °C under nitrogen for 1.5 h. Water (100 ml) was added to the cooled solution, and the precipitated solid was collected by filtration. Crystallization from ethanol gave 1.54 g (80%) of pure 2-cyanodibenzofuran (**474**), m.p. 141-143 °C.

7.12.8. The benzidine rearrangement: Preparation of 3,3'-dimethoxybenzidine (505) [81]

546 → 504 → 505

Zinc dust (500 g) was added gradually to a mechanically stirred solution of 2-nitro anisole (**546**, 250 g), methanol (1000 ml), and water (250 ml) containing dissolved sodium hydroxide (365 g). The mixture was heated under reflux, and after 3 h, water (250 ml) was added, and the heating continued. After a further 7 h, the product was filtered, and the residue was washed with water, mixed with water, and filtered through copper gauze, which removed the zinc and zinc oxide from the bulk of the coarsely granular hydrazo-compound. Further quantities were obtained by diluting the aqueous-methanolic filtrate and, by drying and extracting the zinc / zinc oxide mixture with ethanol, the whole separation was conducted as rapidly as possible in order to avert undue atmospheric oxidation. The yield was 179 g (90%) and the product **504** was purified by crystallization from ethanol (2x). In the first crystallization, 10% aqueous ammonium sulfide, equivalent to 10 ml per litre of ethanol, was added, and the solution filtered from sulfur, which separated during cooling. The second crystallization was carried out in an atmosphere of nitrogen, and a few drops of ammonium sulfide were added. The faintly yellow substance **504** had m.p. 105 °C.

Hydrogen chloride was passed into an ethanolic (1000 ml) solution of hydrazo-compound **504** (100 g) contained in a vessel, from which the air had been displaced by nitrogen. The dihydrochloride was precipitated from the orange solution by diethyl ether as colourless crystals, which, after being washed with diethyl ether containing a small proportion of ethanol (5-10%), had m.p. 272 °C, and on basification with ammonia gave 75 g (75%) of pure 3,3'-dimethoxybenzidine (**505**), m.p. 138 °C.

7.13. Conclusion

In this Chapter, several specific, but important aryl-aryl bond forming reactions have been decribed. Whereas the Meyers approach is a widely used general method for synthesis of biaryl-2-carboxylic acids, the free-radical arylations of liquid arenes with arylhydrazines or arylboronic acids are modern alternatives to the efficient phase transfer Gomberg-Bachmann-Hey reaction, described in the Chapter 2. The Motherwell synthesis of biaryls is also a versatile method for synthesis of 2-hydroxy- or 2-amino-substituted biphenyls. Several methods for the homo-couplings of organometallics are useful, and in certain instances more effective alternatives to analogous reactions starting directly from the corresponding aryl halides or sulfonates (Chapter 3). Synthesis of biaryls involving arylbismuth and arylantimony reagents are powerful and perspective methods. Aryllead(IV) tricarboxylates are also valuable reagents for the small-scale preparation of biaryls derived form electron-rich arenes. The benzidine rearrangement offers a simple and convenient way for obtaining the respective biaryls derived from readily available anilines, without any expensive catalyst. Diaryliodonium salts are very effective electrophilic counterparts in the cross-coupling reactions with a number of arylmetallic reagents affording the biaryls in excellent yields. In this Chapter, the reactions with aryltrifluoroborates, arylbismuthum(V) and aryltellurium reagents have been described, providing an interesting and powerful contribution to the synthetic methodology.

7.14. References

1. W. B. Motherwell and A. M. K. Pennell, J. Chem. Soc., Chem. Commun. (1991) 877.
2. M. L. E. N. da Mata, W. B. Motherwell and F. Ujjainwalla, Tetrahedron Lett. 38 (1997) 137.
3. M. L. E. N. da Mata, W. B. Motherwell and F. Ujjainwalla, Tetrahedron Lett. 38 (1997) 141.
4. E. Bonfand, L. Forslund, W. B. Motherwell and S. Vázquez, Synlett (2000) 475.
5. A. Studer and S. Amrein, Synthesis (2002) 835.
6. A. Couture, E. Deniau, P. Grandclaudon and C. Hoarau, J. Org. Chem. 63 (1998) 3128.
7. A. S. Demir, Ö. Reis and E. Özgül-Karaaslan, J. Chem. Soc., Pekin Trans. 1 (2001) 3042.
8. A. S. Demir, Ö. Reis and M. Emrullahoğlu, Tetrahedron 58 (2002) 8055.
9. A. S. Demir, Ö. Reis and M. Emrullahoğlu, J. Org. Chem. 68 (2003) 578.
10. X.-C. Li, H. Sirringhaus, F. Garnier, A. B. Holmes, S. C. Moratti, N. Feeder, W. Clegg, S. J. Teat and R. H. Friend, J. Am. Chem. Soc. 120 (1998) 2206.
11. V. G. Nenajdenko, D. V. Gribkov, V. V. Sumerin and E. S. Balenkova, Synthesis (2002) 124.
12. H. Gilman and M. Lichtenwalter, J. Am. Chem. Soc. 61 (1939) 957.
13. A. Inoue, K. Kitagawa, H. Shinokubo and K. Oshima, Tetrahedron 56 (2000) 9601.
14. A. McKillop, L. F. Elsom and E. C. Taylor, Tetrahedron 26 (1970) 4041.
15. E. C. Taylor, H. W. Altland and A. McKillop, J. Org. Chem. 40 (1975) 2351.
16. S. M. H. Kabir and M. Iyoda, Synthesis (2000) 1839.
17. K. M. Hossain, T. Shibata and K. Takagi, Synlett (2000) 1137.
18. P. Rosa, N. Mézailles, F. Mathey and P. Le Floch, J. Org. Chem. 63 (1998) 4826.
19. H. C. Brown and C. H. Snyder, J. Am. Chem. Soc. 83 (1961) 1002.
20. S. W. Breuer and F. A. Broster, Tetrahedron Lett. (1972) 2193.
21. M. Moreno-Mañas, M. Pérez and R. Pleixats, J. Org. Chem. 61 (1996) 2346.
22. D. J. Koza and E. Carita, Synthesis (2002) 2183.
23. M. S. Wong and X. L. Zhang, Tetrahedron Lett. 42 (2001) 4087.
24. J. D. Wilkey and G. B. Schuster, J. Org. Chem. 52 (1987) 2117.
25. H. Sakurai, C. Morimoto and T. Hirao, Chem. Lett. (2001) 1084.
26. S. Yamaguchi, S. Ohno and K. Tamao, Synlett (1997) 1199.
27. E. Piers, J. G. K. Yee and P. L. Gladstone, Org. Lett. 2 (2000) 481.
28. M. Iyoda, T. Kondo, K. Nakao, K. Hara, Y. Kuwatani, M. Yoshida and H. Matsuyama, Org. Lett. 2 (2000) 2081.

29. R. L. Beddoes, T. Cheeseright, J. Wang and P. Quayle, Tetrahedron Lett. 36 (1995) 283.

30. G. Harada, M. Yoshida and M. Iyoda, Chem. Lett. (2000) 160.

31. S.-K. Kang, T.-G. Baik, X. H. Jiao and Y.-T. Lee, Tetrahedron Lett. 40 (1999) 2383.

32. S.-K. Kang, T.-H. Kim and S.-J. Pyun, J. Chem. Soc., Perkin Trans. 1 (1997) 797.

33. Y. Nishihara, K. Ikegashira, F. Toriyama, A. Mori and T. Hiyama, Bull. Chem. Soc. Jpn. 73 (2000) 985.

34. R. A. Kretchmer and R. Glowinski, J. Org. Chem. 41 (1976) 2662.

35. L. Buzhansky and B.-A. Feit, J. Org. Chem. 67 (2002) 7523.

36. J. Bergman, Tetrahedron 28 (1972) 3323.

37. J. Bergman, R. Carlsson and B. Sjöberg, Org. Synth. 57 (1977) 18.

38. T. Arnault, D. H. R. Barton and J.-F. Normant, J. Org. Chem. 64 (1999) 3722.

39. A. I. Meyers and E. D. Mihelich, J. Am. Chem. Soc. 97 (1975) 7383.

40. A. I. Meyers, R. Gabel and E. D. Mihelich, J. Org. Chem. 43 (1978) 1372.

41. J. A. Findlay, A. Daljeet, P. J. Murray and R. N. Rej, Can. J. Chem. 65 (1987) 427.

42. A. D. Patten, N. H. Nguyen and S. J. Danishefsky, J. Org. Chem. 53 (1988) 1003.

43. M. A. Rizzacasa and M. V. Sargent, J. Chem. Soc., Perkin Trans. 1 (1988) 2425.

44. M. Rizzacasa and M. V. Sargent, J. Chem. Soc., Chem. Commun. (1989) 301.

45. A. B. Hughes and M. V. Sargent, J. Chem. Soc., Perkin Trans. 1 (1989) 1787.

46. T. Hattori, N. Hayashizaka and S. Miyano, Synthesis (1995) 41.

47. D. Goubet, P. Meric, J.-R. Dormoy and P. Moreau, J. Org. Chem. 64 (1999) 4516.

48. J. Clayden and M. Julia, J. Chem. Soc., Chem. Commun. (1993) 1682.

49. A. N. Nesmejanov, B. A. Sazonova and A. B. Gerasimenko, Dokl. Akad. Nauk SSSR 147 (1962) 634.

50. A. Varvoglis, Synthesis (1984) 709.

51. M. Xia and Z. Chen, Synth. Commun. 30 (2000) 63.

52. M. Xia and Z. Chen, Synth. Commun. 29 (1999) 2457.

53. S.-K. Kang, H.-C. Ryu and J.-W. Kim, Synth. Commun. 31 (2001) 1021.

54. S.-K. Kang, S.-W. Lee, M.-S. Kim and H.-S. Kwon, Synth. Commun. 31 (2001) 1721.

55. A. M. Echavarren, B. Gómez-Lor, J. J. González and Ó. de Frutos, Synlett (2003) 585.

56. D. E. Ames and A. Opalko, Synthesis (1983) 234.

57. D. E. Ames and A. Opalko, Tetrahedron 40 (1984) 1919.

58. S. Pogodin, P. U. Bidermann and I. Agranat, J. Org. Chem. 62 (1997) 2285.

59. D. D. Hennings, S. Iwasa and V. H. Rawal, J. Org. Chem. 62 (1997) 2.

60. T. Harayama, T. Akiyama, H. Akamatsu, K. Kawano, H. Abe and Y. Takeuchi, Synthesis (2001) 444.

61. T. H. M. Jonckers, B. U. W. Maes, G. L. F. Lemière, G. Rombouts, L. Pieters, A. Haemers and R. A. Dommisse, Synlett (2003) 615.

62. M. S. McClure, B. Glover, E. McSorley, A. Millar, M. H. Osterhout and F. Roschangar, Org. Lett. 3 (2001) 1677.

63. S. Oi, S. Fukita, N. Hirata, N. Watanuki, S. Miyano and Y. Inoue, Org. Lett. 3 (2001) 2579.

64. H. C. Bell, J. R. Kalman, J. T. Pihney and S. Sternhell, Aust. J. Chem. 32 (1979) 1521.

65. H. C. Bell, J. R. Kalman, G. L. May, J. T. Pinhey and S. Sternhell, Aust. J. Chem. 32 (1979) 1531.

66. R. A. Abramovitch, D. H. R. Barton and J.-P. Finet, Tetrahedron 44 (1988) 3039.

67. J. Morgan and J. T. Pinhey, J. Chem. Soc., Perkin Trans. 1 (1990) 715.

68. S. Saito, T. Kano, Y. Ohyabu and H. Yamamoto, Synlett (2000) 1676.

69. S.-K. Kang, S.-C. Choi and T.-G. Baik, Synth. Commun. 29 (1999) 2493.

70. D. H. R. Barton, J. P. Kitchin, D. J. Lester, W. B. Motherwell and M. T. Barros Papoula, Tetrahedron 37 (1981) 73.

71. D. H. R. Barton, N. Y. Bhatnagar, J.-P. Finet and W. B. Motherwell, Tetrahedron 42 (1986) 3111.

72. D. H. R. Barton, N. Ozbalik and M. Ramesh, Tetrahedron 44 (1988) 5661.

73. T. Ohe, T. Tanaka, M. Kuroda, C. S. Cho, K. Ohe and S. Uemura, Bull. Chem. Soc. Jpn. 72 (1999) 1851.

74. P. Jacobson, Justus Liebigs Ann. Chem. 428 (1922) 76.

75. D. H. Smith, J. R. Schwartz and G. W. Wheland, J. Am. Chem. Soc. 74 (1952) 2282.

76. G. S. Hammond and W. Grundemeier, J. Am. Chem. Soc. 77 (1955) 2444.

77. D. A. Blackadder and C. Hinshelwood, J. Chem. Soc. (1957) 2898.

78. D. A. Blackadder and C. Hinshelwood, J. Chem. Soc. (1957) 2904.

79. D. V. Banthorpe and E. D. Hughes, J. Chem. Soc. (1962) 3308.

80. D. V. Banthorpe, E. D. Hughes, C. Ingold, R. Bramley and J. A. Thomas, J. Chem. Soc. (1964) 2864.

81. C. K. Ingold and H. V. Kidd, J. Chem. Soc. (1933) 984.

82. D. V. Banthorpe, J. Chem. Soc. (1962) 2407.

83. D. V. Banthorpe, J. Chem. Soc. (1962) 2413.

84. H. J. Shine and J. P. Stanley, J. Org. Chem. 32 (1967) 905.

85. H. Beyer and H.-J. Haase, Chem. Ber. 90 (1957) 66.

86. G. Wittig and J. E. Grolig, Chem. Ber. 94 (1961) 2148.

87. T. Pyl, H. Lahmer and H. Beyer, Chem. Ber. 94 (1961) 3217.

88. H. J. Shine, H. Zmuda and K. H. Park, J. Am. Chem. Soc. 103 (1981) 955.

89. K. H. Park and J. S. Kang, J. Org. Chem. 62 (1997) 3794.

90. P. W. Jeffs, J. F. Hansen and G. A. Brine, J. Org. Chem. 40 (1975) 2883.

91. G. Bringmann and J. R. Jansen, Tetrahedron Lett. 25 (1984) 2537.
92. N. S. Narasimhan and I. S. Aidhen, Tetrahedron Lett. 29 (1988) 2987.
93. M. D'Auria, E. De Luca, G. Mauriello and R. Racioppi, Synth. Commun. 29 (1999) 35.
94. R. Gompper, G. Seybold and B. Schmolke, Angew. Chem. 80 (1968) 404.
95. C. F. Wilcox and S. S. Talwar, J. Chem. Soc. C (1970) 2162.
96. D. Del Mazza and M. G. Reinecke, J. Org. Chem. 53 (1988) 5799.
97. U. N. Rao and E. Biehl, J. Org. Chem. 67 (2002) 3409.
98. M. Frid, D. Pérez, A. J. Peat and S. L. Buchwald, J. Am. Chem. Soc. 121 (1999) 9469.
99. G. R. Newcome, J. Gross and A. K. Patri, J. Org. Chem. 62 (1997) 3013.
100. H.-T. Shih, H.-H. Shih and C.-H. Cheng, Org. Lett. 3 (2001) 811.

8. SYNTHESIS OF AXIALLY CHIRAL BIARYLS

8.1. Introduction

The sterical demands are the main reason of restricted rotation about the aryl-aryl linkage (biaryl axis) in biaryls substituted at four-, three-, or in extremely bulky cases di-, *ortho*-positions. The consequence of this phenomena is that two possible atropisomers can not interconvert under ambient conditions. These structures contain the chiral axis and are chiral compounds. Since a number of naturally occuring biaryl compounds are chiral, the atropisomer-selective synthesis of biaryls have been the challenging area for last two decades [1-3]. From the numerous excellent papers, several approaches to the synthesis of atropisomerically-enriched biaryls have been developed. Herein, some diastereoselective reactions as well as certain important enantioselective (atropisomer-selective) approaches to chiral biaryls are described. Beside a great number of natural products [1,2], many chiral ligands are axially chiral biaryls [3,4]. Several 1,1'-binaphthyl-based ligands are crucial for valuable industrial enantioselective catalytic processes and as chiral selectors in certain chiral stationary phases [3-5]. Binaphthol (**4**) is the most important starting material for the preparation of a wide variety of metal complexes acting as chiral catalysts, or in the synthesis of related atropisomeric ligands such as BINAP (**547**) [4].

4   547

Apart from enantio- and diastereoselective synthesis, binaphthols as the widely used chiral auxiliaries have been obtained in optically pure form by several method including resolution with cinchonidine [3], 1,2-diamino-cyclohexane [6], proline [7], resolution of its cyclic phosphate ester with cinchonine [8], resolution of cyclic borate ester with proline [7], as well as enantioselective enzyme-catalysed (bovine pancreas acetone powder) hydrolysis of di-pentanoic ester of racemic binaphthol [6]. Among wide-spread applications, binaphthol (**4**) has also been used as chiral auxiliary for the resolution of other racemic substances. Thus the cyclic borate esters are effective reagents for the resolution of diastereomeric amino alcohols. In this manner, racemic amino alcohol **548** (1 eq.) was reacted with (*R*)-(+)-1,1'-bi-2-naphthol (**4**, 1 eq.) and boric acid (0.5 eq.) in THF affording the diastereomeric borate **549**. From the latter, optically pure **548** was obtained in 42% yield and 99% diastereomeric excess (d.e.) after acidification and extraction [9], Scheme 1.

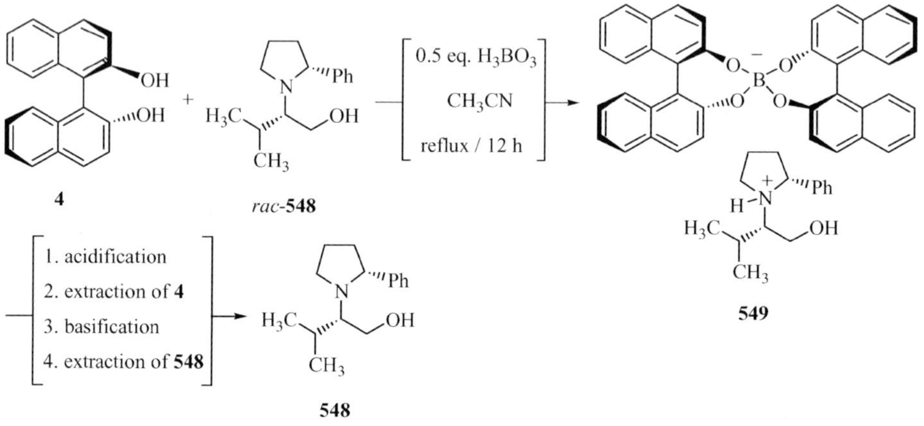

Scheme 1

Additionally, optically pure binaphthol (**4**) has been used as starting material in the synthesis of several efficient chiral crown-ethers, capable to undergo the formation of diastereomeric complexes with only one enantiomer of racemic alkylammonium salts, e.g. amino acids, thus providing an excellent racemate resolving devices. To extend this approach to more general host-guest chemistry, the chiral-host can be effective resolving agent for racemic guest-molecule [10]. In a number of papers covering this interesting field, binaphthol and its derivatives have been investigated. The absolute configuration at chiral axis have been determined either by measurement of the circular dicroism (CD), or, of course, by X-ray crystal structure determination [1,11]. In the older literature an alternative method for determination of the absolute configuration is described. The method is based on the polarizability theory of optical activity, but also a measured value of optical rotation is needed [12]. However, the first two above mentioned methods are far more important and reliable. Alternatively, the absolute configurations of the given enantiomerically enriched samples of axially chiral biaryls can be determined by various conversions to compounds of established configurations. In this Chapter, several interesting diastereoselective and enantioselective reactions for the synthesis of axially chiral biaryls, that have appeared in last 23 years, are discussed. Some of these reactions are atropisomer-selective versions of already described reactions, e.g. Ullmann, Suzuki-Miyaura reaction or palladium-catalysed arylation of arenes with aryl halides, etc.

8.2. Diastereoselective synthesis of biaryls

Diastereoselective synthesis include all reactions and approaches that are based on the atropisomer-selective formation of chiral biaryl axis under the influence of chirality already present in the given molecule. The first employed diastereoselective reaction involved the intramolecular Ullmann coupling of symmetrical and unsymmetrical aryl halides bounded on the certain selected chiral auxiliary at proper intermolecular distances ("chiral bridge"). This was reported by Miyano's group through a few excellent papers [13-16]. Once again, optically pure binaphthol (**4**) was elegantly used as chiral auxiliary, acting as "chiral bridge" for two molecules of identical aryl bromide, namely 1-bromonaphthyl-2-carboxylic acid (**550**). In this manner, binaphthol (**4**) and two equivalents of acid **550** were converted to diester **551**. The latter was subjected to the Ullmann coupling reaction with copper bronze in refluxing DMF affording the biaryl **552** in 36% yield with very high diastereoslectivity, accompanied with formation of reductive dehalogenation (see Chapter 2) side-product **553** [13]. In this case, the chiral (*S*)-binaphthol-moiety induced an axial dissymmetry of the (*S*)-configuration into the newly formed 1,1'-binaphthyl linkage to produce the **552** with no detectable amount of opposite diastereomer. Reductive ester cleavage of **552** with lithium aluminum hydride in refluxing ether gave enantiomerically pure biaryl **554** indicating virtually complete diastereoselectivity [13], Scheme 2. Concerning the reactivity of dibromide **551**, it is noteworth that an electron-withdrawing ester group in an *ortho*-position to the bromine facilitates the Ullmann coupling reaction. In contrast, the analogous benzylic ether **555** was remained unreacted after several hours refluxing in DMF with activated copper bronze.

(*S*)-**555**

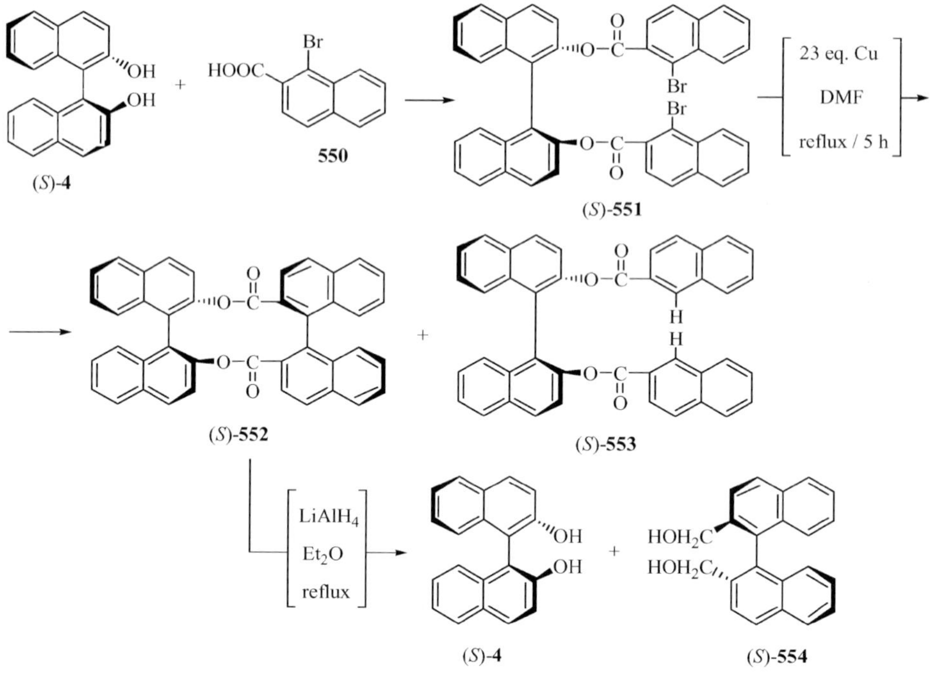

Scheme 2

Miyano's approach was effective also in the synthesis of (R)-6,6'-dinitrodiphenic acid (**556**) from 3-nitro-2-halobenzoic acids **557** or **558**, *via* diesters **559** and **560**, derived from (R)-binaphthol (**4**), in moderate yield and good optical purity, 80-85% d.e. [14].

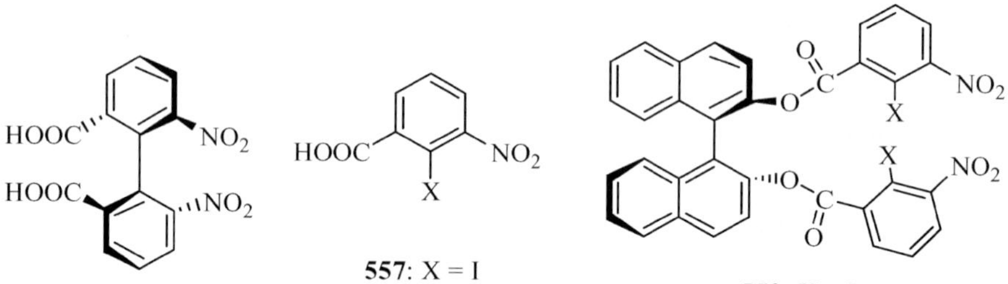

557: X = I

558: X = Cl

559: X = I

560: X = Cl

(R)-556

Beside reductive dehalogenation side-products, the Miyano's method led to the formation of dimer such as **561**, as a product of intermolecular Ullmann coupling of compound **551**, and open-chain dimer **562**, as a product of both intermolecular coupling of **551** and (followed by) reductive dehalogenation [15].

(S,S,S,S)-**561** (S,R,S)-**562**

Another very effective application of the Miyano's method is the synthesis of unsymmetrical diphenic acids form variously substituted 2-halobenzoic acids, using the (R)-1,1'-bi-2-naphthol (**4**) as "chiral bridge", in good yields and very high diastereoselectivity, more than 99% d.e. [16]. For example, (R)-**4** was acylated with 2-iodo-4,6-dichlorobenzoyl chloride (**563**) to give the monoester **564** in 85% yield. The latter was further reacted with 3-nitro-2-iodobenzoyl chloride (**565**) to produce the diester **566** in 93% yield. The intramolecular Ullmann cyclization of unsymmetrical diester **566**, effected at slightly higher dilution factor, gave the respective biaryl **567** in 80% yield, respectively. Finally, base-catalysed saponification of biaryl **567** ester functions led to the formation of unsymmetrical diphenic acid (R)-**568** in 90% yield, and enantiomeric purity (e.e.) more than 99%, with regeneration of optically pure chiral auxiliary, (R)-**4** [16], Scheme 3.

Miyano's synthesis of axially chiral diphenic acid-derived biaryls is an important and powerful access to this class of compounds providing good overall yields, high enantiomeric purity of thus obtained products, and high-yielding regeneration of readily available chiral auxiliary, binaphthol (**4**). Similar approach to the synthesis of axially chiral biaryls was realized with chiral oxazoline moiety in the *ortho*-position to the bromine producing good yield of coupling product and very high diastereoselectivity. In this fashion, chiral oxazoline **569**, derived form *t*-leucinol, underwent an intermolecular diastereoselective Ullmann reaction to afford the biaryl **570** in 77% yield and >97% d.e. [17], Scheme 4.

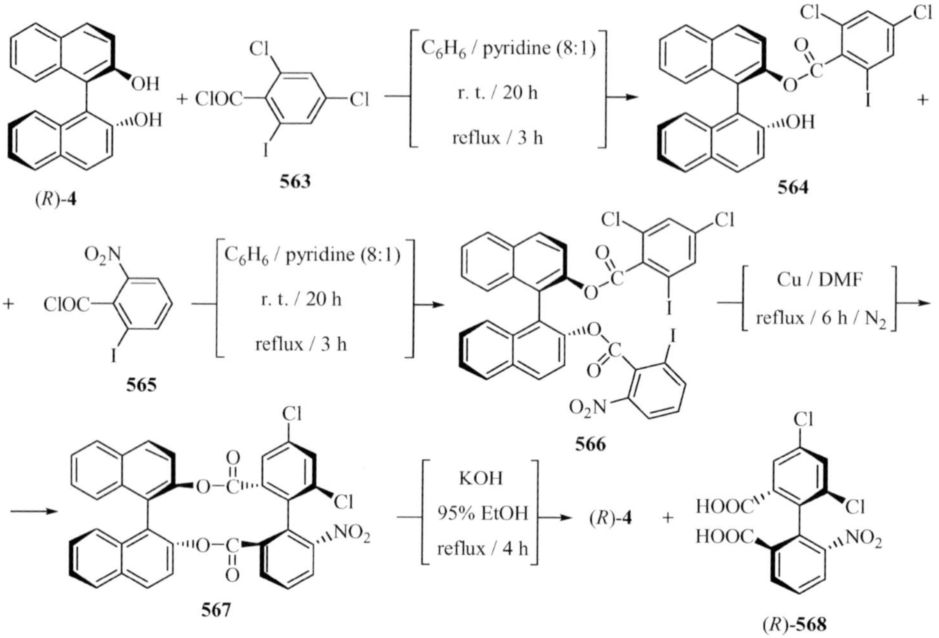

Scheme 3

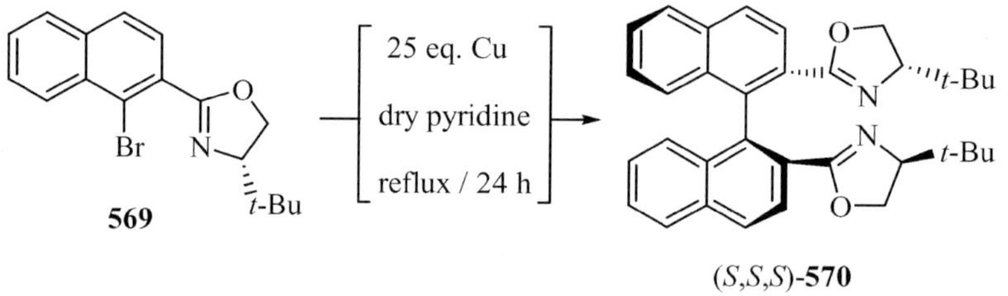

Scheme 4

Apart from the Ullmann reaction, diastereoselective coupling has been carried out using *in situ* generated arylstannane with subsequent Stille reaction. Thus compound **571**, was converted to bistriflate **572**, followed by subsequent protection of amino group to give **573** in 88% overall yield. The latter was reacted with hexamethyldistannane, Me$_6$Sn$_2$, catalysed by Pd(PPh$_3$)$_4$ to give the arylstannane **574** which was further cyclized, by an intramolecular Stille reaction giving **575**, and deprotected to the respective biaryl **576** in 64% yield [18], respectively, Scheme 5.

Beside Ullmann and Stille reaction, successful alternative was provided by the Lipshutz synthesis of biaryls *via* higher order cuprate reagents. Similarly to the Miyano's approach, among *ortho*-halophenols (usually bromo), 1-bromo-2-naphthol (**577**) is bounded to, a readily available tartaric acid derivative, 1,4-dibenzyloxy-2,3-

butanediol (**578**) by the Mitsunobu reaction to give the diether **579**. The latter was subjected to the Lipshutz reaction, *via* organocuprate **580** to afford **581** in 78% yield. The chiral auxiliary was removed by oxidative cleavage with *N*-bromosuccinimide (NBS) to produce enantiomerically pure (*S*)-binaphthol (**4**) in 86% yield [19], respectively, Scheme 6.

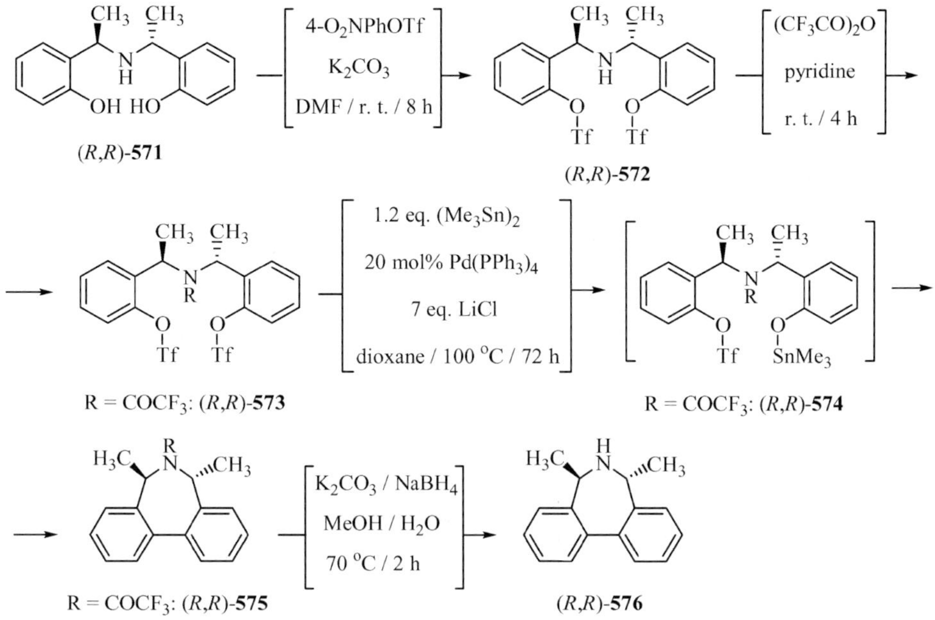

Scheme 5

Using other chiral 1,2-diols such as 1,2-diphenyl-1,2-ethanediol or 2,3-butanediol, the Lipshutz approach allows to rich very high diastereoselectivity [19]. Among reactions that have been used in the diastereoselective synthesis of unsymmetrical axially chiral biaryls, beside Ullmann and Lipshutz reactions, are the Suzuki-Miyaura and Meyers synthesis of biaryls.

Diastereoselective version of the Suzuki-Miyaura reaction was accomplished using planar chiral tricarbonylchromium-complexed aryl bromide with arylboronic acids to form the respective biaryls with very high diastereomeric excess (d.e.) For example, compound **582** underwent the Suzuki-Miyaura reaction with boronic acid **583** to give the biaryl **584** in 88% yield [20], Scheme 7.

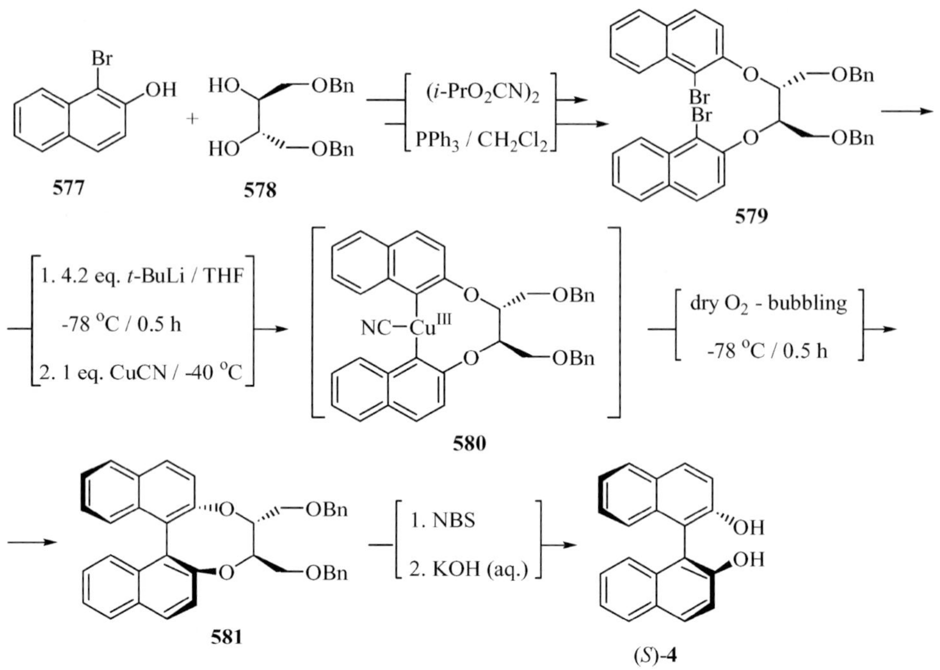

Scheme 6

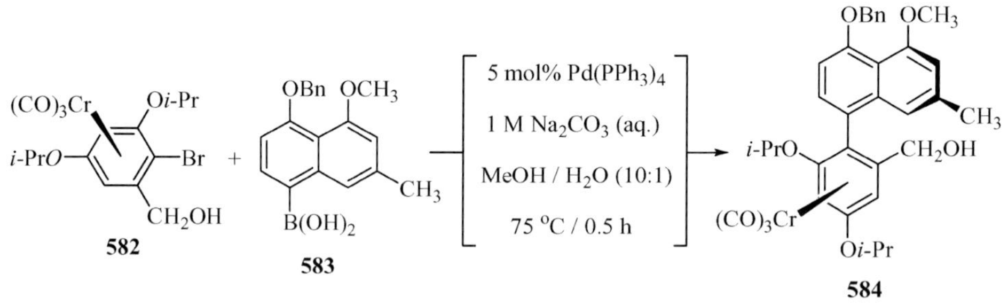

Scheme 7

However, in the Suzuki-Miyaura reaction of relatively simple, enantiomerically pure, both naphthyl bromides and naphthylboronic acids gave diastereomeric mixtures, indicating rather low diastereoselectivity [21]. More important access to axially chiral biaryls is the Meyers approach in its diastereoselective version [22-25]. It was found that oxazolines, derived from readily available amino alcohol **585** [22], underwent the Meyers synthesis of biaryls giving the expected biaryls in high d.e.'s [23-25]. Thus 1-methoxynaphthyl-2-carboxamide (**586**) was activated with triethyloxonium tetrafluoroborate to **587**, which reacted with **585** to give the oxazoline **588**. The latter is subjected to the Meyers synthesis using 2-methoxynaphthylmagnesium bromide (**589**, two equivalents to **588**) furnishing the binaphthyl **590** in 71% yield, which upon subsequent removal of chiral auxiliary gave the respective binaphthyl (*R*)-**591** in 65% yield and 96% e.e. [23], Scheme 8.

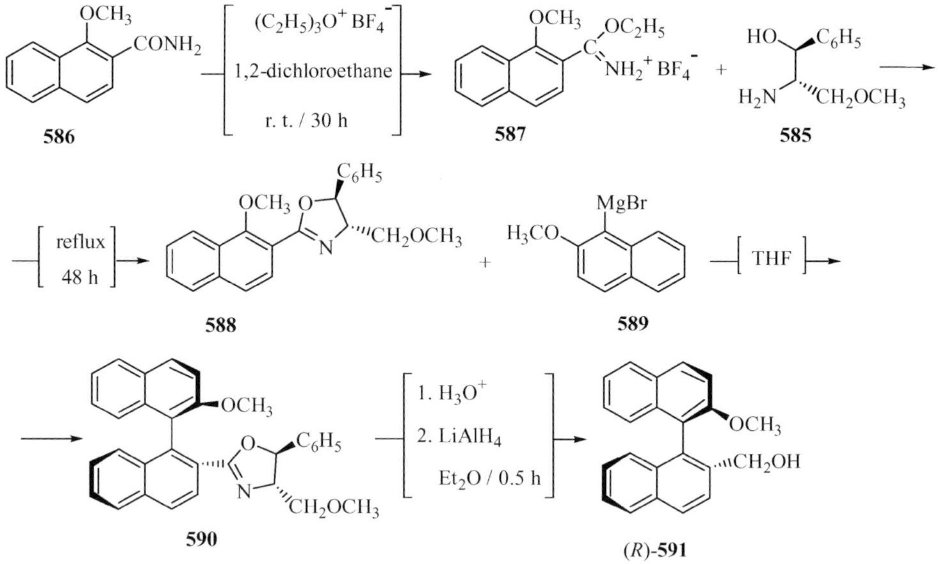

Scheme 8

Diastereoselective Meyers synthesis have been also efficiently employed at tri-*ortho*-substituted biphenyls [24], and various natural products, e.g. (-)-steganone. For instance, compound **592** was converted to the corresponding Grignard reagent **593**, which subsequently reacted with chiral oxazoline **594** to afford the respective biphenyl **595**, isolated after single recrystallization (EtOAc) as pure diastereomer in 65% yield [25], Scheme 9.

Somewhat between diastereoselective and enantioselective approaches is the Cram's synthesis involving the oxazolines with chiral alkoxide leaving groups [26]. In this manner, bromine in oxazoline **596** was substituted with sodium alkoxides, derived from readily available natural alcohols such as menthol (**597**), fenchyl alcohol (**598**), borneol (**599**), quinine (**600**), and quinidine (**601**) to give the respective chiral oxazolines **602**-**606**, Scheme 10. The Meyers reaction of oxazolines **602**-**606** and 1-naphthylmagnesium bromide (**608**) was effected at low temperatures (-42 °C) affording the expected biaryl **609** with respective chiral induction.

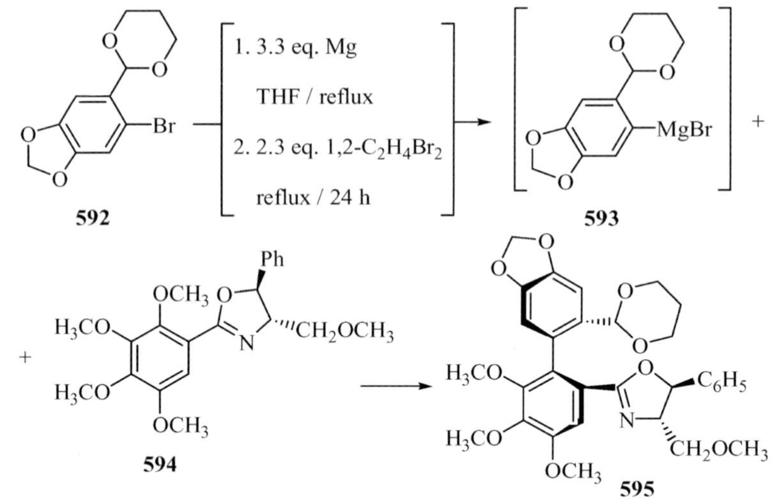
1. 3.3 eq. Mg
THF / reflux
2. 2.3 eq. 1,2-$C_2H_4Br_2$
reflux / 24 h
592
593
594
595

Scheme 9

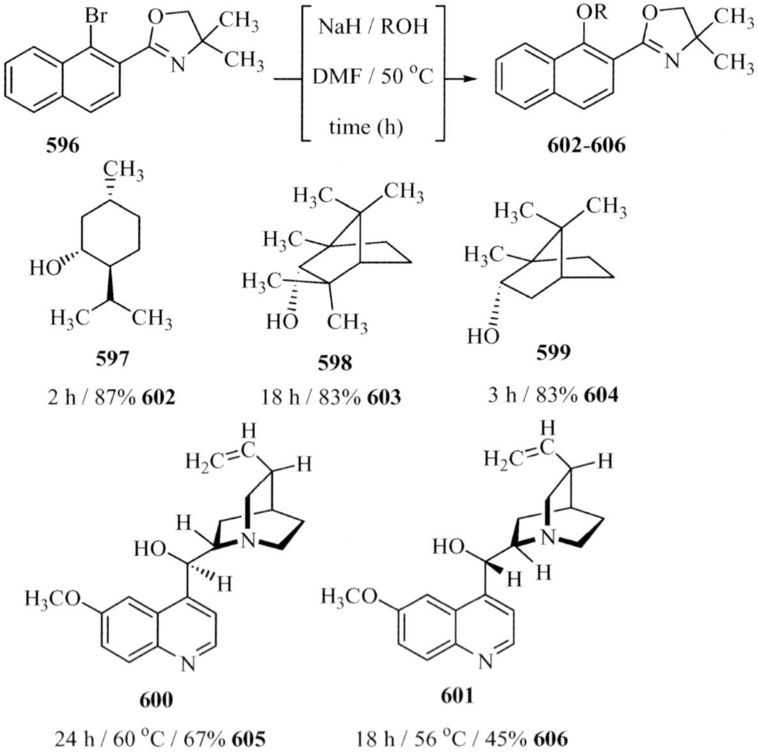
NaH / ROH
DMF / 50 °C
time (h)
596
602-606
597
2 h / 87% 602
598
18 h / 83% 603
599
3 h / 83% 604
600
24 h / 60 °C / 67% 605
601
18 h / 56 °C / 45% 606

Scheme 10

The efficacy and the chiral leaving group-induced enantioselectivity in the Meyers synthesis of axially chiral biaryls of some selected examples are given in the Table 1.

Table 1. The yields and enantioselectivity in the Meyers synthesis of biaryl **608** from oxazolines bearing the chiral leaving group [26]

Leaving group	Time (h)	Yield (%)	e.e. (%)	Config.
l-menthyl (**602**)	1	80	67	*S*
bornyl (**603**)	1	83	10	*R*
α-fenchyl (**604**)	5	78	45	*S*
quininyl (**605**)	1	12	80	*S*
quinidinyl (**606**)	0.75	15	81	*R*

Certain heterocycles, e.g. pyridines or quinolines, bearing of an electron-withdrawing group such as oxazoline, undergo the Michael-type nucleophilic 1,4-addition accompanied with loss of aromaticity to give the new C-C bond. Thus formed dihydropyridine or benzodihydropyridine can be oxidatively aromatized with conservation of chirality, primary induced by an influence of chiral oxazoline moiety. In this manner, Meyers and coworkers [27] described the Michael-type addition of 1-naphthyllithium (**609**) to the oxazoline **610** at low temperature to form **611** in 90% yield. The latter was oxidatively aromatized to the naphthylquinoline **612** in 87% yield with 88:12 ratio of two diastereomers. Diastereoselectivity in this reaction remained on the same level as obtained by the nucleophilic addition of **609** to **610** indicating the virtually complete conservation of chirality, from sp^3-type in the compound **611** to the axially chiral compound **612**, Scheme 11.

Another important access to the axially chiral biaryls derived from electron-rich substrates is the oxidative coupling of arenes in diastereoselective version (OCA reactions, Chapter 6) [28-31]. The axial chirality in these reactions is induced by the chirality already present in the starting materials. These reactions have been conducted by using $K_3[Fe(CN)_6]$ [28], VOF_3 [29], $VOCl_3$ [30], $Tl(OCOCF_3)_3$ [30], $PhI(OAc)_2$ [30], $Mn(acac)_3$ [31], and certain other oxidative coupling reagents [31]. For example, chiral tetrahydronaphthol (*S*)-**613** was dimerized to (*S,S*)-*trans*-**614** in 62% yield with 66% optical purity [28], respectively, Scheme 12.

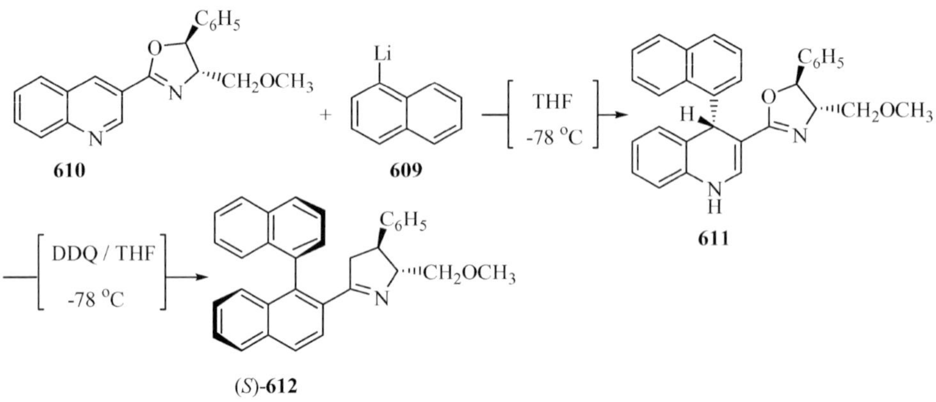

610   **609**

611

(S)-612

Scheme 11

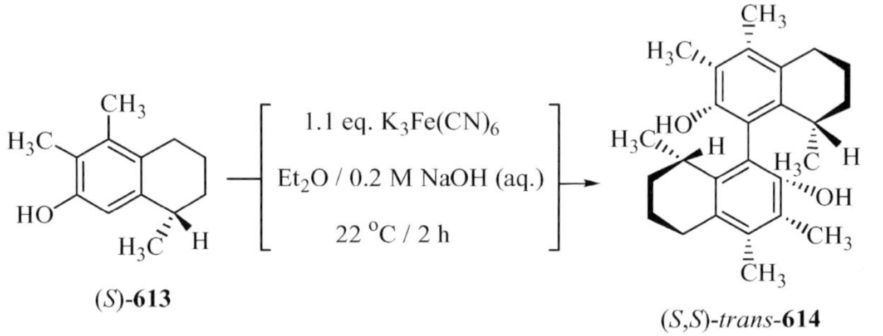

(S)-613

(S,S)-trans-614

Scheme 12

Generally, diastereoselectivity in the oxidative coupling reactions is strongly dependent on the structure of the starting chiral arene, but, in a number of examples the results were good to excellent. This princip has been extended to achiral substrates. These are converted to derivatives of suitable chiral auxiliaries from the natural chiral pool which then undergo the oxidative coupling reaction in the diastereoselective manner, whereas upon subsequent removal of chiral carrier the atropisomerically enriched biaryls were obtained. In this fashion, 2,7-dihydroxynaphthalene (**615**) was bounded to the steroidal bile-acid template obtained from methyl 7-deoxycholate (**616**) *via* activation through the **617** (65%). Thus obtained derivative **618** was oxidatively coupled with manganese(III)-acetylacetonate in hot acetonitrile affording the binaphthyl **619** in 35% yield. After removal of bile-acid chiral auxiliary by saponification, (*S*)-binaphthyl (**620**) was produced in 77% yield and d.e. >99% [31], respectively, Scheme 13.

Additional useful access to axially chiral biaryls is the diastereoselective intramolecular palladium-catalysed arylation of arenes with aryl halides. Bringmann's group has employed this reaction in the key-step of the synthesis of naphthyl isoquinoline alkaloids, e.g. (-)-ancistrocladine [32]. Thus compound **621** was transformed to diastereomeric biaryls **622** and **623** in ratio 3 : 1, Scheme 14.

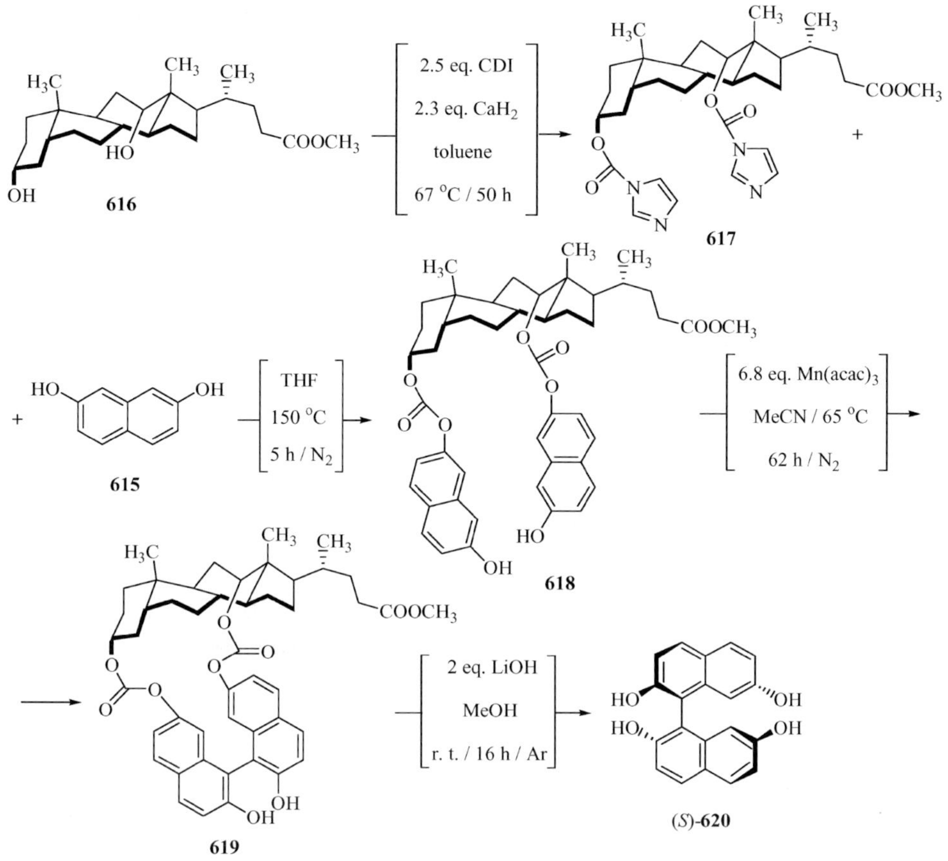

Scheme 13

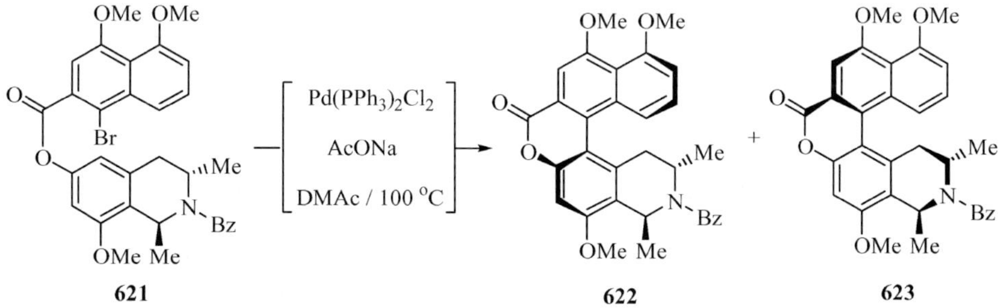

Scheme 14

Lactones such as **622** and **623** in the "axially racemic" form are "axially-prostereogenic" and with suitable nucleophiles such as alkoxides, including the chiral ones, undergo the atropdiastereoselective ring-opening reaction to afford diastereomers with remarkable selectivity [33]. In this manner, lactone **624** is reacted with potassium isopropoxide in isopropanol at room temperature for 10 sec to form the respective atropisomeric esters **625** and **626** in molar ratio 20.9 : 1. However, the same reaction after 22 h gave the latter products in ratio 1 : 1.1 indicating the kinetically controlled atrop-diastereoselective ring-opening reaction pathway [33], Scheme 15.

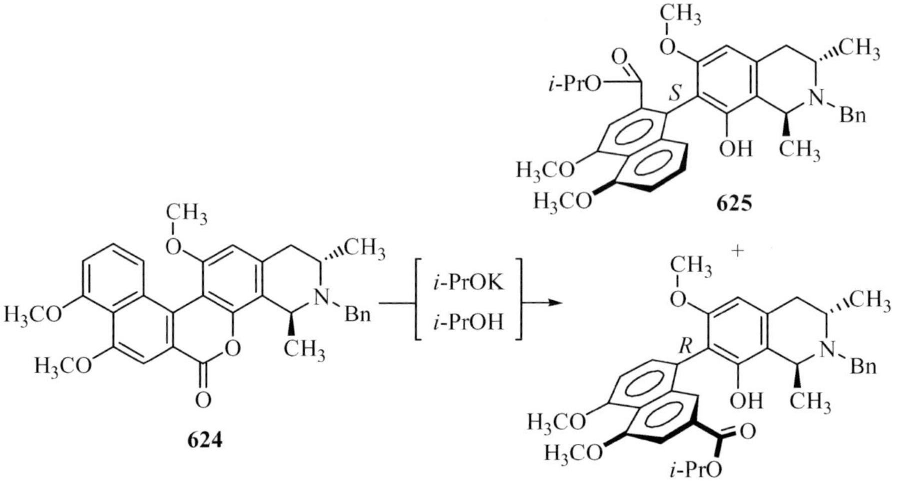

Scheme 15

Diastereoselective synthesis of biaryls is important field in organic chemistry due to a number of natural products and a significant number of drug candidates, chiral ligands and auxiliaries that are axially chiral biaryls. One can expect great results in this chemistry, concerning both new reactions and approaches, as well as more effective methodology.

8.3. Enantioselective synthesis of biaryls

Enantioselective synthesis of axially chiral biaryls include all approaches where the starting material is achiral and the chirality is induced by an influence of chiral catalyst or reagent. Among enantioselective reactions that have been successfully used in the synthesis of axially chiral biaryls are the Kharasch reaction [34-36], Suzuki-Miyaura reaction [37,38], oxidative phenolic coupling mediated by copper complexes with chiral amines [39-43], or ruthenium complexes with chiral salenes [44], and various methods involving desymmetrization (atropisomerically-selective reactions) either *via* the Kharasch reaction [45,46] or by enzymatic procedures [47], as well as some specific enantioselective reactions [48]. Hayashi's group has found that nickel complex of diphenylphosphinoferrocene **627** acts as efficient and highly enantioselective catalyst for the Kharasch cross-coupling reaction of 2-substituted naphthylmagnesium bromides with 2-bromonaphthalenes to chiral binaphthyls in good to excellent yields and enantiomeric excesses (e.e.), generally in range 70-95%. For example, reaction of 2-methylnaphthylmagnesium bromide (**628**) and 2-methyl-1-bromonaphthalene (**629**) led to the formation of expected binaphthyl **630** in 69% yield and 95% e.e. [34], respectively, Scheme 16.

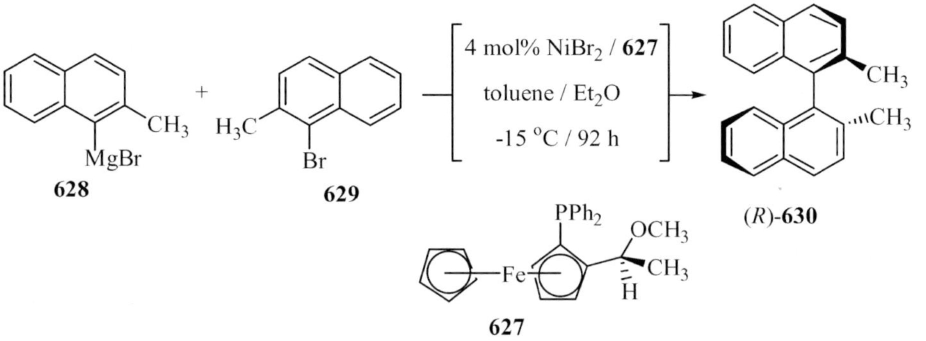

Scheme 16

Hayashi's method has been effective in the synthesis of various 2,2'-disubstituted-1,1'-binaphthalenes [34,36], 1,1':5',1''- and 1,1':4',1''-ternaphthalenes [35] providing high yields and enantioselectivity.

The enantioselective Suzuki-Miyaura reaction (SM) can be accomplished by using the chiral ligands **213** [37] or **631** [38], whose palladium complexes catalysed the cross-coupling reaction of sterically hindered iodides and bromides with also quite encumbered arylboronic acids furnishing the axially chiral, sterically demanded biaryls in moderate to good yields with low to moderate enantioselectivity.

Cammidge's group [37] reported the first asymmetric SM reaction of 1-iodo naphthalene (**632**) with 2-methylnaphthalene-1-boronic acid (**633**) catalysed by the complex *in situ* generated from palladium(II) chloride and the chiral phosphine ligand (*S*)-**213** resulting with formation of biaryl (*R*)-**634** in 44% yield and 63% e.e., Scheme 17.

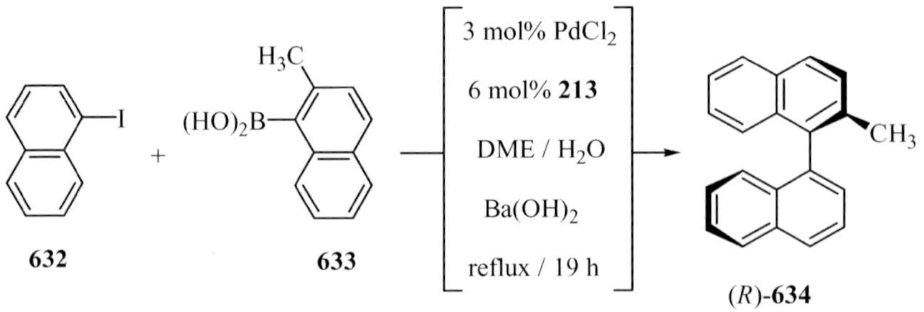

632 633 (*R*)-**634**

Scheme 17

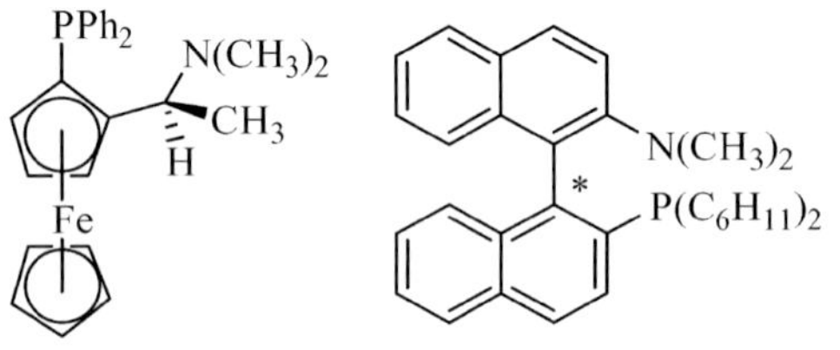

213 **631**

Buchwald's binaphthyl **631** was proved as more effective ligand, providing good to high enantioselectivity (70-92% e.e.). Thus 2-phenylnaphthylboronic acid (**635**) was reacted with 2-nitro-iodobenzene (**18**) to give the respective biaryl **636** in 86% yield and 73% e.e. [38], Scheme 18.

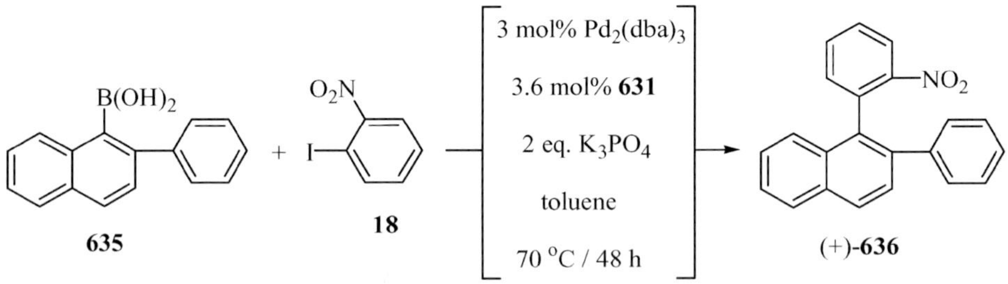

635 **18** (+)-**636**

Scheme 18

The enantioselective Suzuki-Miyaura reaction is apparently very promising area, whereas above shown examples are only the begining of development of this very powerful approach.

Copper(II)-amine complexes are widely used and very effective phenolic oxidative coupling reagents. Since the reaction proceeds within the copper coordinative sphere, chiral amine ligand does induce the enantioselective oxidative couplings of phenols such as 2-naphthol [39-41], or 9-phenanthrol [42,43] to form respective atropisomeric biaryls in high yields and good enantiomeric excesses. Chiral amines **637-641**, derived from readily available amino acid proline, in the presence of copper(II) chloride have profound effect on the enantioselectivity in the catalytic oxidative couplings with oxygen as ultimate oxidant [40,41]. For example, diamine **637** and copper(II) chloride in refluxing dichloromethane efficiently catalysed the oxidative coupling of naphthol **393** in an oxygen atmosphere giving binaphthol **642** in 78% yield with 70% e.e. [40], Scheme 19. Except copper(II) complexes of proline-based 1,2-diamines **637-641**, chiral 1,3-diamines like naturally occuring alkaloid sparteine (**643**) exhibit moderate (e.e.'s up to 47%), but noteworth effect transfering the chirality during the oxidative phenolic couplings with oxygen under this catalytic conditions [41].

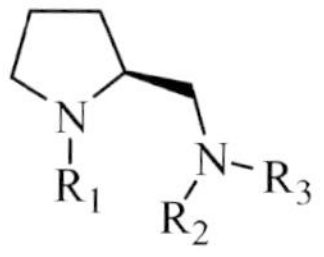

637: R_1 = H, R_2 = C_6H_5, R_3 = C_2H_5

638: R_1 = H, R_2 = 4-MeOC$_6$H$_4$, R_3 = CH$_3$

639: R_1 = H, R_2 = 4-CF$_3$C$_6$H$_4$, R_3 = CH$_3$

640: R_1 = H, R_2 = C_6H_5, R_3 = $C_6H_5CH_2$

641: R_1 = H, R_2 = 2-naphthyl, R_3 = CH$_3$

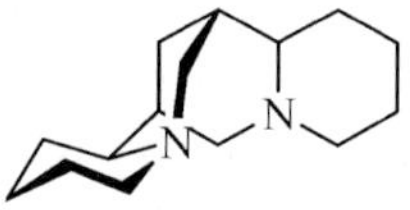

643

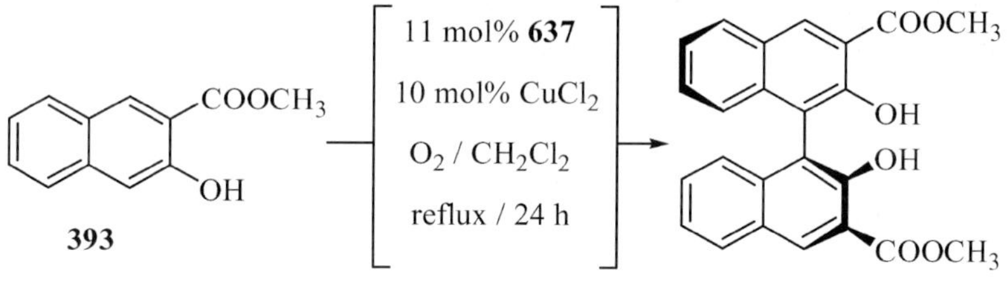

Scheme 19

Moreover, the enantiomeric excess can rich up to 96% with almost quantitative yield of binaphthols by using copper(II) chloride complex of (*S*)-(+)-amphetamine (**644**) employing the molar ratio: CuCl$_2$: **644** : 2-naphthol = 2 : 8 : 1. In this manner, (*S*)-(-)-binaphthol (**4**) was produced in 98% yield and 96% e.e. [39], Scheme 20.

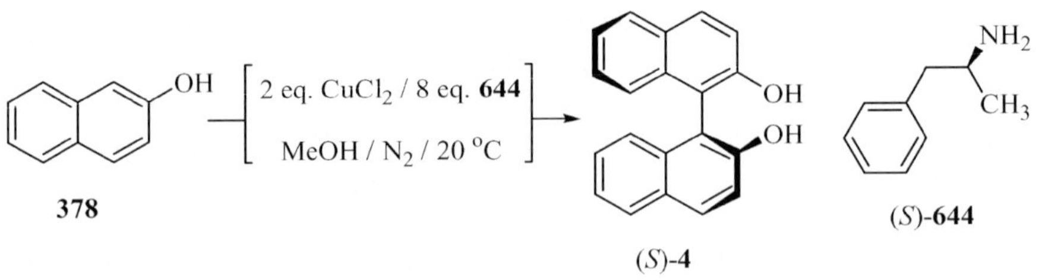

Scheme 20

However, this stoichiometric copper-mediated reaction is not the enantioselective oxidative coupling, but stereoselective crystallization as shown by Brussee and coworkers [39]. Even though the enantiomerically pure (S)-(-)-binaphthol (**4**) is found to be optically stable in refluxing dioxane / water mixture (100 °C) for 24 h, the above mentioned CuCl$_2$ / **644** complex is able to affect the racemization even at room temperature [39]. More interestingly, an equimolar mixture of (R)-(+)-amphetamine (**644**), copper(II) chloride, and racemic binaphthol upon standing in methanolic solution (25 °C / 20 h / N$_2$) gives the precipitate which, after isolation and extraction, affords the mixture of (S)-(-)- and (R)-(+)-binaphthols in the molar ratio of 89 : 11. The latter process exhibits strong dependence on the reaction temperature increasing the enantiomeric purity from 5% e.e. up to 96% e.e. within the very small temperature range, 10 to 20 °C [39]. Further excellent stoichiometric reagent for enantioselective phenolic oxidative coupling is copper(II) nitrate trihydrate in the presence of 1,2-diphenylethylamine (**645**). Highly enantioselective dimerization of 9-phenanthrol (**646**) to the respective biphenanthrol **647** was realized using this reagent in 86% yield and 98% optical purity [43], respectively, Scheme 21.

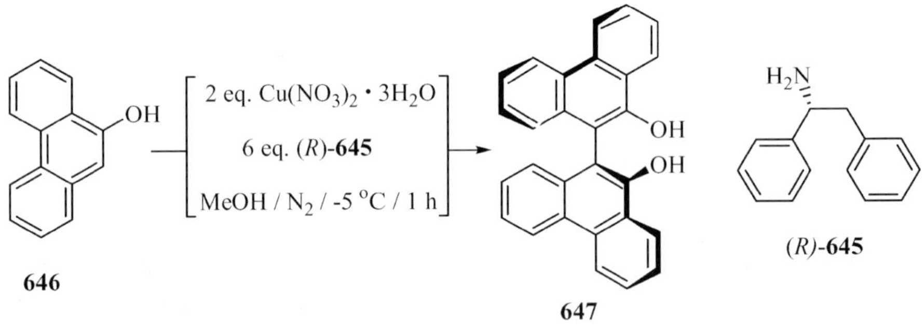

Scheme 21

This chiral copper(II) complex was also effective in the oxidative cross-couplings between two electronically different polycyclic phenols, generally giving good yields and high enantioselectivity [42].

Apart from copper-based complexes, Katsuki's group reported that ruthenium complex of chiral salen **648** readily effects the enantioselective oxidative couplings of 2-naphthols in air under very mild reaction conditions and irradiation of visible light [44]. In this fashion, 6-bromo-2-naphthol (**649**) was converted to binaphthol **650** in 82% yield and 68% e.e. [44], Scheme 22.

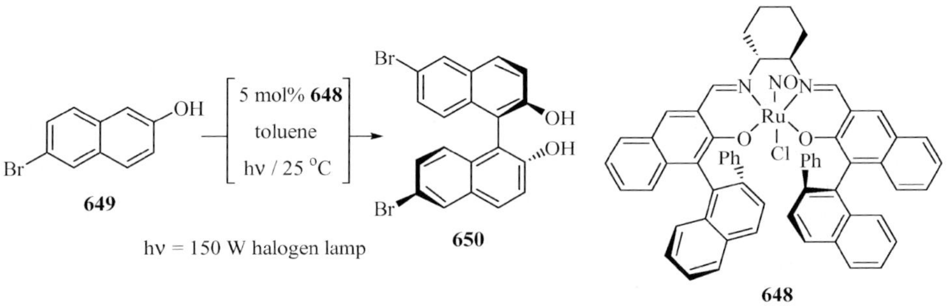

Scheme 22

An interesting approach to the synthesis of axially chiral biaryls was introduced by Hayashi's group [45,46]. They found that complexes of chiral phosphines such as phephos (**651**) with palladium(II) chloride exhibit a profound catalytic activity in the cross-coupling reactions of aryl Grignard reagents with enantiotopic, axially prochiral bistriflates furnishing the respective chiral biaryls of good to excellent enantioselectivity. Prochiral bistriflate **652** was reacted with phenylmagnesium bromide (**221a**) to give the biaryl **653** in 87% yield and 93% e.e. [45], respectively, Scheme 23.

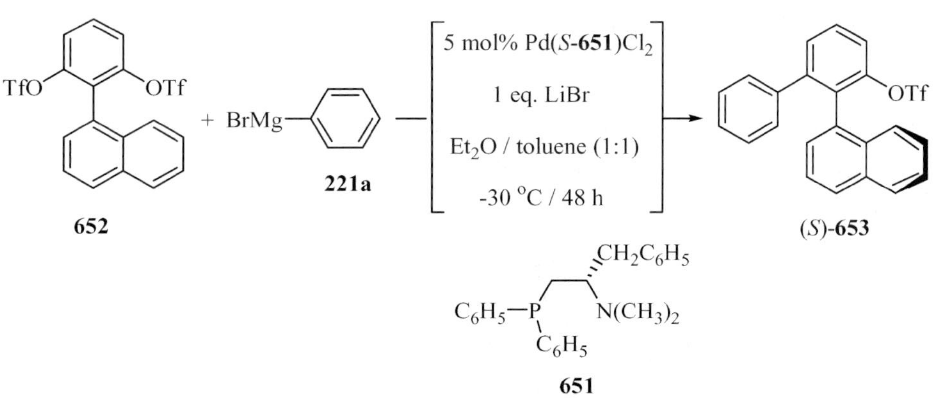

Scheme 23

Chiral ligand **651** is obtained from the appropriate natural amino-acid phenylalanine, whereas the corresponding derivatives of valine or leucine proved to be slightly less effective [46]. Axially prochiral, enantiotopic, biaryl-2,6-diols have been converted to the respective chiral compounds *via* enzymatic desymmetrization. Thus *Pseudomonas cepacia* lipase (PCL) catalysed the atropisomerically-selective hydrolysis of diacetate **654** to give monoacetate **655** in 67% yield and 96% e. e. [47], Scheme 24.

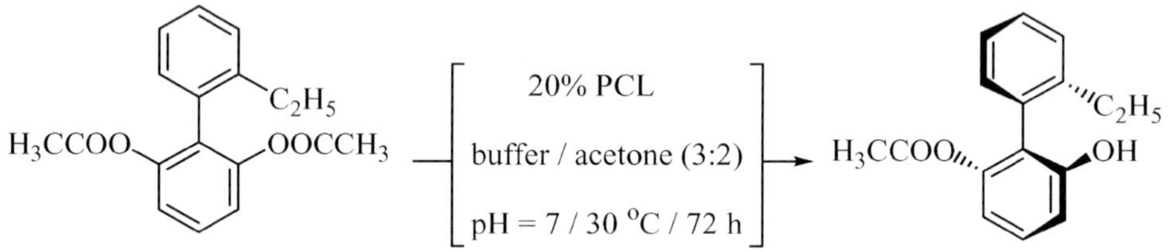

Scheme 24

Somewhat specific but interesting access to axially chiral 2-naphthols is the asymmetric coupling of phenols with aryllead(IV) tricarboxylates, developed by Yamamoto's group [48]. Pinhey's group has found that the base such as pyridine accelerates the *C*-arylation of phenols with aryllead(IV) triacetates, see Chapter 7. Yamamoto's group probed the effect of naturally occuring alkaloid brucine (**656**) as chiral base in this reaction. The latter, acting as chiral ligand for lead, resulted a significant chiral induction during, for example, arylation of 2-naphthol lithium salt (**378a**) with 2-isopropylphenyllead(IV) triacetate (**657**) furnishing the chiral biaryl **658** in 86% yield and 77% e.e. [48], Scheme 25.

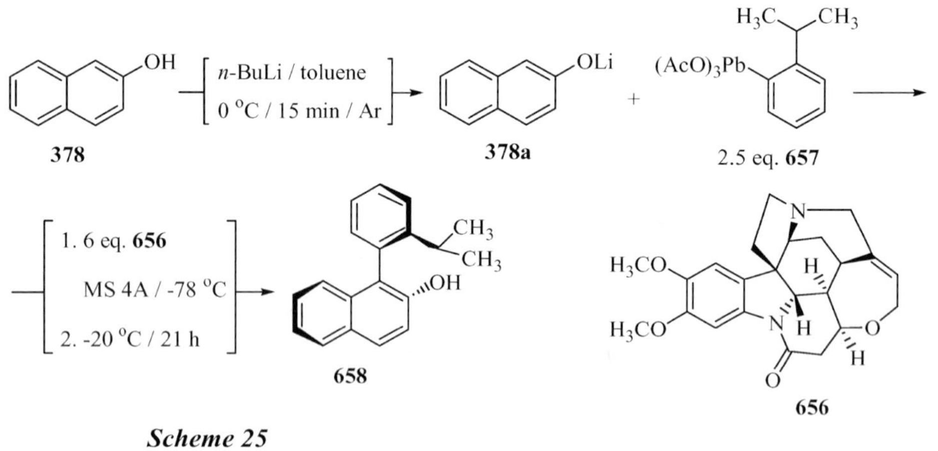

Scheme 25

Simple unsubstituted phenols are arylated at both *ortho*-positions, whereas sterical encumbrances in both reactants enhance the enantio- and diastereoselectivity which were moderate to high [48]. However, the use of over-stoichiometric amount of extremely poisonous brucine makes this method rather unatractive.

Enantioselective as well as diastereoselective synthesis of axially chiral biaryls is the subject of a rapidly growing interest, due to its role as a crucial motif in a great number of drug candidates, chiral polymers, chiral auxiliaries, catalysts, etc.

8.4. Selected synthetic procedures

8.4.1. Diastereoselective Miyano synthesis of biaryls: Synthesis of (S)-2,2'-bis(hydroxymethyl)-1,1'-binaphthyl (554) [13,14]

To a stirred ice-cooled solution of (*S*)-binaphthol (**4**, 1.43 g, 5 mmol) in benzene (100 ml) and pyridine (10 ml) was slowly added a slightly excess amount of 1-bromo-2-naphthoic chloride (**659**, 3.23 g, 12 mmol). The reaction mixture was diluted with benzene (50 ml), and then 2 M HCl (50 ml) was added. The aqueous layer was extracted with benzene (3×30 ml). The combined organic phases were washed successively with 2 M aqueous hydrochloric acid, 1 M sodium sulfite, and water, then dried over Na_2SO_4 in the presence of activated charcoal. The filtrate was evaporated under a reduced pressure, and the crude diester **551** was purified by column chromatography, m.p. 180-182 °C, $[\alpha]_D^{22}$ +34.7° (*c* = 0.922, acetone) [13].

In a 200-ml, two-necked, round-bottomed flask, copper powder (2.0 g, 0.0315 mol, 36 eq.) was pretreated for activation just prior to use (see Chapter 2). Dry DMF (30 ml) was added to the copper powder under a nitrogen atmosphere, and the resulting suspension was heated to reflux temperature. The solution of (S)-**551** (0.65 g, 0.864 mmol) in dry DMF (30 ml) was added dropwise over 8 h-period into the refluxing suspension of activated copper powder. After the addition had been completed, the reaction mixture was stirred under reflux for another 2 h. The reaction mixture was allowed to cool to room temperature, than it was diluted with benzene (50 ml), and filtered. Solids were washed with benzene. Combined organic phases were washed successively with 2 M HCl and water (3×), and dried over Na_2SO_4 in the presence of activated charcoal. Evaporation of the solvent gave a pale yellow product (S,S)-**552**, 0.48 g (93.8 %), $[\alpha]_D^{22}$ -201° (c = 1.03, benzene) [14].

To a solution of (S,S)-**552** (0.30 g, 0.506 mmol) in dry diethyl ether (40 ml) was added lithium aluminum hydride (0.20 g, 5.27 mmol, 10 eq.) and the mixture was refluxed with stiring for 4 h. The reaction mixture was quenched with cautious addition of ethyl acetate and then water. The water (20 ml) was added and organic layer was separated. Aqueous layer was extracted with diethyl ether (3×10 ml). Combined organic extracts were washed with water, dried over Na_2SO_4, filtered and evaporated to dryness. Preparative TLC on silica gel with chloroform / ethyl acetate (4:1) afforded 80 mg (50.3 %) of pure (S)-**554**, $[\alpha]_{546}^{22}$ -61.1° (c = 0.973, acetone), and 127 mg of recovered (S)-binaphthol (**4**), $[\alpha]_D^{22}$ -35.0° (c = 1.06, THF) [14].

8.4.2. Diastereoselective Lipschutz synthesis of biaryls via higher order cyanocuprates: Preparation of binaphthyl 581 [19]

To a dried three-necked round-bottom flask (100 ml) was added copper(I) cyanide (180 mg, 2 mmol) followed by dried THF (35 ml). The mixture was cooled to -78 °C under argon. In a separate round-bottom flask (50 ml), dibromodiether **579** (1.31 g, 2 mmol) was dissolved in THF (35 ml). The flask was cooled to -78 °C, and *tert*-butyl lithium (4.98 ml, 8.4 mmol, 4.2 eq, 1.7 M solution in pentane) added dropwise to give a clear yellow solution. The reaction mixture was stirred at this temperature for 0.5 h.

Thus obtained bislithium compound was transferred with a cannula to the flask, containing the CuCN suspension. The mixture was warmed to -40 °C with gentle stirring. A clear yellow solution of the higher order cuprate **580** was obtained. This was recooled to -78 °C. The argon flow was then stopped and dry oxygen (passed through a trap at -78 °C) was bubbled through the reaction mixture for 0.5 h, whereupon the reaction mixture turned dark. The mixture was allowed to warm to 0 °C, and the oxygen flow was continued for 1.5 h. The reaction was then quenched with a solution of methanol and concentrated aqueous $NaHSO_3$ (2 ml). The mixture was warmed to room temperature and poured into a solution of 10% NH_3 in concentrated NH_4Cl (100 ml). After it had been stirried for 0.5 h, the organic layer was separated. The aqueous phase was extracted with diethyl ether (3x), and the combined organic layers were subsequently washed with 5% HCl, saturated $NaHCO_3$, and brine. The organic phase was dried over $MgSO_4$ and filtered. The solvent was evaporated to dryness. The crude product, upon preparative chromatography over silica gel using *n*-hexane / ethyl acetate (9:1), R_f= 0.48, gave 720 mg (73%) of diastereomerically pure **581** as a white crystalline product, EI-MS: $C_{36}H_{29}O_2$(M+1) calcd: 493.21674, found: 493.21676.

8.4.3. Diastereoselective Cram synthesis of biaryls using a chiral leaving group: Preparation of binaphthyl 608 [26]

A solution of 1-bromonaphthalene (0.70 ml, 5 mmol) in dry THF (2 ml) was cooled to -78 °C, and *sec*-BuLi (3.8 ml, 5 mmol, 1.3 M solution in cyclohexane) was added. The yellow heterogeneous mixture was stirred for 1 h at -78 °C, and then was warmed to -42 °C. To the yellow solution of **609**, a solution of chiral oxazoline **602** (0.105 g, 0.28 mmol) in dry THF (4 ml) was added, and the resulting reaction mixture was stirred at -42 °C for 1 h. To this was added a saturated aqueous NH_4Cl solution (5 ml), and the mixture was warmed to 25 °C. The aqueous phase was neutralized with 10% HCl and washed with diethyl ether. The organic phases were combined, washed with water and brine, dried ($MgSO_4$), and concentrated. Thick layer chromatography on silica gel with dichloromethane / *n*-pentane (9.8 : 0.2) as an eluent gave 84 mg (80 %) of pure **608**, $[\alpha]_{589}^{25}$ +88.4° (c = 3.30, THF), 67% e.e.

8.4.4. Enantioselective Hayashi synthesis of biaryls: Preparation of (R)-2,2'-dimethyl-1,1'-binaphthyl (630) [34]

To a mixture of catalyst **627** (0.34 g, 0.8 mmol), anhydrous nickel bromide (87 mg, 0.4 mmol), and 1-bromo-2-methylnaphthalene (**629**, 2.92 g, 13 mmol) was added methylmagnesium bromide (5 ml, 1 mmol, 0.2 M) in diethyl ether, and the mixture was refluxed for 10 min. The orange solution turned dark brown. The Grignard reagent **628*** (10 mmol), which is an orange slurry prepared in diethyl ether (15 ml) and diluted with toluene (15 ml), was added at -5 °C. The reaction mixture was stirred at -15 °C for 92 h and hydrolyzed with diluted hydrochloric acid. The organic layer and ether extracts from the aqueous layer were combined, washed with saturated NaHCO$_3$ and then water, dried over anhydrous magnesium sulfate, filtered and evaporated to dryness. The residue was chromatographed on a silica gel column with *n*-hexane as an eluent to give 1.91 g (68 %) of (*R*)-(-)-2,2'-dimethyl-1,1'-binaphthyl (**630**), $[\alpha]_D^{22}$ -37.1° (c = 1.0, CHCl$_3$), 95% e.e.

*The Grignard reagent **628** was prepared by adding a solution of 1-bromo-2-methylnaphthalene (**629**) in dry diethyl ether to magnesium ribbons under ultrasonic irradiation and was diluted with toluene (toluene / diethyl ether, 1 : 1) to produce a yellow slurry whose concentration was 0.3-0.4 M [34].

8.4.5. Preparation of (S)-(-)-1,1'-binaphthyl-2,2'-diol (4) [39]

To a solution of (S)-(+)-amphetamine (**644**, 3.25 g, 24 mmol) in methanol (20 ml) was added a solution of copper(II) chloride (0.81 g, 6 mmol) in methanol (10 ml), and the solution was stirred at room temperature for 0.5 h under a nitrogen atmosphere. Then, a solution of 2-naphthol (**378**, 0.43 g, 3 mmol) in methanol (10 ml) was added and the resulting mixture diluted with methanol (20 ml). The reaction mixture was stirred at 25 °C for 20 h under nitrogen. The precipitated complex was filtered off. The collected complex was destroyed with 4 M HCl (40 ml), and diluted with additional water (100 ml). The product was crystallized from the solution and was isolated by filtration, and dried *in vacuo* to give 0.42 g (98 %) of pure (S)-**4**, m.p. 208-210 °C, 96% e.e. The whole work-up procedure, including filtration was performed under a nitrogen atmosphere.

8.4.6. *Enantioselective phenolic oxidative coupling mediated by Cu (NO₃)₂ in the presence of chiral 1,2-diphenylethylamine: Preparation of (S)-10,10'-dihydroxy-9,9'-biphenanthryl (647) [43]*

646   (S)-**647**   (R)-**645**

To a solution of (R)-(-)-1,2-diphenylethylamine (**645**, 11.84 g, 0.06 mol) and copper(II) nitrate trihydrate (4.83 g, 0.02 mol) in methanol (60 ml) cooled to -5 °C was added a solution of **646** (1.94 g, 10 mmol) in methanol (20 ml) under a nitrogen atmosphere. After stirring at -5 °C for 1 h, the reaction mixture was quenched with 2 M aqueous hydrochloric acid (100 ml) and the product was extracted with diethyl ether (3 × 50 ml). The combined organic phases were dried (NaSO₄), filtered and evaporated to dryness. The residue was purified by preparative chromatography over silica gel to provide 1.66 g (86 %) of pure (S)-**647**, m.p. 234-236 °C, 98% optical purity.

8.5. Conclusion

In this Chapter, several diastereoselective and enantioselective methods for synthesis of biaryls containing a chiral axis are described. The most versatile methods for diastereoselective synthesis of symmetrical and unsymmetrical biaryls are intramolecular homo-coupling of dihalides (Miyano-Ullmann and Lipschutz reactions), triflates (Stille reaction), or related substrates containing the chiral induction group ("chiral bridge"). The latter can be part of starting molecule or chiral auxiliary which is, after generation of a new aryl-aryl bond, subsequently removed by saponification (e.g. esters), hydrogenation (e.g. benzyl ethers), etc. The most important diastereoselective synthesis of biaryls derived from phenols is the oxidative coupling with various manganese(III) and copper(II) complexes. The enantioselective version of the copper(II) salts-mediated oxidative couplings in the presence of chiral amines is essential access to chiral binaphthols and related structures. Specially interesting approach to the synthesis of axially chiral biaryls is the enantioselective Suzuki-Miyaura cross-coupling reaction where promising results have been obtained.

8.6. References

1.   G. Bringmann, R. Walter and R. Weirich, Angew. Chem., Int. Ed. Engl. 29 (1990) 977.
2.   G. Bringmann, M. Breuning and S. Tasler, Synthesis (1999) 525.
3.   L. Pu, Chem. Rev. 98 (1988) 2405.
4.   R. Noyori, Chem. Soc. Rev. 18 (1989) 187.
5.   F. Mikeš and G. Boshart, J. Chromatography 149 (1978) 455.
6.   G. Bringmann, R. Walter and R. Weirich, Stereoselective Synthesis, vol. 1: Biaryls, Eds. G. Helmchen, R. W. Hoffmann, J. Mulzer and E. Schaumann, Thieme, Stuttgart, 1996.
7.   M. Periasamy, L. Venkatraman and K. R. J. Thomas, J. Org. Chem. 62 (1997) 4302.
8.   J. Reeder, P. P. Castro, C. B. Knobler, E. Martinborough, L. Owens and F. Diederich, J. Org. Chem. 59 (1994) 3151.
9.   M. Periasamy, N. S. Kumar, S. Sivakumar, V. D. Rao, C. R. Ramanathan and L. Venkatraman, J. Org. Chem. 66 (2001) 3828.
10. D. J. Cram, Angew. Chem., Int. Ed. Engl. 27 (1988) 1009.
11. L. Huang, Y.-K. Si, G. Snatzke, D.-K. Zheng and J. Zhou, Coll. Czech. Chem. Commun. 53 (1988) 2664.
12. D. D. Fitts, M. Siegel and K. Mislow, J. Am. Chem. Soc. 80 (1958) 480.

13. S. Miyano, M. Tobita, M. Nawa, S. Sato and H. Hashimoto, J. Chem. Soc., Chem. Commun. (1980) 1233.
14. S. Miyano, S. Handa, K. Shimizu, K. Tagami and H. Hashimoto, Bull. Chem. Soc. Jpn. 57 (1984) 1943.
15. S. Miyano, S. Handa, M. Tobita and H. Hashimoto, Bull. Chem. Soc. Jpn. 59 (1986) 235.
16. S. Miyano, H. Fukushima, S. Handa, H. Ito and H. Hashimoto, Bull. Chem. Soc. Jpn. 61 (1988) 3249.
17. M. B. Andrus, D. Asgari and J. A. Sclafani, J. Org. Chem. 62 (1997) 9365.
18. L. A. Saudan, G. Bernardinelli and E. P. Kündig, Synlett (2000) 483.
19. B. H. Lipshutz, F. Kayser and Z.-P. Liu, Angew. Chem., Int. Ed. Engl. 33 (1994) 1842.
20. T. Watanabe, M. Shakadou and M. Uemura, Synlett (2000) 1141.
21. H.-F. Chow and C.-W. Wan, J. Org. Chem. 66 (2001) 5042.
22. A. I. Meyers, G. Knaus, K. Kamata and M. E. Ford, J. Am. Chem. Soc. 98 (1976) 567.
23. A. I. Meyers and K. A. Lutomski, J. Am. Chem. Soc. 104 (1982) 879.
24. A. I. Meyers and R. J. Himmelsbach, J. Am. Chem. Soc. 107 (1985) 682.
25. A. I. Meyers, J. R. Flisak and R. A. Aitken, J. Am. Chem. Soc. 109 (1987) 5446.
26. J. M. Wilson and D. J. Cram, J. Org. Chem. 49 (1984) 4930.
27. A. I. Meyers and D. G. Wettlaufer, J. Am. Chem. Soc. 106 (1984) 1135.
28. B. Feringa and H. Wynberg, J. Org. Chem. 46 (1981) 2547.
29. K. Tomioka, T. Ishiguro and K. Koga, Tetrahedron Lett. 21 (1980) 2973.
30. M. A. Schwartz and P. T. K. Pham, J. Org. Chem. 53 (1988) 2318.
31. A. K. Bandyopadhyaya, N. M. Sangeetha and U. Maitra, J. Org. Chem. 65 (2000) 8239.
32. G. Bringmann, J. R. Jansen and H.-P. Rink, Angew. Chem., Int. Ed. Engl. 25 (1986) 913.
33. G. Bringmann and H. Reuscher, Angew. Chem., Int. Ed. Engl. 28 (1989) 1672.
34. T. Hayashi, K. Hayashizaki, T. Kiyoi and Y. Ito, J. Am. Chem. Soc. 110 (1988) 8153.
35. T. Hayashi, K. Hayashizaki and Y. Ito, Tetrahedron Lett. 30 (1989) 215.
36. S. L. Colletti and R. L. Halterman, Tetrahedron Lett. 30 (1989) 3513.
37. A. N. Cammidge and K. V. L. Crépy, Chem. Commun. (2000) 1723.
38. J. Yin and S. L. Bushwald, J. Am. Chem. Soc. 122 (2000) 12051.
39. J. Brussee, J. L. G. Groenendijk, J. M. Koppele and A. C. A. Jansen, Tetrahedron 41 (1985) 3313.
40. M. Nakajima, K. Kanayama, I. Miyoshi and S.-i. Hashimoto, Tetrahedron Lett. 36 (1995) 9519.

41. M. Nakajima, I. Miyoshi, K. Kanayama, S.-i. Hashimoto, M. Noji and K. Koga, J. Org. Chem. 64 (1999) 2264.
42. K. Yamamoto, H. Yumioka, Y. Okamoto and H. Chikamatsu, J. Chem. Soc., Chem. Commun. (1987) 168.
43. K. Yamamoto, H. Fukushima and M. Nakazaki, J. Chem. Soc., Chem. Commun. (1984) 1490.
44. R. Irie, K. Masutani and T. Katsuki, Synlett (2000) 1433.
45. T. Hayashi, S. Niizuma, T. Kamikawa, N. Suzuki and Y. Uozumi, J. Am. Chem. Soc. 117 (1995) 9101.
46. T. Kamikawa and T. Hayashi, Tetrahedron 55 (1999) 3455.
47. T. Matsumoto, T. Konegawa, T. Nakamura and K. Suzuki, Synlett (2002) 122.
48. S. Saito, T. Kano, H. Muto, M. Nakadai and H. Yamamoto, J. Am. Chem. Soc. 121 (1999) 8943.

SUBJECT INDEX